3rd International Conference on Mechanical, Manufacturing and Process Plant Engineering (ICMMPE 2017)

IOP Conference Series: Materials Science and Engineering
Volume 328

Batu Ferringhi, Malaysia
22 - 23 November 2017

Editors:

Mokhtar Awang
Farazila Yusof

ISBN: 978-1-5108-5985-2
ISSN: 1757-8981

Printed from e-media with permission by:

Curran Associates, Inc.
57 Morehouse Lane
Red Hook, NY 12571

Some format issues inherent in the e-media version may also appear in this print version.

Printed by Curran Associates, Inc. (2018)

For permission requests, please contact the Institute of Physics
at the address below.

Institute of Physics
Dirac House, Temple Back
Bristol BS1 6BE UK

Phone: 44 1 17 929 7481
Fax: 44 1 17 920 0979

techtracking@iop.org

Additional copies of this publication are available from:

Curran Associates, Inc.
57 Morehouse Lane
Red Hook, NY 12571 USA
Phone: 845-758-0400
Fax: 845-758-2633
Email: curran@proceedings.com
Web: www.proceedings.com

Table of contents

Volume 328

3rd International Conference on Mechanical, Manufacturing and Process Plant Engineering (ICMMPE 2017)

22–23 November 2017, Batu Ferringhi, Penang, Malaysia

Accepted papers received: 28 February 2018
Published online: 16 March 2018

Preface

Papers

IOP Conf. Series: Materials Science and Engineering **328** (2018) 011001 doi:10.1088/1757-899X/328/1/011001

Preface:

This volume presents a compilation of research works covering the fields of mechanical, manufacturing and plant engineering. All manuscripts in this volume were presented during the 3rd International Conference on Mechanical, Manufacturing and Process Plant Engineering 2017 (ICMMPE 2017) which was held at Parkroyal Hotel, Penang, Malaysia on 22-23 November 2017. The editor(s) of the proceeding would like to express the utmost gratitude and thanks to all reviewers in the technical team for making this volume a success.

Editors:

Ir. Dr. Mokhtar Awang

Dr. Farazila Yusof

IOP Conf. Series: Materials Science and Engineering **328** (2018) 011002 doi:10.1088/1757-899X/328/1/011002

Peer review statement

All papers published in this volume of *IOP Conference Series: Materials Science and Engineering* have been peer reviewed through processes administered by the proceedings Editors. Reviews were conducted by expert referees to the professional and scientific standards expected of a proceedings journal published by IOP Publishing.

IOP Conf. Series: Materials Science and Engineering **328** (2018) 012001 doi:10.1088/1757-899X/328/1/012001

Effect of post heat treatment on the microstructure and microhardness of diffusion coupled gray cast iron and low carbon steel

F Ayob[1, 2], P Hussain[2], M Awang[2], M R Zoolfakar[1], A Abdullah[1], B A Baharudin[1], and A A Abdul Rahim[1]

[1] Universiti Kuala Lumpur, Lumut, Malaysia
[2] Universiti Teknologi Petronas, Seri Iskandar, Malaysia

E-mail: fauzuddin@unikl.edu.my

Abstract. Experimental studies carried out on the joining of grey lamellar graphite cast iron to low carbon steel by diffusion welding techniques had successfully produced a few diffusion couples, despite facing some earlier failures. Although faced with some success, analysis on the diffusion couples showed voids and incomplete bond/joint at the interface of the joints while the tensile strength value was found to be low. In furtherance, the diffusion couples were then subjected to a post-weld heat treatment (PWHT) that allowed for further diffusion process to take place to remove the voids and complete the bond/weld. The results revealed microvoids and interface lines at the interfaces of the specimens treated at temperatures of 800^0C, and becoming less visible at 900^0C. At an elevated temperature of 1000^0C, on longer treatment times, apparently these microvoids and interface lines had disappeared, and the bond/weld at the interfaces of the diffusional welded couples seemed to be more complete. Thus, heat treatment time and temperature were found to have a strong influence on the structure, thickness and hardness of the diffusion layer produced.

1. Introduction

Grey cast iron and mild steel/ low carbon steel are two of the most common materials being used by various industries. Grey cast irons have found wide acceptance based on a combination of out-standing castability, excellent machinability, economics, and unique properties [1]. While mild steel/ low carbon steel is commonly used due to its good strength properties and relatively easy to fabricate such as by welding, it is usually used as structural materials.

At present, joining of cast iron to mild steel is generally performed by mechanical methods, such as bolting which adds to product weight, increases the number of process operations, and raises costs. An important contribution would be made to manufacturing industry through a capability being available for joining of cast iron to mild steel by welding [2].

Generally, cast iron which carries a good proportion of carbon, sulphur and phosphorous, is poorly weldable by fusion welding because it tends to crack in the vicinity of the weld zone, both during and after welding process [3]. While fusion welding of cast iron alone remains just as difficult to weld, in the context of cast iron being welded together with dissimilar material such as mild steel/ low carbon steel having entirely different properties such example their thermal expansion coefficients, their welding may result in distortion as well as crack formation[2].

Diffusion welding is a joining process between materials wherein the principal mechanism for joint formation is a solid state diffusion. Coalescence of the faying surface is accomplished through the application of pressure at elevated temperature. There is no melting and only limited macroscopic deformation or relative motion of the parts occurs during welding [4]. Diffusion welding is an attractive joining technique for the manufacture of precision apparatus and complex structures [5]. Diffusion welding is mainly used for joining of difficult to weld materials or when standard fusion welding methods cannot be used [6].

In previous works [7], preliminary studies on experiments conducted on joining of grey cast iron to mild steel/ low carbon steel by diffusion welding method had showed initial failures of few specimens. However, upon investigating and recognizing of each equipment's unique characteristics, some adjustments were made on the parameters and procedures of subsequent experiments that were found to finally produce significant visually good joints of diffusion couples.

Presently, diffusion couples produced are being subjected to the scanning electron microscopy (SEM) metallographic examination and energy dispersive spectrometry (EDS) to check for composition, weight concentration and tensile testing. Despite successful production of the diffusion couples, analysis had revealed voids and incomplete bond/joint at the interface of the joints, while the tensile strength value was found to be low.

Heat treatment allows the change in the microstructure and properties of investigated diffusional welded cast iron within a wide range of parameters [8]. Post heat treatment was found to have a strong influence on the micro-structure and micro-hardness of friction stir processed NiAL Bronze (NAB) alloy [9]. Furthermore heat treatment time and temperature were found to have a strong influence on the structure, thickness, inter-diffusion coefficients, composition and hardness of the diffusion layer of diffusional welded couples produced [10].

In furtherance of this study, the diffusional welded couples produced (although being of low quality in nature) were then subjected to a post weld heat treatment (PWHT) with the objectives, namely: (1) To remove voids and complete the bond/weldment by allowing further diffusion process to take place; (2) To investigate the effects of PWHT on the diffusion layers' microstructure, thickness and hardness values.

2. Material and method

2.1 Material

Types, grades, composition and the mechanical properties of the parent materials used in this study are as shown in Table-1 and Table-2.

Table-1.Chemical composition and mechanical properties of grey cast iron

	Chemical Composition (wt%)				
	C	**Si**	**Mn**	**P**	**S**
Grey Cast Iron (ASTM A48C 35)	3.43	2.21	0.62	0.073	0.069
	Tensile Strength (MPa)		**Elongation (%))**	**Hardness (HB)**	
	250		-	175	

Table-2.Chemical composition and mechanical properties of low carbon steel

	Chemical Composition (wt%)				
	C	**Si**	**Mn**	**P**	**S**
BS 449 grade 250	0.19	0.10	0.46	0.011	0.031
	Tensile Strength (MPa)		**Elongation (%)**	**Hardness (HB)**	
	408		35	195	

2.2 Method

With regards to the preliminary study [7], to date, five diffusion couples of grey cast iron to low carbon steel have been successfully produced by diffusion welding techniques. The equipment and the processes used to produce these diffusion couples were also described in the previous paper [7].

In the following study before conducting post weld heat treatment, few of these diffusion couples were then subjected to a scanning electron microscopy (SEM) metallographic examination and energy dispersive spectrometry (EDS) for a composition and weight concentration analysis and a tensile testing. For the tensile testing, the subsize test specimen was prepared as per ASTM standard's E8M-11. Room temperature tensile testing was carried out using a computer controlled Instron testing machine with a crosshead velocity of 0.15mm/minute.

A post weld heat treatment was conducted on a few specimens of the diffusion couples. The main parameters/variables that constitute the PWHT are temperature and time, and performed under atmospheric pressure. The temperatures selected are based on optimum recommended temperatures for diffusion welding process of the appropriate materials as this treatment is meant to allow for further diffusion process to take place. Hence the temperatures and times being chosen were 800^{0}C, 900^{0}C, 1000^{0}C and 2, 4, 8 hours respectively.

An apparatus was designed for the PWHT; the heating system is done through radiation consisting of a coil resistance furnace as the heat source, with appropriate switches and the instrumentation needed to set measure and monitor the desired treatment temperature and time.

After completing the PWHT, the specimens were then prepared for a metallographic examination. Photographs of the prepared metallographic specimens, in the vicinity of diffusion zones, along the bonding interface were then taken by optical microscope (OM). The metallographic specimens were also used for hardness testing. In this test, the micro-hardness tester of the Vickers hardness testing machine was employed. The hardness was measured across the bonding interface.

3. Results and discussion

3.1 Before PWHT

Figure-1 shows one of the successfully joined diffusion couples. The micrographs of few samples exhibit fairly good bonding along the interface of the welded joint with microvoids still present. Figure-2 shows one of the micrographs. From the micrographs, diffusion layer was observed on the low carbon steel side while on the cast iron side close to the interface, partial dissolution of graphite flakes was visible. From the energy dispersive spectrometry (EDS) results, as per Figure-3 and Table-3, a rapid diffusion of carbon occurs from the cast iron to the low carbon steel. From the tensile test, the strength value observed was indeed very low and the fracture was brittle in nature and suspected the joint/weld was incomplete. Figure-4 shows one of the fractured specimens.

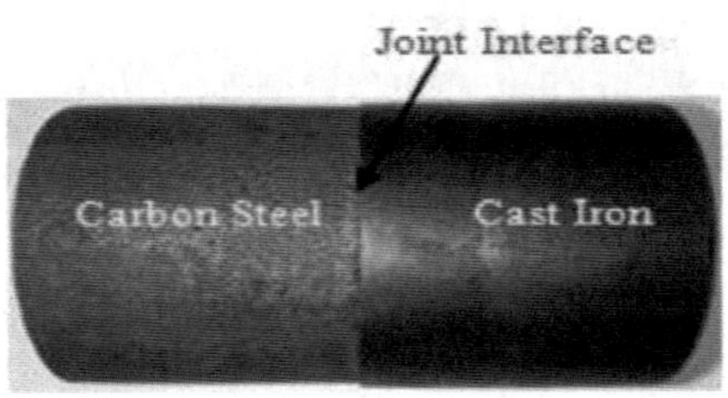

Figure-1.Jointed diffusion couple

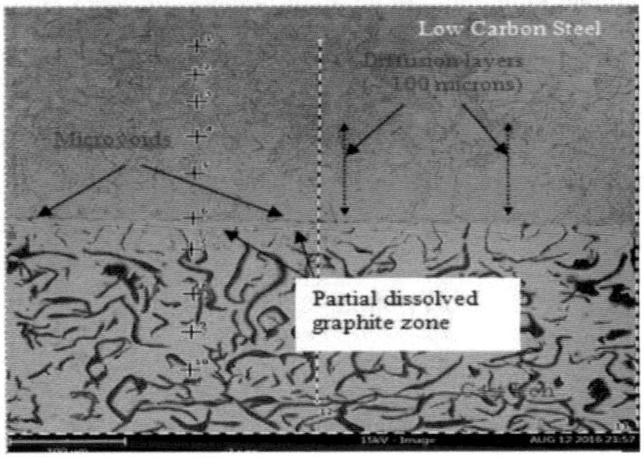

Figure-2.SEM Microphotograph of a diffusion couple (500x)

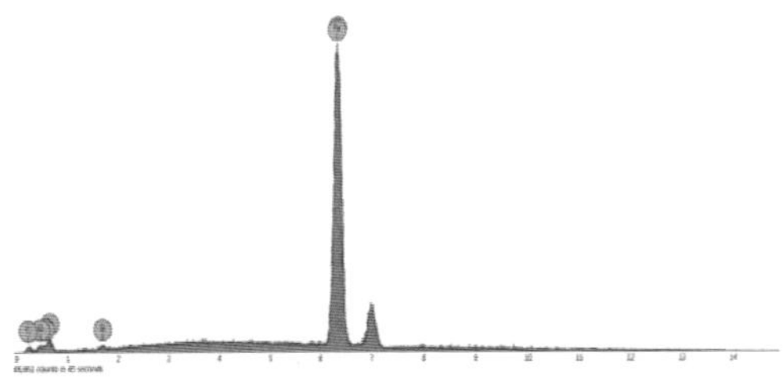

Figure-3.EDS graphic in interface region point 6, of Figure-2.

Table-3. EDS Composition and Weight Concentration at point 6, of Figure-2.

Element	Weight Concentration
Iron	96.5
Oxygen	0.7
Carbon	2.5
Silicon	0.3

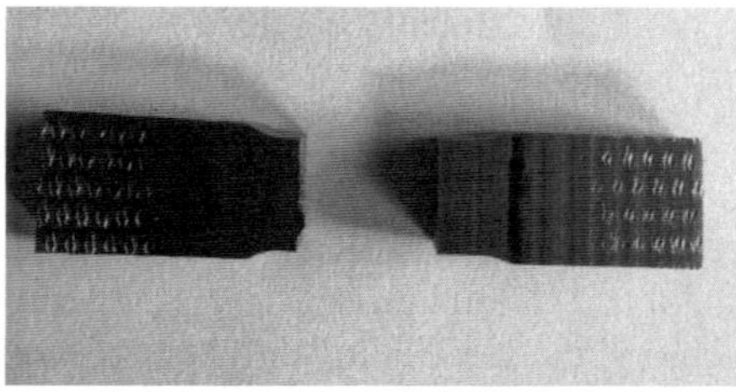

Figure-4.View of a fractured diffusion couple

3.2 After PWHT

3.2.1. Microstructure of diffusion layer. OM microphotographs of the diffusion zone of the diffusion couples between cast iron and low carbon steel after PWHT are as shown in Figure-5 and Figure-6. These photographs show the microstructures of the diffusion layer after a series of heat treatments, conducted under a combination of temperature of 800^{0}C, 900^{0}C and 1000^{0}C, and time at 2 hours, 4 hours and 8 hours respectively. Figure-7 shows the typical analysis of the interfaces and the diffusion layers of the heat treated diffusion couples.

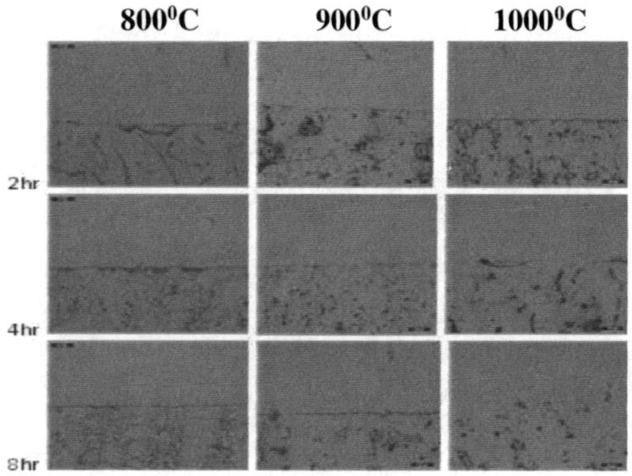

Figure-5.OM microphotographs of specimens after PWHT, before etching (500x)

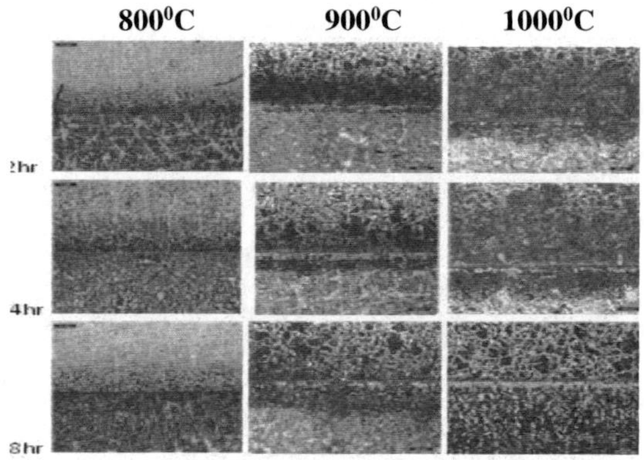

Figure-6.OM microphotographs of specimens after PWHT, after etching (50x)

IOP Conf. Series: Materials Science and Engineering **328** (2018) 012001 doi:10.1088/1757-899X/328/1/012001

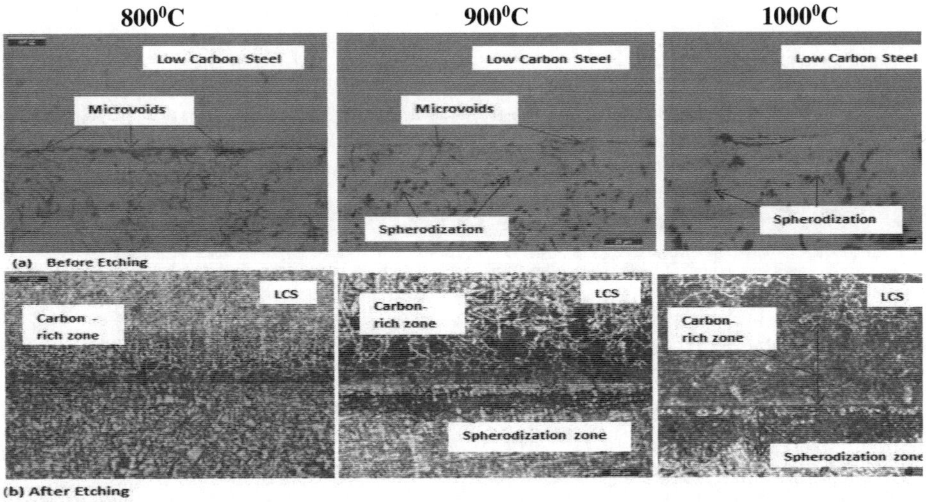

Figure-7.OM microphotographs' typical analysis of the interfaces and the diffusion layers of specimens treated at 800^0C, 900^0C and 1000^0C for 4 hours.

From the observations carried out on the micrographs of all the specimens (Figure-5 and Figure-6), the following inferences apply to all the PWHT diffusion couples:

1) Microvoids and interface lines are very much visible at the interfaces of the specimens treated at temperature of 800^0C, and becoming less visible at 900^0C, while at 1000^0C with longer treatment time, not only they almost disappear but the bond/weld also seems to be more complete. This phenomenon of elimination of the interfacial defects (voids) and the original interface at higher diffusion temperatures were also observed in the diffusion bonding of cast iron to medium carbon steel [11].
2) Further diffusion process has taken place with the increase of the diffusion layer thickness corresponding with increase of PWHT times and temperatures.
3) Spheroidization or dissolution of graphite flakes is observed on the cast iron sides close to the interface of specimens treated at 900^0C and 1000^0C but is not observed on specimens treated at 800^0C. This spherodization phenomenon was also observed by the researchers [1], in their study on the effect of high heating and cooling rate on interface of diffusion bonded gray cast iron to medium carbon steel.
4) A rapid diffusion of carbon and silicon from the grey cast iron to the low carbon steel are expected to take place.
5) As a result of the above, diffusion layers of spheroidization zone and carbon rich zone have formed near the interface of cast iron and low carbon steel sides respectively on specimens treated at 900^0C and 1000^0C, while at 800^0C only carbon rich zone has formed. For all specimens, these diffusion layers thicknesses increase with the increase of PWHT times and temperatures.

3.2.2. Thickness of diffusion layer. As stated above, examination of the microstructures in the microphotographs indicated that further diffusion process has taken place during PWHT and the increase of the diffusion layers thickness formed in each specimen of the diffusion couple, were very much influenced by the heat treatment temperature and time.
Changes in thickness with respect to time as observed from the microphotographs are shown in Figure-8, as time dependence of diffusion layer thickness produced for all diffusion couples.

IOP Conf. Series: Materials Science and Engineering **328** (2018) 012001 doi:10.1088/1757-899X/328/1/012001

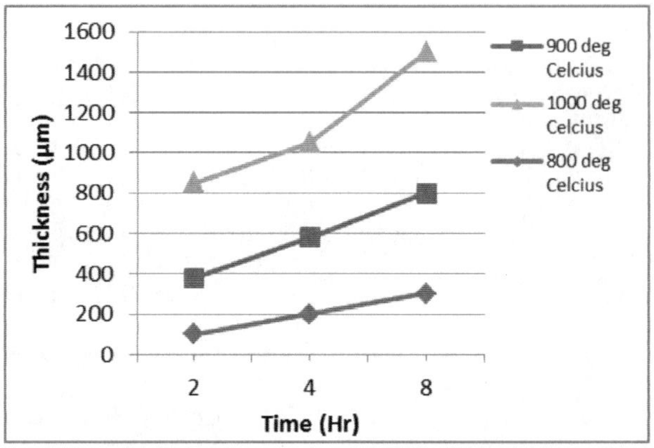

Figure-8. Time dependence of diffusion layer thickness

Some essential characteristics which could be derived from this form of time relationship of diffusion layer thickness are as follow:

1) A steady increase in thickness is observed as diffusion times consecutively increase from 2 to 8 hours was registered for all diffusions of 800^{0}C, 900^{0}C and 1000^{0}C temperatures.
2) A linear form of correlation between diffusion layer thickness and time could be deduced especially for 800^{0}C and 900^{0}C treatments.

The effect of heat treatment temperature on the diffusion layer produced was partly clearly displayed in the time related graph as mentioned above. Figure-9 shows directly the temperature dependence of the diffusion layer thickness, for all diffusion couple specimens.

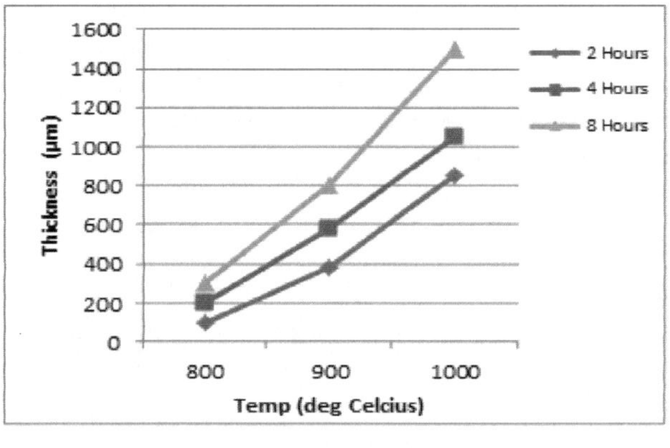

Figure-9. Temperature dependence of diffusion layer thickness

In this form, a further temperature role could be:

1) For all specimens, as the temperature rise from 800^{0}C to 1000^{0}C, a steady increase in diffusion layer thickness was observed.
2) A near quadratic form of correlation between diffusion layer thickness and temperature could be deduced especially for 8 hours treatment.

3.2.3. Microhardness across diffusion layer. Figure-10 (a), (b) and (c) show micro Vickers hardness gradients across the bonding interface of grey cast iron and low carbon steel diffusion couples after 2, 4 and 8 hours at 800^{0}C, 900^{0}C and 1000^{0}C annealing heat treatment; of which microstructures are shown in the Figure-5 and Figure-6 and the related characteristics are as described above.

IOP Conf. Series: Materials Science and Engineering **328** (2018) 012001 doi:10.1088/1757-899X/328/1/012001

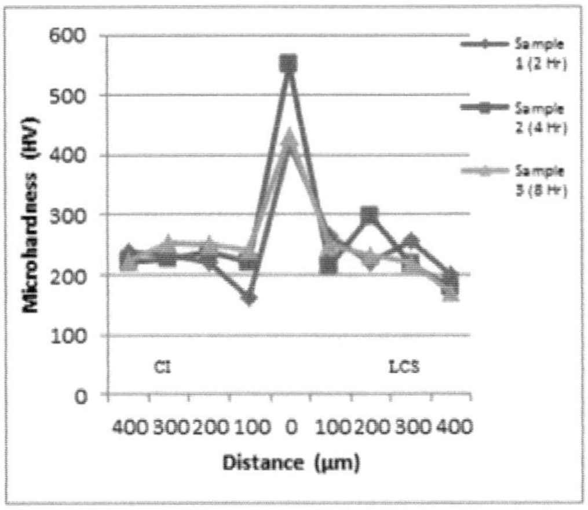

Figure-10(a). Microharness profile across the interface of diffusion couples after PWHT (800^{0}C).

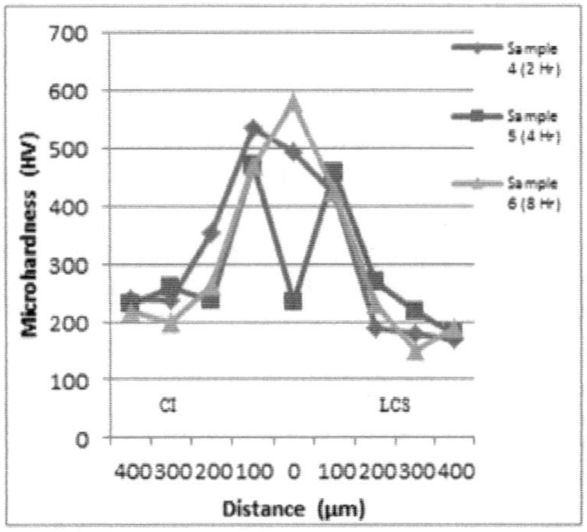

Figure-10(b). Microharness profile across the interface of diffusion couples after PWHT (900^{0}C).

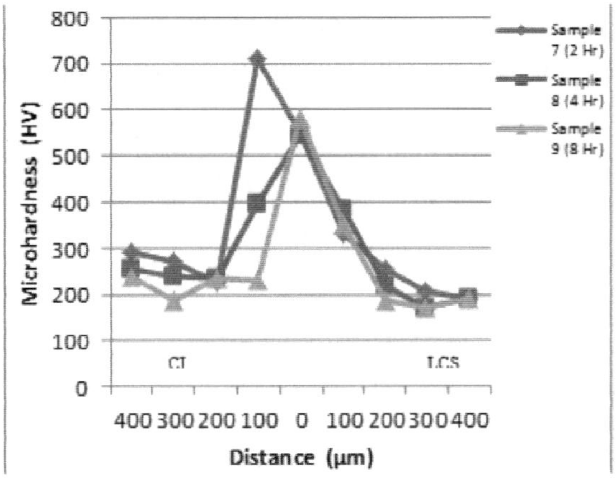

Figure-10(c). Microharness profile across the interface of diffusion couples after PWHT (1000^{0}C).

From these hardness test results Figure-10 (a), (b) and (c), a few specific comments can be deduced:

1) The hardness of the diffusion layer produced in all diffusion couple's specimens is much higher than the hardness of the respective base metals at all treatment time and temperatures.
2) The maximum hardness was observed close to the interface on the both sides of cast iron and low carbon steel. This corresponds to the area of diffusion layers of spheroidization zone and carbon rich zone, especially for 900^{0}C and 1000^{0}C treatments.
3) The above high hardness value can be related to the presence of high amount carbides. According to the researchers [12], increasing hardness value is due to higher the temperature and longer times available for carbon to diffuse.

4. Conclusion

A few diffusion couples of grey cast iron and low carbon steel were successfully produced by diffusion welding techniques after earlier failures. The micrographs examination of few samples of these diffusion couples exhibit fairly good bonding along the interface of the welded joint with microvoids still present. The strength value observed was indeed very low and the fracture was brittle in nature and it is suspected that the joint/weld was incomplete.

Specimens of these successful diffusional welded joints (despite being of low quality joint) were then being subjected to post heat treatment to allow further diffusion process to take place in the diffusion couples.
From the results and the discussions of the metallographic examinations and hardness testing of the specimens after post heat treatment as described above, some conclusions could be derived as follow:
(1) Temperature was observed as the main variable that controls the bond formation and it is necessary to have diffusion above the A3 temperature to eliminate the interfacial voids and for the bond interface to disappear as well for a complete bond.
(2) Heat treatment time and temperature were found to have a strong influence on the structure, thickness, composition and hardness of the produced diffusion layer.
(3) New phase or phases may have been formed in the diffusion layers. The natures of these phases are yet to be determined but they could be assumed to have the following properties: high hardness.
(4) Graphite flakes act as a source of carbon to the austenitic matrix that surrounds it. Their dissolution instigates the break-up and spheroidization.

Reference

[1] B. Kurt, N. Orhan and A. Hasçalık 2007 Effect of high heating and cooling rate on interface of diffusion bonded gray cast iron to medium carbon steel, *Materials & Design*, Volume 28, p. 2229-2233.
[2] N. Fujii, H. Suzuki, K. Yasuda, J. Takahashi 2006 "Comparison of strength characteristics of Cast Iron/Mild Steel welds produced by various welding processes" *Welding International*, 20 (2) 127-136.
[3] Tadashi Momono, Fauzuddin Ayob, Hiroshi G. Katayama, Ichizo Sakura, Kunihiko Matsuo 1990 "Diffusion Bonding of Cast iron to Steel under Atmospheric Pressure", *Casting Science and Technology*, The Japan Foundry men Society, Sap. 25-26, P. 72.
[4] N.F. Kazakov 1985 "An Outline of Diffusion Bonding in Vacuum", Diffusion Bonding of materials, *Pergamon Press*, P. 11.
[5] Airu Wang, Osamu Ohashi and Kenji Ueno 2006 Effect of surface asperity on diffusion bonding. *Material Transactions*, Vol. 47, No.1, pp179 to 184.
[6] Ladislav Kolarik, Jiri Janovec, Marie Kolarikova, Pavel 2015 Influence of Diffusion Welding Time on Homogenous Steel Joints, *Procedia Engineering* 100,1678-1685.
[7] Fauzuddin bin Ayob, Patthi bin Hussain, Aziz bin Abdullah and Abdul Aziz bin Abdul Rahim 2017 Preliminary Study on Diffusion Welding of Marine Grades Grey Cast Iron to Low Carbon Steel, *ARPN Journal of Engineering and Applied Sciences*, Vol. 12, No. 10.
[8] Awfa Abdul-Rassol Abdullah and Rasha Rahman Rawhdan 2014 " Study on Bond Strength of Bonded Gray Cast Iron," *Journal of Al-Nahrain University*, Vol.17 (2), pp.94-100.
[9] Yuting Lv, Liqiang Wang , Xiaoyan Xu and Weijie Lu 2015 Effect of Post Heat Treatment on the Microstructure and Microhardness of Friction Stir Processed NiAl Bronze (NAB) Alloy, *Metals*, 1695-1703; doi:10.3390/met5031695.
[10] Fauzuddin Ayob, Abdul Aziz Abdul Rahim and Syajaratunnur Yaakup 2015 Diffusion Bonding of Nickel to Iron and Steels to Examine The Influences of The Process Parameters and Carbon Contents on Joints Mechanical and Metallurgical Properties, *Marine Frontier@Unikl, Mimet's Technical Bulletin* Volume 6 Edition 2.
[11] F. A. Calvo, A. Ureña, J. M. Gomez de Salazar, F. Molleda 1989 Diffusion bonding of grey cast iron to ARMCO iron and a carbon steel, *Materials Science*, Volume 24, Issue 11, pp 4152-4159.
[12] Dr. Awfa Abdul-Rassol Abdullah and Rasha Rahman Rawhdan 2014 " Microstructure Characteristics in the Interface Zone of Gray Cast Iron Solid State Bonds" *Eng. & Tech. Journal* ,Vol.32,Part (B), No.1.

Synthesizing the Nanocrytalline Cobalt-Iron Coating Through The Electrodeposition Process With Different Time Deposition

Nik Rozlin Nik Masdek[1,a], Mohd Sorfian Hafiz Mansor[2,b], Zuraidah Salleh[3,c], Koay Mei Hyie[4,d]

Universiti Teknologi MARA, Faculty of Mechanical Engineering, 40450 Shah Alam, Selangor Darul Ehsan, Malaysia.

[a]*nikrozlin@yahoo.com*, [b]*pianhafiz@yahoo.com*,

Abstract. In the engineering world, electrodeposition or electroplating has become the most popular method of surface coating in improving corrosion behavior and mechanical properties of material. Therefore in this study, CoFe nanoparticle protective coating has been synthesized on the mild steel washer using electrodeposition method. The electrodeposition was conducted in the acidic environment with the pH value range from 1 to 2 with the controlled temperature of 50°C. The influence of deposition time (30, 60, 90 minutes) towards characteristic and properties such as particle size, surface morphology, corrosion behavior, and microhardness were studied in this investigation. Several results can be obtained by doing this experiment and testing. First, the surface morphology of Cobalt Iron (CoFe) on the electrodeposited mild steel washer are obtained. In addition, the microhardness of the mild steel washer due to the different deposition time are determined. Next, the observation on the difference in the grain size of CoFe that has been electrodeposited on the mild steel plate is made. Last but not least, the corrosion behavior was investigated. CoFe nanoparticles deposited for 30 minutes produced the smallest particle size and the highest microhardness of 86.17 and 236.84 HV respectively. The CoFe nanoparticles also exhibit the slowest corrosion rate at 30 minutes as compared to others. The crystalline size also increases when the time deposition is increased. The sample with 30 minute depositon time indicate the smallest crystalline size which is 15nm. The decrement of deposition time plays an important role in synthesizing CoFe nanoparticles with good corrosion resistance and microhardness. CoFe nanoparticles obtained at 30 minutes shows high corrosion resistance compared to others. In a nutshell, it was observed that the decrement of deposition time improved mechanical and corrosion properties of CoFe nanoparticles.
Keywords— Corrosion, time deposition, electrodeposition, CoFe, mild steel

1.Introduction

Surface and coating technology is a technology that has been used in many engineering applications. The surface coating is used to modify and improve the surface properties of materials for protection in demanding contact conditions or in aggressive environment or for enhanced functional performance. In other words, the surface coating is used to increase the lifetime of components exposed to corrosiveconditions. High mechanical and corrosion protective properties of nanoparticle alloy coating are needed for the design and operation of devices, machine, and structural systems in extreme environments [1].

Electrodeposition is exceptionally versatile and the valuable application keeps being invented. The added value of this method is cost-effective and less time-consuming if compared to physical methods such as sputtering or evaporating. Moreover, by controlling deposition parameters such as deposition time, it is possible to achieve a remarkable change in the chemical composition, mechanical and corrosion properties of electrodeposited materials [2].

In this investigation, the effect of the time deposition on the corrosion behavior and microhardness of CoFe deposited mild steel washer were investigate.

2. Literature Review

A. Factors that Influenced The Rate of Corrosion

The influencing factors of the corrosion can be categorized into two types of factors which are the external factor and internal factor. The external factor is more likely related to the properties of the environment, for instance, temperature, pressure, chemical and the concentration, flow rate, the presence of the stray electric current and the presence of the dissimilar metals. Those are the parameters that can influence the rate of the corrosion. Meanwhile for the internal factor, it is commonly related to the metal properties itself. The examples of the properties are, chemical content, imperfections, structure, presence of stress and heat treatment. Therefore to combat this particular problem, surface modification of the metal material is necessary to protect them against various types of degradation [3].

B. The Effect of The Time Deposition on The Electrodeposited Material's

The previous study was done by using different time deposition (30mins, 90mins, 120mins,) in order to investigate the effect of the Cobalt-Nickel-Iron nanoparticles coated on stainless steel substrate. The crystallite sizes of Co-Ni-Fe nanoparticles were slightly increased when the deposition time was longer. One of the purposes of the electrodeposition process is to prevent the corrosion. In term of the corrosion rate, it can be observed that the increment of the deposition time will increase the rate of corrosion. In this case, Co-Ni-Fe prepared at 30 mins deposition time showed the lowest corrosion rate compared to the 60 mins and 90 mins of deposition time [4]. It can be concluded that the smaller the particle size the higher the corrosion resistance of the material. If we look into the microhardness properties, it was found that the microhardness for the 90 mins deposition time is lower compared to those of 30mins and 60mins. Hence, the lower the deposition time, the higher the microhardness. If we relate this phenomenon to the particle size, we can say that smaller particle size will create a greater number of particle boundaries. These particles boundaries actually act as the barrier to stop the dislocation motion. In term of surface roughness, it was found that surface roughness is increased if the deposition time is increased [4].

C. Nanoparticle Materials

Nanoparticle materials are the material that have extremely small particle in heir microstructure. The size of nanoparticle material could range from 1 to 100 nanometers in size. In order to enhance the mechanical and physical behavior of nanoparticle material, we need to change the particle into nano size. In term of the corrosion rate, the corrosion rate will decrease with the decrement of the particle size. Small particle size has a high volume friction fraction of particle boundaries. High volume boundaries will result in better physical and mechanical properties. Other than that, the hardness also has similar relation to the particle size. The material is said to be harder if it has smaller particle size. When we talk about the hardness, the parameter that influence the level of hardness of a certain material is the density of particle boundaries. Smaller particle size has high density of the particle boundaries which is good to disrupt the propagation of dislocations [5].

3. Methodology

In this experiment, the specimen used was a mild steel washer with a dimension of approximately 2.5 cm in diameter. The sulphate bath was prepared in order to conduct the electrodeposition process of the CoFe nanoparticles on the mild steel washer. Figure 1 shows the mild steel bolt that is used in this investigation.

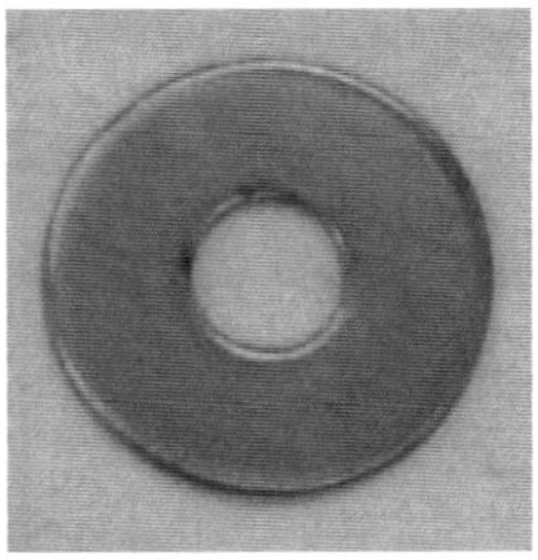

Fig. 1. Mild steel washer used in this experiment

The experiment begins with the preparation of the electrolyte bath before electrodeposition process takes place. The bath is a mixture of cobalt sulphate, iron sulphate, boric acid, saccharine and also sodium chloride. Table 1 shows the mass of the compound in the bath composition. According to the previous study, boric acid and saccharine were added in the electrolyte as a pH buffer and also functioning as grain refinement agent [5]. All the substances are mixed together to form a solid chemical compound. The mixing agent used in this experiment is the distilled water and the mixing process is conducted in the beaker and the volume of distilled water is 500ml. Then, the plating bath was constantly stirred using a magnetic stirrer to ensure proper mixing of the electrolyte solution. The temperature was controlled at 50°C while the current density was maintained at 1.1 A/cm^2. The electrodeposition process starts with the immersing the prepared mild steel washer (cathode) into the bath parallel with the graphite rod electrode (anode).

Each experiment was carried out with a freshly prepared solution. In this experiment, the manipulated parameter is the time deposition while others parameters were kept constant. The time deposition used were 30, 60, 90 minutes. After completing the electrodeposition process, the specimen was rinsed with distilled water and then dried room temperature. The characteristic test and corrosion behavior study were made after the electrodeposition process is done.

The equipments used to investigate the characterization of CoFe coating were ULTIMA IV FD 3668N, X-ray diffractometer (XRD), and the microstructure of the CoFe coating is synthesized by Scanning Electron mIcroscope (SEM). The hardness of the coating sample was measured with a MITUTOYO

MVK-H1, Vickers microhardness tester. Meanwhile, to study the corrosion behavior, Salt Spray (fog) Test was conducted according to ASTM B117.

TABLE 1. Bath composition of CoFe deposition

Bath composition	Mass of compound (g)
Cobalt Sulphate	7.027
Iron Sulphate	2.78
Sodium Chloride	8.244
Boric Acid	2.0
Saccharine	0.6836

4. Results and Discussion

D. Morphology and Microstructure

The morphology and microstructural characterizations were observed by using Scanning Electron Microscope (SEM). Figure 2 shows the grain structure of the CoFe nanoparticles at different time deposition. From the finding, it can be observed that all the three samples of electrodeposited mild steel washer produced the dendritic structure, but still has the differences observed between samples. Based from the microstructure, it can be observed that the CoFe coating prepared at 30 minutes produced a compact and full structure. The coating prepared at 30 minutes also present smaller size and lower in agglomeration [6]. It also shows no obvious voids existing on this microstructure. Next in the CoFe coating prepared at 60 minutes, the structure is also almost compact and full but there were some voids existed and it can be seen on the microstructure. It is also noted that the particle size was larger and almost combined with each other to form agglomerates [5]. The previous study clarified that the longer the deposition time, the larger the particle size [2]. Lastly, for the CoFe prepared at 90 minutes, the presence of the voids increases. It also can be noted that the particle size of this sample was larger. The presence of the voids on the microstructure of the sample is believed to come from the uncoated CoFe nanoparticle. It can be indicated that the coating is not fully covered during the electrodeposition process. The formation of the voids might be due to the existence of the oxide in the space between grains and also may be due to the agglomeration process covering the sample surface. By relating the microstructure and the corrosion resistance, it can be concluded that the presence of voids and the larger the particle size that contribute to the formation of the agglomeration will decrease the corrosion resistance [4].

The particles size for CoFe coating is determined from the measurement of the particles was taken by measuring the average grain size of at least 5 grains. The particle size of CoFe coating for 30mins, 60mins, and 90mins are 3.29μm, 6.72μm, and 7.68μm respectively. Based on the observation, the smallest grain size was found in the sample with 30 mins of deposition time. Meanwhile, the largest grain size is obtained from the sample with 90 mins of deposition time. Based on previous study [6], coating with nanocrystalline grain size often demonstrated excellent chemical, mechanical and physical properties. These characteristic are resulted from the reduction in its grain size in addition to the existence of a huge numbers of grain boundaries in the microstructure. To summarize this finding, the grain size is increased when the time of deposition increase.

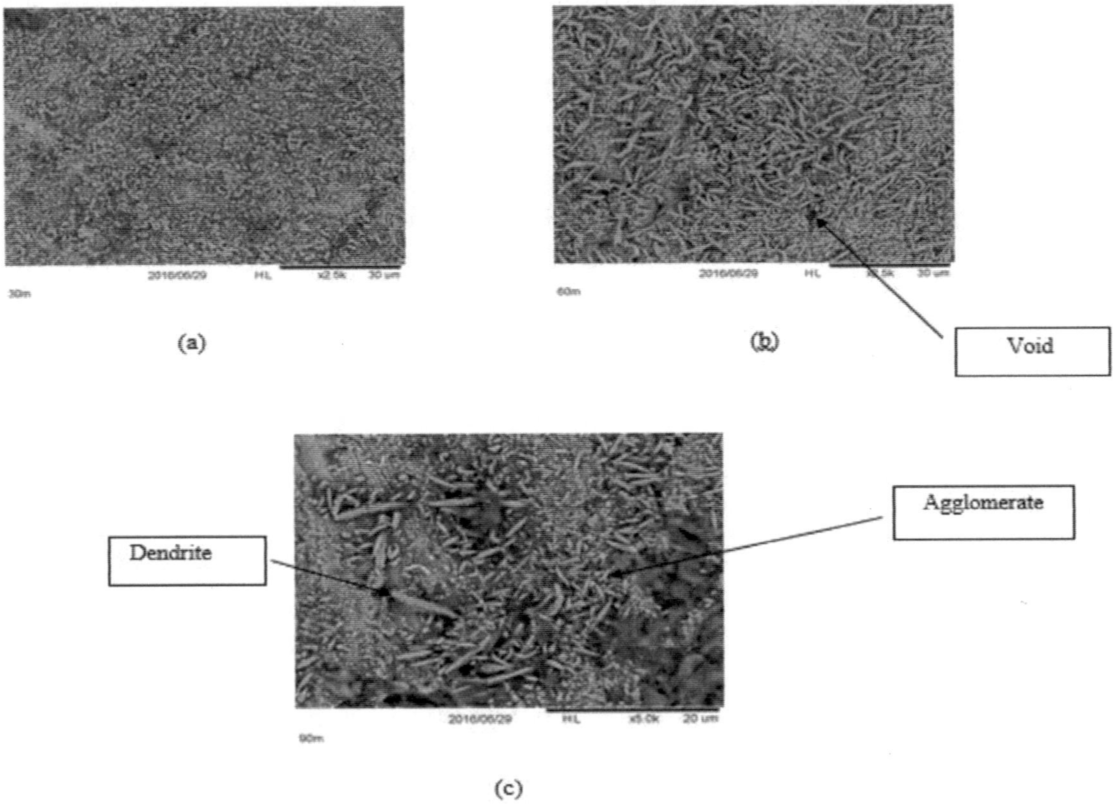

Fig. 2. SEM micrograph of CoFe at (a) 30mins (b) 60mins (c) 90min

E. Microhardness

Microhardness test was conducted by taking the average measurement by using Vickers microhardness. Five measurements were performed on the different spots on the sample and the average measurement was calculated. Figure 3 shows the effect of the microhardness of the CoFe coating based on different deposition time. From figure 3, it can be seen that the highest value of microhardness was showed by the sample with deposition time of 30mins. Meanwhile, the average microhardness of CoFe nanoparticles prepared at 90 minutes was found to be the lowest than the samples prepared on 30 and 60 minutes.

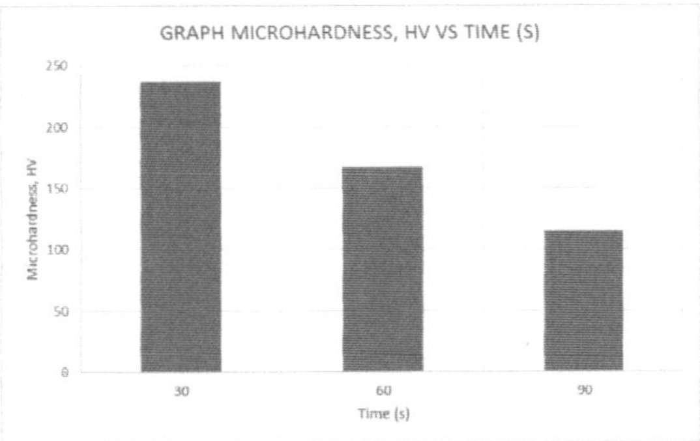

Fig. 3. The relationship between time deposition and microhardness

By referring to the finding above, it can be concluded that the microhardness of the CoFe nanoparticle decreased with the increment of the deposition's time. As we can see, the final value of the microhardness of the CoFe nanoparticles at 30, 60, 90 minutes are 236.84 HV, 166.74 HV, and 118.64 HV respectively. The increment and decrement in the nanoparticles hardness are related to the particle size of nanocrystalline material itself. Particles size plays a major role that contributed to the microhardness behavior of the CoFe coated mild steel washer. Figure 4 shows the correlation between the microhardness and particle size of the CoFe nanoparticles prepared at different deposition time. The microhardness of the CoFe nanoparticles prepared at 30 minutes seems to be higher compared those with 60 mins and 90 mins of deposition time. The highest microhardness of the CoFe nanoparticles deposited at 30 minutes is believed to be affected by the formation of the numerous grain boundaries in the microstructure. The formation of the grain boundaries has caused the reduction of particle size [7]. Previous study clarified that the decreasing in the particle size will result in the increasing of the microhardness of the nanocrystalline material [8]. In addition, the stoppage of dislocation motion or the changing dislocation direction occurs with the formation of grain boundaries as it acts as hindrance. This will result in the hardening of the material and the plastic deformation in the material can be prevented due to the restricted motion of dislocation in the nanoparticles. As the conclusion, the CoFe nanoparticles coated prepared at 30 minutes of deposition time has the greatest value of hardness compared to the other samples.

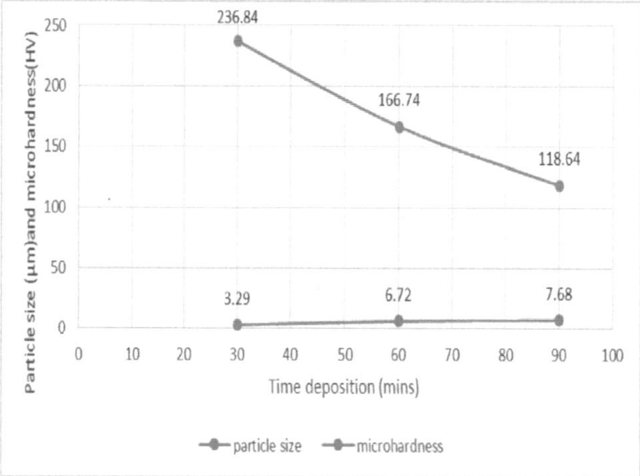

Fig. 4. The correlation between microhardness and particles size.

F. Phase and Crystallographic Structure

XRD measurements of as-synthesized Co-Fe coatings were carried out from 2θ angle of 10° to 90° angle using Cu radiation. Figure 5 to 7 shows the XRD patterns of the electrodeposited Co-Fe coatings at different deposition times of 30, 60, and 90 minutes.

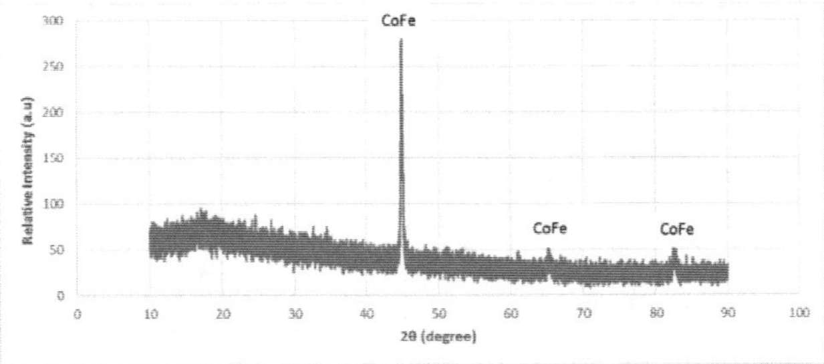

Fig. 5. The XRD pattern for CoFe nanoparticles (30minutes)

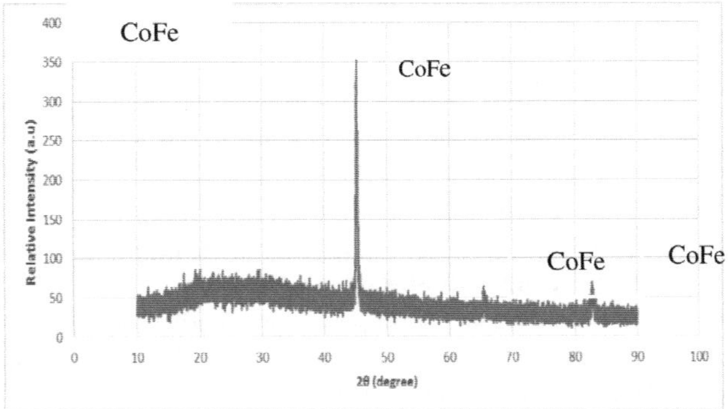

Fig. 6. The XRD pattern for CoFe nanoparticles (60minutes)

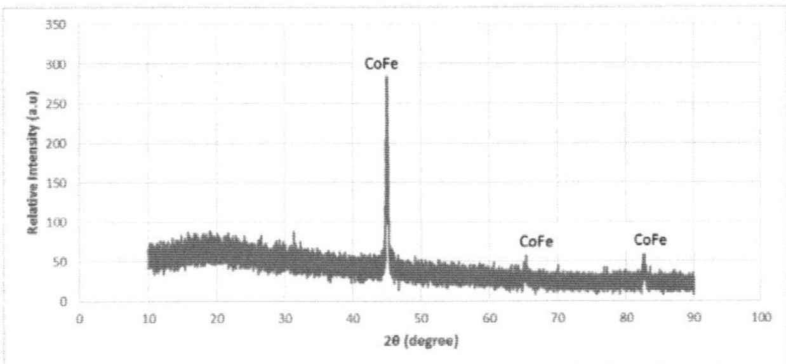

Fig. 7. The XRD pattern for CoFe nanoparticles (90minutes)

The XRD spectrum for the Co-Fe phase reveals the characteristic peaks at angles of 45°, 65°, and 82°. The Co-Fe phase is identified as the BCC crystal structure with JCPDF No. 030657519 [4,14]. From the figures above, it can be seen that there are differences between the peaks or the relative intensity. Based on the previous study, it clarifies that the difference between the XRD peaks intensity is related to the morphology. Morphology is fundamentally different among the samples due to the difference of crystal face. This study also stated that, crystal morphology changes based on the reaction conditions such as pH, additives and reaction time [9]. The calculation of the crystalline size is made by using the Debye Scherrers' equation. Theoretically, the crystallite size can be calculated using Debye Scherrers's equation assuming that there is no crystal defect and voids [10, 11]. Based on the calculation, the values obtained for 30, 60, 90 minutes of the deposition time are 15nm, 22nm, and 22.3nm respectively. Table 2 summarized the value of crystalline size of the CoFe nanoparticles regard to its deposition times. From the result, it can be concluded that the increment of the crystalline size is occurred with the increment of the time deposition [12].

TABLE 2. Crystallite size of CoFe nanoparticles.

Time deposition (minutes)	Crystallite size (nm)
30	15
60	22
90	22.3

G. Salt Spray (Fog) Test

The Salt Spray (fog) Test is an accelerated corrosion test. All the specimens were placed in enclosed salt spray chamber and being exposed to a continuous indirect spray of salt water solution. This corrosion test was conducted according to the ASTM B117 standard by dissolving 265g sodium chloride (NaCl) into 5000 ml of distilled water. The sample is placed in the chamber for 24 hours before the samples were taken out. Figure 8 shows the result of the corrosion formation after 24 hours of the salt spray (fog) test. The first sample, which is the 30 minute of deposition time, demonstrates a small area of corrosion. The light brown color corrosion only takes place at the side part of the washer. The center part of the washer shows no obvious corrosion takes place and the coating is still intact with the sample. For the second sample, which is 60 minutes of deposition time, the sample shows several light reddish brown areas of corrosion. The corrosion formation did not covered all the surface of the sample as there certain small area on the sample is still intact with the coating. Meanwhile, for the third sample which is 90 minutes of deposition time, it can clearly be seen that the washer demonstrates a large reddish brown area of corrosion compared to the other two samples. The corrosion fully forms on the surface of the sample. Based on this result, after these three samples undergo the salt spray (fog) test, the 30 minutes deposited washer shows the best corrosion resistance compared to others samples. Light brown area of corrosion is barely found on the surface of the sample. The CoFe coating also still remains after 24 hours of the salt spray (fog) test. Therefore it can be concluded that, the area of the corrosion increase with the increment of deposition times [13]. Other than that, it can also be concluded that the corrosion resistance is increased as the time deposition decreases [14].

Sample	After Salt Spray (fog) Test
CoFe (30 minutes)	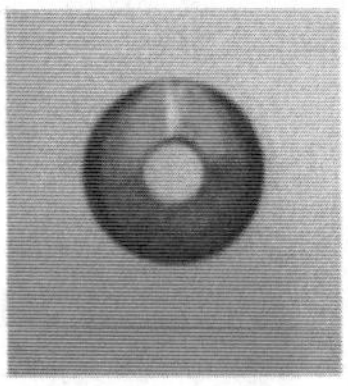
CoFe (60 minutes)	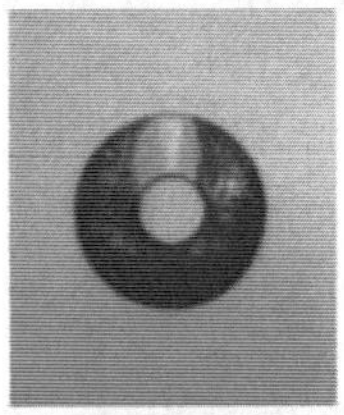
CoFe (90 minutes)	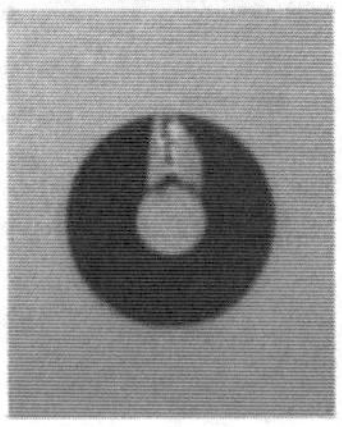

Fig. 8. Sample after the salt spray (fog) test

5. Conclusion

The synthesizing of nanocrystalline Cobalt-Iron (CoFe) coating through the electrodeposition process with the different deposition time is successfully done. The various deposition times (30-90 minutes) have a huge effect on the surface morphology, corrosion behaviour and microhardness of the CoFe nanoparticles. It can be seen that from the result, has influenced the microhardness of CoFe nanoparticles. The sample with the lowest deposition time which is 30 minutes showed the highest average of microhardness (236.84 HV) and lowest particles size (3.29μm). In addition, the crystallite size of the CoFe coating for the sample prepared on 30 minutes of deposition time gives the smallest size (15nm). Meanwhile, the corrosion behavior of the CoFe nanoparticle was accelerated with the increment of the deposition time. From this study, it can be concluded that the deposition time plays a major role in the synthesizing process of the CoFe nanoparticles. From this study also, it can be said that the most efficient deposition time in order to produce the CoFe nanoparticles with magnificent corrosion resistance and microhardness is 30 minutes. Therefore, it is important to synthesize the CoFe nanoparticles at a shorter time in order to obtain a remarkable corrosion behavior and hardness.

Acknowledgment

The authors gratefully acknowledge the help of the Research Management Centre (RMC) and Ministry of Higher Education (MOHE) Malaysia in providing the Fundamental Research Grant Scheme (Project Number: 600-RMI/FRGS 5/3 (80/2015), 600-RMI/RAGS 5/3 (37/2015), and 600-RMI/DANA 5/3/LESTARI (67/2015) research grants and Universiti Teknologi MARA for their support.

References

[1] N.M Nik Rozlin, Akram M. Alfantazia, "Nanocrystalline cobalt–iron alloy: Synthesis and characterization", (Material Science and Engineering)Feb. 2012, pp. 388-394

[2] Ken Osborne, " Electroplating, Feb 2012.

[3] V. Chawla, D. Puri, S. Prakash, and B. Sidhu, "Hard Coatings on ASTM-SA213-T-22 Boiler Steel," vol. 5, no. 3, pp. 247–253, 2011.

[4] Nor Azrina Resali, Koay Mei Hyie , M. N. Berhan, Zuraidah Salleh, Salmiah Kasolang, "Cobalt-Nickel-Iron Nanoparticles Coated on Stainless Steel Substrate", (Procedia Engineering),2013, pp. 30-36

[5] N. A. Resali, K. M. Hyie, W. N. R. Abdullah, M. a. a. Ghani, and a. Kalam, "The Effect of Bath Ph on the Phase Formation of Ternary Co-Ni-Fe Nano-Coatings," Appl. Mech. Mater., vol. 391, pp. 9–13, Sep. 2013.

[6] K. M. Hyie, A. Ahmad, N. A. Resali, M. F. Munir, C. S. Li, and S. Saidin, "Corrosion Study of Electrodeposited Co-Ni-Fe Protective Coating on Electroless Nickel Immersion Gold (ENIG) Flexible Printed Circuit," Procedia Technol., vol. 15, pp. 793-798, 2014.

[7] K. M. Hyie, N. A. Resali, W. N. R. Abdullah, and W. T. Chong, "Synthesis and characterization of nanocrystalline pure cobalt coating: Effect of pH," Procedia Eng., vol. 41, no. Iris, pp. 1627–1633, 2012.

[8] N. A. Resali, W. N. R. Abdullah, K. M. Hyie, A. Kalam, N.R. Nik Roselina, C.M. Mardziah , "The Influence of Bath Concentration on Particle Size of Cobalt Nanoparticles", Applied Mechanics and Materials, Vol. 393, pp. 140-145, Sep. 2013.

[9] Suryanarayana, "The structure and properties of nanocrystalline materials: Issues and concerns," Jom, vol. 54, no. 9, pp. 24–27, Sep. 2002.

[10] W. N. R. Abdullah, K. Mei Hyie, N. A. Resali, and W. T. Chong, "Effect of Time Depositions on Electrodeposited Cobalt-Iron Nanocoating," Adv. Mater. Res., vol. 576, pp. 565–568, 2012.

[11] B. Y. Yoo, S. C. Hernandez, D. Y. Park, and N. V. Myung, "Electrodeposition of FeCoNi thin films for magnetic-MEMS devices," Electrochim. Acta, vol. 51, no. 28, pp. 6346–6352, 2006.

[12] Mikiyasu Inoue, Izumi Hirasawa 'The relationship between crystal morphology and XRD peak intensity on CaSO4.2H2O", pp. 169-175, 2013

[13] K. M. Hyie, N. A. Resali, and W. N. R. Abdullah, "Study of Alloys Addition to the Electrodeposited Nanocrystalline Cobalt," Adv. Mater. Res., vol. 486, pp. 108-113, Mar. 2012.

[14] Y. J. Lee, S. C. Wu, J. H. Chen, M. T. Yeh, K. M. Lin, and H. C. Lin, "Effects of Cold-Spray Coatings on the Corrosion Properties of AZ80 Magnesium Alloy," Key Eng. Mater., vol. 573, pp. 43–48, 2014.

[15] Y. J. Lee, S. C. Wu, J. H. Chen, M. T. Yeh, K. M. Lin, and H. C. Lin, "Effects of Cold-Spray Coatings on the Corrosion Properties of AZ80 Magnesium Alloy," Key Eng. Mater., vol. 573, pp. 43–48, 2014.

[16] K. M. Hyie, W. Normimi, R. Abdullah, and N. A. Resali, "Electrodeposited Cobalt and Its Alloys Nanocoating : A Brief Review," pp. 43–47, 2012.

[17] N. A. Resali, W. N. R. Abdullah, K. M. Hyie, A. Kalam, N.R. Nik Roselina, C.M. Mardziah , "The Influence of Bath Concentration on Particle Size of Cobalt Nanoparticles", Applied Mechanics and Materials, Vol. 393, pp. 140-145, Sep. 2013.

IOP Conf. Series: Materials Science and Engineering **328** (2018) 012003 doi:10.1088/1757-899X/328/1/012003

Additive Manufacturing Technology for Biomedical Components: A review

Haizum Aimi Zaharin [a, b] *, Ahmad Majdi Abdul Rani [b], Turnad Lenggo Ginta [b], Farooq I Azam[b]

a Centre for Intelligent Signal and Imaging Research, Universiti Teknologi Petronas, Malaysia

b Department of Mechanical Engineering, Universiti Teknologi Petronas, Malaysia

haizum_15001997@utp.edu.my

Abstract. Over the last decades, additive manufacturing has shown potential application in ranging fields. No longer a prototyping technology, it is now being utilised as a manufacturing technology for giant industries such as the automotive, aircraft and recently in the medical industry. It is a very successful method that provides health-care solution in biomedical sectors by producing patient-specific prosthetics, improve tissues engineering and facilitate pre-operating session. This paper thus presents a brief overview of the most commercially important additive manufacturing technologies, which is currently available for fabricating biomedical components such as Stereolithography (SLA), Selective Laser Sintering (SLS), Selective Laser Melting (SLM), Fused Deposition Modelling (FDM) and Electron Beam Melting (EBM). It introduces the basic principles of the main process, highlights some of the beneficial applications in medical industry and the current limitation of applied technology.

1. Introduction

AM is the process of joining materials, usually layer upon layer, to make objects from 3D model data, as opposed to subtractive manufacturing technologies [1]. This technology started originally in 1988 for producing prototype model and was known as rapid prototyping (RP). Within three decades, RP has evolved from simple 3D prototyping to complex structure known as rapid manufacturing (RM). As improvements to output quality from this technique and looking at the basic principle of fabricating parts using additive of material approach, the term RP and RM are used interchangeably with the term additive manufacturing [2]. Lately, AM technologies is progressively employed in medical field and slowly taking over the conventional method of producing patient-specific medical device. With further evolution of bio-cell printing AM, this technology has a potential to print the whole body organs within the next 20 years [3]. AM principle involves the process of cutting 3D drawing into 2D slices, where the data set is derived from original CAD data. 2D slices will be printed layer upon layer until a solid 3D model is formed. Essentially, AM is based on two-stage process: 1) virtual stage that includes modelling and simulating, requires a computer and 3D modelling software (CAD), 2) physical stage which is manufacturing, occupies layering material and machine equipment [4]. AM equipment reads the data from produced CAD files and defines the data by depositing successive layer of material in layers, to build 3D physical object [5].

IOP Conf. Series: Materials Science and Engineering **328** (2018) 012003 doi:10.1088/1757-899X/328/1/012003

2. Image Developing and Processing for Additive Manufacturing

Bio-model based on additive manufacturing technique designates one-to-one scale of the human anatomical part acquired from 3D medical imaging [6]. The source of the image should be cleared to produce a good medical model. Currently, image data comes from computed tomography scan (CT scan), magnetic resonance (MRI) or other imaging modalities. MRI scanner delivers the energy produced by water molecules with the pulse given by radio waves to generate volumetric data set. Due to the bone nature which is nearly void of water, bone image is left out as ingenerated data. Meanwhile, CT scan uses x-rays to generate images of human body and displays better contrast between bones and soft tissue. Subject to the bio-model to be formed, MRI images are more suitable to deal with soft tissues while CT images are usually used as the source image when dealing with bones. Next, unwanted images is executed and the region of interest is extracted using image processing software. There are few imaging software packages that are reliable in processing medical data, repositioning the fracture and partly designing the new model of medical prosthesis. For instance, MIMICS and 3-Matics software developed by Materialise [7]. This software is used in the production of hearing aids. The medical images in DICOM (Digital Imaging and Communication in Medicine) format (.dcm) will be loaded into the software and selected data set can be opened in DICOM viewer software as 2D and 3D view. Model image can be reconstructed, smoothed leaned and repaired by modifying mathematical algorithm developed by the software. The edited version of model image is exported and saved into supported 3D printing format which are, Object file (.OBJ) or Standard Triangulation Language (.STL) file formats. Data is further modified and optimized by Computer-aided Design (CAD) software where the region of interest is selected and generated as 3D model. This data is sliced using RP Tools and sent to AM machine for fabrication [6], [8]. The usage of integrated systems to transfer DICOM into CAD had been studied [9] and successful results were achieved in the design and modification process using CAD.

3. Stereolitography

Stereolithography (SLA) is a form of three-dimensional printing technology also known as photo-solidification or resin printing. This technique is used to construct 3D parts using a layer by layer concept through a process called photopolymerization. In photopolymerization process, energy source usually optical light is used to scan over a vat of light-curable resin causing the molecules to link and solidify specific region on the liquid surface. The depth of material is increased gradually by moving the floor of vat steadily downward. Successively, cured resin in layers and solid object is formed. The surrounding resin does not provide mechanical stability; thus, a support structure is required for the forming of the overhanging layers (Figure 1). The fabricated 3D object is taken off by draining and washing-off excess resin. SLA part are semi-transparent and very good in allowing visual access especially to body cavities, for example marrow spaces, sinuses and neurovascular channel [10]. However, the model structure has considerably poor mechanical properties. It needs to go through post-curing process under the exposure of ultraviolet light to improve mechanical properties of the structure [11]. Typical material use in SLA are liquid photopolymer and composites.

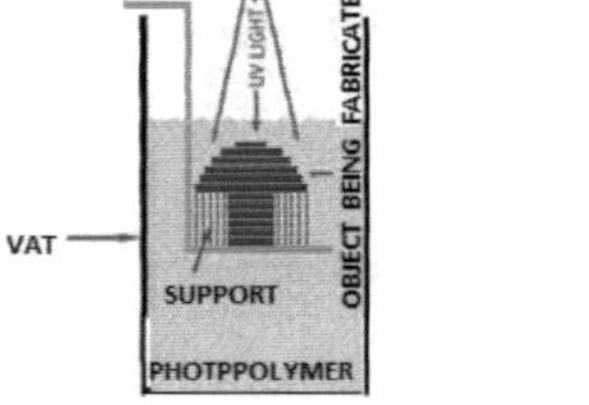

Figure 1. Stereolithography process diagram [12]

SLA products are reported to be very useful as an aid to diagnosis, teaching and surgical planning purpose [13]–[15]. With help from SLA technology, surgical duration and perioperative blood loss might decrease, while complication rate rremainthe same [16]. Surgeons also claimed that the preoperative planning is very effective [10] and minimizes the exposure of radiation during surgeries [17]. The SLA prototype is also used as a physical model to assist surgeons in better understanding of a complex fracture [18]. Besides that, SLA product can be used to fabricate patient-specific implant components. It was reported that there was no harmful effect shown under the dorsal skin of rats after one-month implantation period using implant prepared by SLA process [19]. SLA implant also has been reported to be an excellent method for treatment planning for maxillofacial trauma compared to conventional milling since it provides detail anatomical model. On the other hand, the disadvantages include the initial expense of SLA model such as cost for image processing and time taken for manufacturing and delivery [10].

4. Selective Laser Sintering (SLS) and Selective Laser Melting (SLM)

Selective Laser Sintering (SLS) is a form of AM which uses high power laser as the source of heat energy to sinter polymer powder on defined region and binding them to becoming three-dimensional solid model. During this process, fine polymer powder is heated with CO_2 laser causing the powders to sinter together. As illustrated in Figure 2, a roller is placed alongside one of the feed beds and the powder is delivered by raising the feed bed steadily. Then, the roller pushes the raised powder across, covering the model platform with a powder layer. After that the laser, directed to the model platform, starts tracing cross section of the designed digital object onto the powdered material. Immediately after the first layer is created, the model platform descending gradually. This process continues with the same procedure until the entire three-dimensional object has been constructed. Unlike SLA, this process is self-supporting, where it allows for parts to be built on other part and unused material. However, SLS disadvantages include rough surface finish and less detailed features as compared to SLA [20]. SLS shows promising application for bone tissue engineering as the porous scaffolds can be computationally designed and fabricated [21]. Numerous biomedical applications of SLS shows promising results in oral, maxillofacial, neurological surgery, orthopaedics and tissue engineering applications [6], [8], [14], [22], [23]. Wide variety of process material for SLS in medical are polymers, metal alloy, biomaterial metal-polymer and metal-ceramics mixture.

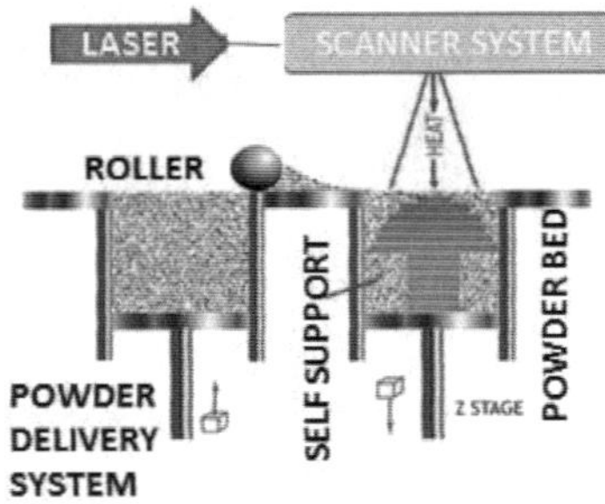

Figure 2. General concept of SLS/SLM [24]

Selective Laser Melting (SLM) basically is a subcategory of SLS. Instead of only fusing the metal powder to bond specific region, the laser is set up to fully melt the material and binding them in layers. Disadvantage of SLM process in comparison with SLS is that, it is not easy to control the production of surface tension because this process melts the material completely [22]. Common process materials for SLM are copper, aluminium, stainless steel and cobalt chrome. Application and advantages of SLM in medical have shown much progress over the years. It was mentioned that SLM maximizes the utilization of material because of the possibility to reuse un-melted powder [23]. In a scientific point of view, SLM can also solve aseptic loosening problem by controlling the mechanical properties of implant. It was employed SLM method to fabricate implants by modifying the mechanical properties that mimics nearly like the natural human bone by developing open porous implants [25].

5. Fused Deposition Modelling (FDM)

Fused Deposition Modelling (FDM) is an AM process that uses a heated nozzle process. FDM consist of build platform, extrusion nozzle and control system. Initially, plastic filament from the coil reel is fed into heated nozzle by driving motor. Once the material is transferred into the heated nozzle, molten material is then extruded onto base plate in layer form based on pre-design toolpath until final 3D product is produced (Figure 3).

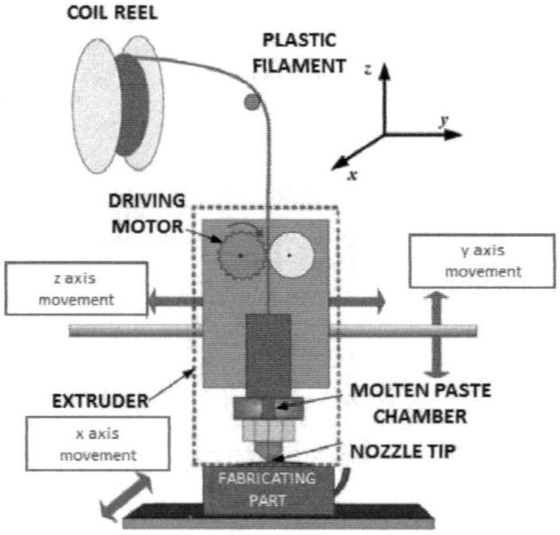

Figure 3. FDM working process [26]

FDM technique builds a functional model from available engineering thermoplastic such as investment casting wax, sulfones, ABS, polycarbonate and elastomers. FDM demands no special ventilation and involves no harmful chemicals by-products. It is also a good method of producing 3D parts with complex shapes. Nevertheless, this method suffers from disadvantages like long build time, poor surface finish and low mechanical strength [26]. It was investigated that FDM produced 3D scaffold with highly controllable channel size and porosity and fully interconnected channel, which can further be studied and modified to improve the mechanical strength of the manufactured model especially in bio-medical [27]. It was also demonstrated that the successful FDM fabrication of patient-specific implant with varying densities for cranial defect and femur part [28].

6. Electron Beam Melting (EBM)

Electron beam melting (EBM), like SLS and SLM use metal powder to fabricate metal parts. In contrary with SLS, EBM and SLM techniques achieved full melting for material to adhere. EBM works by melting metal powder with an electron beam in vacuum at a very high temperature (up to 1000°C). As a result, the EBM products have better mechanical properties (i.e.: strength, elasticity, fatigue, chemical composition, microstructure) than cast products and comparable with wrought material [29]. Since EBM takes place under vacuum condition, it is suitable to work with reactive materials that are highly reactive to oxygen such as titanium [30]. Vacuum environment in EBM also provides benefits such as maintaining a high-quality electron beam as collision between electrons and gas molecules is avoided, facilitating outgas impurities incorporated in the metal powder, and keeping good thermal insulation. EBM process starts with generated electron in electron beam column at the top of the vacuum chamber. High speed electrons are then accelerated in electrical field through two controlled magnetic fields. Stigmators and focus coils work as magnetic lens which help to focus beam into specified diameter while deflection coils redirect the focus beam to aimed region on the building table (Figure 4). Typical material used by EBM for medical field are commercial pure Titanium, Ti-6Al-4V, stainless steel, magnesium alloys, and nickel alloys. Currently, main advantage of EBM is on titanium near-net-shape parts for medical implants. Few papers presented that EBM had been found to be a suitable method for manufacturing of patient specific porous implants [31]–[33]. EBM was utilised to modify the mechanical properties of titanium hip stem to reduce stress shielding effect [35]. The study showed promising result to fabricate custom and complex design of orthopaedic implants, although building non-stochastic lattice structure with correct lattice struts orientation to achieve a good stress distribution is a real challenge.

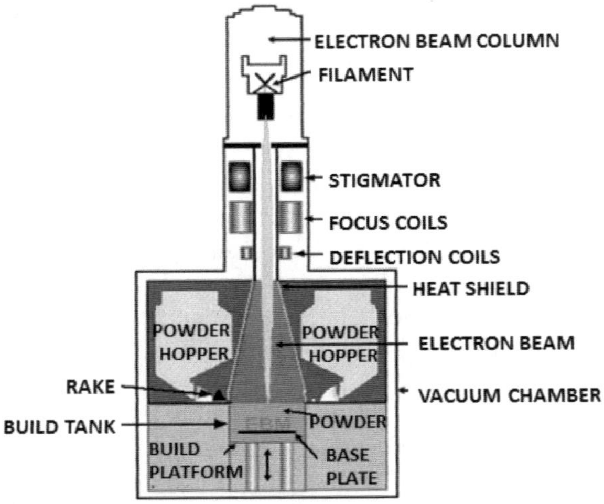

Figure 4. EBM working principle scheme [36]

7. Conclusion

AM technology in combination with medical imaging technique has shown to have a strong prospective for biomedical applications. This technique has successfully facilitated and improved the quality, duration of operating procedure and overcomes the problem of patient compromising the proper match of implant using individualised-tailor-made implant. Although AM technology has been greatly developed and improved, many limitations remain to be addressed. High cost of AM machines and materials, number of materials available, mechanical properties, size and accuracy of final parts produced are still the main challenges in adoption of AM in biomedical industry. However, AM is developing rapidly and is showing great potential in the future of biomedical manufacturing industry.

References

[1] ASTM International, "F2792-12a - Standard Terminology for Additive Manufacturing Technologies," *Rapid Manuf. Assoc.*, pp. 10–12, 2013.

[2] I. Gibson, D. Rosen, and B. Stucker, *Additive Manufacturing Technologies*. 2015.

[3] Royal Academy of Engineering, "Additive Manufacturing : Opportunities and Constraints," in *Additive Manufacturing: Opportunities and Constraints*, 2013, no. May 2013, p. 21.

[4] D. Eyers and K. Dotchev, "Technology review for mass customisation using rapid manufacturing," *Assem. Autom.*, vol. 30, no. 1, pp. 39–46, 2010.

[5] A. Rosochowski and A. Matuszak, "Rapid tooling: the state of the art," *J. Mater. Process. Technol.*, vol. 106, no. 1–3, pp. 191–198, 2000.

[6] J. W. Choi and N. Kim, "Clinical application of three-dimensional printing technology in craniofacial plastic surgery," *Arch. Plast. Surg.*, vol. 42, no. 3, pp. 267–277, 2015.

[7] Materialise, "AM software system and service provider." [Online]. Available: www.materialise.com.

[8] M. Moayedfar, "MANUFACTURE OF MEDICAL ORTHOPAEDIC IMPLANTS USING," vol. 11, no. 1, pp. 177–180, 2016.
[9] F. S. Gayzik, D. P. Moreno, C. P. Geer, S. D. Wuertzer, R. S. Martin, and J. D. Stitzel, "Development of a full body CAD dataset for computational modeling: A multi-modality approach," *Ann. Biomed. Eng.*, vol. 39, no. 10, pp. 2568–2583, 2011.
[10] D. B. Powers, W. A. Edgin, and L. Tabatchnick, "Stereolithography: a historical review and indications for use in the management of trauma.," *J. Craniomaxillofac. Trauma*, vol. 4, no. 3, pp. 16–23, 1998.
[11] F. P. W. Melchels, J. Feijen, and D. W. Grijpma, "A review on stereolithography and its applications in biomedical engineering," *Biomaterials*, vol. 31, no. 24, pp. 6121–6130, 2010.
[12] flickr, "Stereolithography." [Online]. Available: https://www.flickr.com/photos/organ_printer/878476150.
[13] C. K. Chua, K. F. Leong, and C. S. Lim, "Rapid Prototyping: Principles and Applications," *Building*, p. 512, 2010.
[14] T. M. Wong *et al.*, "Or thopaedic Surger y The use of three-dimensional printing technology in orthopaedic surgery : A review," vol. 25, no. 1, pp. 1–7, 2017.
[15] J. Poukens, J. Haex, and D. Riediger, "The use of rapid prototyping in the preoperative planning of distraction osteogenesis of the cranio-maxillofacial skeleton.," *Comput. Aided Surg.*, vol. 8, no. 3, pp. 146–154, 2003.
[16] M. Yang *et al.*, "Application of 3D Rapid Prototyping Technology in Posterior Corrective Surgery for Lenke 1 Adolescent Idiopathic Scoliosis Patients," *Medicine (Baltimore).*, vol. 94, no. 8, p. e582, 2015.
[17] G. a Brown, K. Firoozbakhsh, T. a DeCoster, J. R. Reyna, and M. Moneim, "Rapid prototyping: the future of trauma surgery?," *J. Bone Joint Surg. Am.*, vol. 85–A Suppl, pp. 49–55, 2003.
[18] V. Bagaria, A. Kuthe, and S. Deshpande, "Use of rapid prototyping and three-dimensional reconstruction modeling in management of complex acetabular fracture," *Curr. Orthop. Pract.*, vol. 20, no. 3, pp. 337–340, 2009.
[19] T. Matsuda and M. Mizutani, "Liquid acrylate-endcapped biodegradable poly(??-caprolactone-co-trimethylene carbonate). II. Computer-aided stereolithographic microarchitectural surface photoconstructs," *J. Biomed. Mater. Res.*, vol. 62, no. 3, pp. 395–403, 2002.
[20] A. Mazzoli, "Selective laser sintering in biomedical engineering," *Med. Biol. Eng. Comput.*, vol. 51, no. 3, pp. 245–256, 2013.
[21] J. M. Williams *et al.*, "Bone tissue engineering using polycaprolactone scaffolds fabricated via selective laser sintering," *Biomaterials*, vol. 26, no. 23, pp. 4817–4827, Aug. 2005.
[22] A. Mazzoli, "Selective laser sintering in biomedical engineering," *Med. Biol. Eng. Comput.*, vol. 51, no. 3, pp. 245–256, 2013.
[23] L. C. Ardila *et al.*, "Effect of IN718 recycled powder reuse on properties of parts manufactured by means of Selective Laser Melting," *Phys. Procedia*, vol. 56, no. C, pp. 99–107, 2014.
[24] flickr, "Selective Laser Syntering." [Online]. Available: https://www.flickr.com/photos/organ_printer/878476066. [Accessed: 20-Jul-2017].
[25] R. Wauthle *et al.*, "Additively manufactured porous tantalum implants," *Acta Biomater.*, vol. 14, no. March, pp. 217–225, 2015.
[26] Y. A. Jin, H. Li, Y. He, and J. Z. Fu, "Quantitative analysis of surface profile in fused deposition modelling," *Addit. Manuf.*, vol. 8, pp. 142–148, 2015.
[27] I. Zein, D. W. Hutmacher, K. C. Tan, and S. H. Teoh, "Fused deposition modeling of novel scaffold architectures for tissue engineering applications," *Biomaterials*, vol. 23, no. 4, pp. 1169–1185, 2002.
[28] D. Espalin, K. Arcaute, D. Rodriguez, F. Medina, M. Posner, and R. Wicker, "Fused deposition modeling of patient- specific polymethylmethacrylate implants," *Rapid Prototyp. J.*, vol. 16, no. 3, pp. 164–173, 2010.

[29] www.arcam.com, "EBM® Electron Beam Melting – in the forefront of Additive Manufacturing." [Online]. Available: http://www.arcam.com/technology/electron-beam-melting/. [Accessed: 10-Jul-2017].
[30] Wikipedia.org, "Electron Beam Melting." [Online]. Available: https://en.wikipedia.org/wiki/Electron_beam_additive_manufacturing. [Accessed: 10-Jul-2017].
[31] X. Zhao *et al.*, "Comparison of the microstructures and mechanical properties of Ti–6Al–4V fabricated by selective laser melting and electron beam melting," *Mater. Des.*, vol. 95, pp. 21–31, 2016.
[32] M. Jamshidinia, L. Wang, W. Tong, and R. Kovacevic, "The bio-compatible dental implant designed by using non-stochastic porosity produced by Electron Beam Melting® (EBM)," *J. Mater. Process. Technol.*, vol. 214, no. 8, pp. 1728–1739, Aug. 2014.
[33] C. Gilbert, S. Pauline, K. Radovan, A. Raed, and A. Khaldou, "Digital Engineering of Bio-Adaptable Dental Implants," in *Implant Dentistry - A Rapidly Evolving Practice*, InTech, 2011.
[34] O. L. A. Harrysson, O. Cansizoglu, D. J. Marcellin-Little, D. R. Cormier, and H. A. West, "Direct metal fabrication of titanium implants with tailored materials and mechanical properties using electron beam melting technology," *Mater. Sci. Eng. C*, vol. 28, no. 3, pp. 366–373, Apr. 2008.
[35] L. Morgan, U. Lindhe, and O. Harrysson, "Rapid Manufacturing with Electron Beam Melting (EBM) – A manufacturing revolution ?," *Solid Free. Fabr. Symp. Proc.*, vol. August, no. 41, pp. 433–438, 2003.
[36] A. Sidambe and A. T., "Biocompatibility of Advanced Manufactured Titanium Implants—A Review," *Materials (Basel).*, vol. 7, no. 12, pp. 8168–8188, Dec. 2014.

IOP Conf. Series: Materials Science and Engineering **328** (2018) 012004 doi:10.1088/1757-899X/328/1/012004

Study on the Effect of Diamond Grain Size on Wear of Polycrystalline Diamond Compact Cutter

A. M. Abdul-Rani[1], Adib Akmal Bin Che Sidid[2], Azri Hamim Ab Adzis[3],

[1]Department of Mechanical Engineering, Universiti Teknologi PETRONAS
[2]Department of Petroleum Engineering, Universiti Teknologi PETRONAS
[3]Mechanical Engineering Section, Universiti Kuala Lumpur, Malaysia France Institute

majdi@utp.edu.my, adib.akmal@yahoo.com, azri@unikl.edu.my

Abstract. Drilling operation is one of the most crucial step in oil and gas industry as it proves the availability of oil and gas under the ground. Polycrystalline Diamond Compact (PDC) bit is a type of bit which is gaining popularity due to its high Rate of Penetration (ROP). However, PDC bit can easily wear off especially when drilling hard rock. The purpose of this study is to identify the relationship between the grain sizes of the diamond and wear rate of the PDC cutter using simulation-based study with FEA software (ABAQUS). The wear rates of a PDC cutter with a different diamond grain sizes were calculated from simulated cuttings of cutters against granite. The result of this study shows that the smaller the diamond grain size, the higher the wear resistivity of PDC cutter.

1. Introduction

In the Oil and Gas industry, a suitable drill bit needs to be selected to get the maximum Rate of Penetration (ROP) and to minimize the cost of drilling. The more popular types of drill bit are the Roller Cone (RC) bit and the Polycrystalline Diamond Compact (PDC) bit. PDC bit advantage over RC bit is its high ROP. RC bit, however, is more wear resistant compared with PDC.

The abrasive properties of the rock will cause the bit to be worn off and reduce its cutting efficiency. Worn PDC bit needs to be replaced with a new one. This will consume a lot of time and money. Therefore, more studies are needed to ensure PDC bit can last longer. The objective of the paper is to analyze the effect of diamond grain size of PDC cutter on wear by using Finite Element Analysis (FEA) model. The simulation models were run to get the stress distribution on the PDC cutters with various grain sizes. The stresses in the cutter then will be converted into wear rate.

2. Literature Review

2.1. Bit Wear

The definition of wear in drilling industry is when there is erosion or loss on the material surface due to the frictional force. Ekberg believed that plastic shear followed by shear fraction are signs of wear [1]. The wear of a PDC bit can be seen on the cutter. Moseley et. al. stated that common PDC cutter wear are due to bond failure between the diamond table and the bit body, chipped cutter, broken cutter, delamination, heat checking and abrasive wear [2].

2.2. Bit Grain Size
Wilmot and Penrose claimed that the grain size of diamond gives a big impact on the abrasion and affects the resistance of a cutter. The smaller the grain size (fine particle), the higher the abrasion resistance will be. The abrasion resistance of PDC is measured in term of G-ratio. It indicates the volume of rock detached per unit diamond consumed. The greater the value of G-ratio, the greater the abrasion resistance will be [3]. The other factor which relate wear to the grain size is the amount of grain boundaries as stated by Gedeon. The grain boundaries have a function to stop the dislocation of the particles. Using smaller the grain size will lead to high ratio of surface area to volume and this will give a higher amount of grain boundaries. Thus, the strength of the material will be higher [4].

2.3. Boundary Strengthening
Material strength is strongly related to the grain size of a material. Hall-Petch equation describes the influence of the grain size on the yield strength of a material [5]. The equation describes that with smaller mean grain size, the yield strength of the material will increase and this will lead to the increase of the material strength. The Hall-Petch equation is as below.

$$\sigma_y = \sigma_o + Kd^{-0.5} \tag{1}$$

Where σ_y is the Yield Strength, σ_o is the PDC stress constant, K is the Hall-Petch constant and d is the mean grain size. The equation rationalized that the deformation caused by the dislocation motion is affected by the propagation of dislocation from a deformed grain to an un-deformed grain where the grain boundary is the wall for the dislocation to occur [6]. The dislocation will occur and pile up at the grain boundary which is the wall and will cause the stress concentration at the wall. This will lead to the deformation of a material. The number of dislocation in the pile-up affects the development of stress concentration at the boundary. However, fine-grained material has a small number of dislocation gathered at the boundary which leads to a small stress concentration at the pile up.

2.4. Analytical Model
A new 3D analytical PDC cutter-rock interaction model by taking priority of the effect of side rake angle, the coefficient friction of rock and back rake angle was developed by Rajabov, et al. and Che et al. They proved that the horizontal force acting on the cutter can be computed if the normal force, angle of the back rake and rock friction coefficient are known. They also believed that the increase of the natural force acting on the cutter will increase the horizontal force. Furthermore, they also stated that as the back-rake angle increase, the horizontal force will also increase. Below is the equation proposed by Rajabov, et al. and Che et al. [7][8].

$$F_H = F_N \left[\frac{1-\mu \tan \alpha}{\tan \alpha + \mu}\right] \tag{2}$$

Where F_H is the horizontal force, F_N is the normal force, μ is friction coefficient and α is the back-rake angle.

Based on Merchant's cutting model, the total stress acting on the cutter can be calculated. However, there are a few assumptions made in Merchant's model by Juneja, Sekhon and Seih such as the tool must have a sharp edge, the cutting edge is a straight-line perpendicular to the direction of the cutter and it generates a place surface as the work moves past it. Furthermore, the workpiece experienced the deformation across the thin shear plane and it is rigid and perfectly plastic. Moreover, the shearing surface is a plane extending upward from the cutting edge and the chip does not flow to side. Finally, the depth of cut is kept constant between work and tool with continuous chip [9].

IOP Conf. Series: Materials Science and Engineering **328** (2018) 012004 doi:10.1088/1757-899X/328/1/012004

The total stress acting on the bit is the summation of normal stress and shear stress or horizontal stress. The higher the contact area on the cutter, the lower the total number of stress acting on the PDC cutter is. The wear model is an analytical model that can compute a number from Von Mises stress acting on the material into the wear rate value respective to the stress acting on the material [10]. The wear model equation is:

$$W = k_2 V \sigma^{\frac{1}{bn'}} \tag{3}$$

Where W is the wear rate, k_2 is an arbitrary wear constant, V is the total volume deformed, σ is the Von Mises stress in MPa, b is the constant which is 5 and n' is the cyclic strain-hardening coefficient. In this model, V which is the total volume deformed is approximately equal to Pi x r^2 x d, where r is the radius of contact area and d is the thickness of the material. For PDC cutter, the value of k_2 is 573 N/mm^2, n' is 0.31, 1/bn' is 6.45 and k_2V is 1.5 x 10^{11} [11].

3. Methodology

3.1. PDC Single Cutter Test

The theoretical basis of the PDC cutter was studied using the general analytical model. A 3D simulation model of rock breaking using a single PDC cutter was established with Finite Element Method (FEA) based on elastoplastic mechanics and rock mechanics. The single cutter FEA models of various grain sizes of PDC cutter were done using ABAQUS to identify the Von Mises stress acting on the cutter and to determine the best grain size. Similar FEA set up was used by Ahmad Majdi et. al. to study the effect of cutter sizes and geometry [12].

3.2. Design Features

Based on Hall-Petch Equation, the Yield Strength of a material can be computed by changing its grain size. Table 1 shows the yield strength of cutters with different grain sizes.

Table 1. Yield Strength of PDC Cutter

Diamond Grain Size (μm)	Yield Strength (MPa)
5	5849.7
15	4286.1
25	3804.6
50	3320.0

3.3. Parameters

A single PDC cutter model was simulated to cut against a granite block. A cylindrical cutter of radius 0.012m and thickness of 0.0035m is chosen with back rake angle kept constant at 10o and the depth of cut is 0.003m. Table 2 shows the PDC cutter and granite parameters.

Table 2. Physical Properties of PDC Cutter and Granite

	PCD	Granite
Unconfined Compressive Strength, UCS (MPa)	N/A	300
Density, ρ (kgm^{-3})	3510	26200
Young's Modulus, E (GPa)	890	70
Poisson Ratio	0.07	0.30
Shear Modulus, G (GPa)	545	N/A
Ultimate Tensile Strength, UTS (MPa)	N/A	256
Shear Yield Stress, τ_{yield} (MPa)	N/A	132
Kinetic Coefficient, μ_k	0.45	0.45

Figure 1 shows the cutter and granite block configuration. There are two forces acting on the cutter which are normal force, F_N and horizontal force, F_H, where F_N is equal to the weight on bit acting on the cutter itself. The cutter is assumed to move in a straight direction with a constant velocity of 0.306 m/s. The objective of the analytical model is to identify the axial stress and shear stress acting on the PDC cutter during the drilling operation.

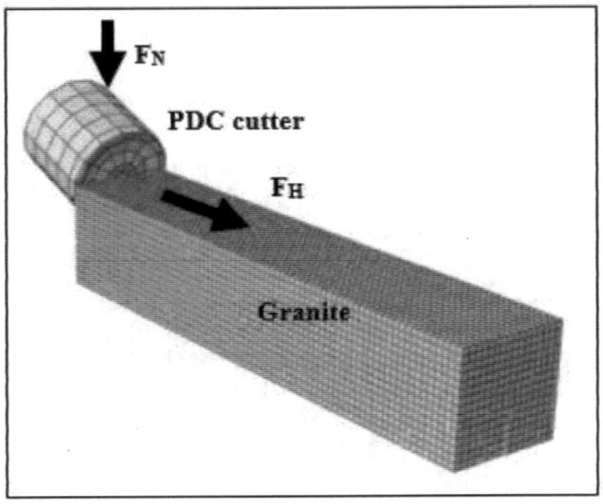

Figure 1. Single Cutter Model

4. Results and Discussion

4.1. PDC Single Cutter Analytical Model

The PDC cutter test data are shown in Table 3. From equation (1), the value of horizontal force, FH acting on the single PDC cutter is 3678.81 N. The total stress acting on the single PDC cutter can be calculated by adding up the axial stress and shear stress.

Table 3. Behaviour of PDC Cutter

Parameters	Value
Surface Area of Cutting Face, A	4.5238x10-4 m^2
Shear Contact Area, A_s	3.2640x10-5 m^2
Horizontal force	3678.81 N
Normal stress, σ	5.517 MPa
Shear stress, τ	112.5895 MPa
Total Stress	118.1065 MPa

4.2. PDC Single Cutter Simulation

A 3D model of PDC cutter and formation was designed by using ABAQUS FEA. The horizontal force and normal force are applied on the PDC cutter which were obtained from the analytical model. The boundary condition of the formation is set as "Encastre" and the PDC cutter is set as "Velocity" with x-axis velocity of 0.306 m/s. The simulation step period is set to be 1.5 step. The tabulated Von Mises stresses acting on the PDC cutter is converted in a graph and is shown in the Figure 2. The graph showed that the smaller the grain size, the lesser the Von Mises stresses acting on the PDC cutter.

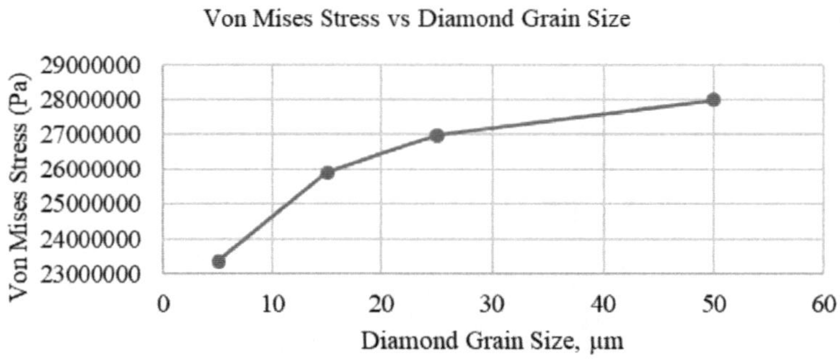

Figure 2. Von Mises Stress vs Diamond Grain Size

4.3. PDC Cutter Wear Rate

The Von Mises Stress recorded from the simulation result then can be converted into wear rate value using equation (2). The wear rate then can be analyzed to determine the best diamond grain size for material properties of PDC cutter. Based on the result shown in Figure 3, it can be concluded that the higher the Von Mises stress, the higher the wear rate of the material will be.

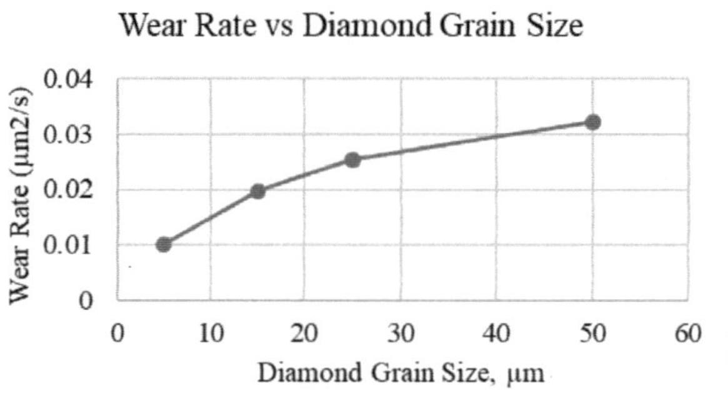

Figure 3. Wear Rate vs Diamond Grain Size

From the Figure 3, it can be observed that the wear rate decreases as the diamond grain size increases. The smaller diamond grain size will have the most grain boundary and naturally will increase the yield strength of the material which makes it more wear resistance. So, the diamond grain size should be kept small to maximize the yield strength of the cutter and maximize the cutter life.

5. Conclusion

This project shows the result of effect of diamond grain size on wear of PDC cutter through Finite Element Analysis method. An analytical model is used to compute the velocity, horizontal force and normal force to be applied in the simulation. The simulated 3D model of single PDC cutter and formation with variable diamond grain sizes. The constant parameters are the back-rake angle, the depth of cut and other material properties. The resultant Von Mises stress distribution from the simulations are converted into wear rate by using the wear model. The result shows that the smaller the diamond grain size, the lesser the stress distribution exerted on the cutter. This is due to the larger grain boundary which leads to lesser dislocation piled up at the boundary. This in turn will cause lesser stress to be exerted on the cutter. In conclusion, this project found that 5μm size of diamond grain has the best properties to reduce cutter wear rate.

6. Recommendation

It is recommended to simulate the model using a longer range of period step. The longer period step will give a better result because the once the time taken reach the optimum time, the stress distribution number will keep constant and stable.

References

[1] Ekberg A 1997 Rolling contact fatigue of railway wheels—a parametric study *Wear* **211(2)** 280-288

[2] Moseley S G, Bohn K P and Goedickemeier M 2009 Core drilling in reinforced concrete using polycrystalline diamond (PCD) cutters: Wear and fracture mechanisms *Int. J. Refractory Metals and Hard Materials* **27(2)** 394-402.

[3] Wilmot G M and Penrose B 2003 Advanced Cutting Structure Improves PDC Bit Performance in Hard and Abrasive Drilling Environments *Society of Petroleum Engineers* vol. SPE no **81167** 27 April 2003 (2003-04-27), - 30 April 2003 (2003-04-30)

[4] Gedeon M March 2010 *Grain Size and Material Structure* (Brush Wellman's Alloy Customer Service Department)

[5] Morris Jr JW 2001 May 1 *The influence of grain size on the mechanical properties of steel.* Lawrence Berkeley National Laboratory

[6] Stawovy MT 1998 Processing of aluminum alloys containing displacement reaction products (Doctoral dissertation).

[7] Rajabov V, Miska SZ, Mortimer L, Yu M, and Ozbayoglu ME January 2012 The effects of back rake and side rake angles on mechanical specific energy of single PDC cutters with selected rocks at varying depth of cuts and confining pressures. In *IADC/SPE Drilling Conference and Exhibition*. Society of Petroleum Engineers.

[8] Che D, Han P, Guo P and Ehmann K 2012 Issues in Polycrystalline Diamond Compact Cutter–Rock Interaction From a Metal Machining Point of View—Part I: Temperature, Stresses, and Forces. *J. Manufacturing Science and Engineering* **134(6)** 064001.

[9] Juneja BL, Sekhon GS and Seih N 2003 *Fundamentals of Metal Cutting and Machine Tool* (India: New Age International.)

[10] Tangena AG 1987 *Tribology of thin film system* (Doctoral thesis, Eindhoven Technical University)

[11] Tangena AG, Wijnhoven PJ Jan 1 1988 The correlation between mechanical stresses and wear in a layered system *Wear* **121(1)** 27-35.

[12] Ahmad Majdi Abdul Rani, Kong Kai Vern, Azri Hamim Ab Adzis, TVVLN Rao 2017 Effect of

Size and Geometry of Polycrystalline Diamond Cutter (PCD) on Wear Rate in Multilayer Formation. *Science International (Lahore)* **29(3)** 529-534

IOP Conf. Series: Materials Science and Engineering **328** (2018) 012005 doi:10.1088/1757-899X/328/1/012005

An In-Depth Review on Direct Additive Manufacturing of Metals

Farooq I Azam* [1]**, Ahmad Majdi Abdul Rani**[1]**, Khurram Altaf**[1]**, T.V.V.L.N Rao**[2]**, Haizum Aimi Zaharin**[1]

1 Mechanical Engineering Department, Universiti Teknologi PETRONAS, Malaysia

2 Department of Mechanical-Mechatronics Engineering, The LNM Institute of Information Technology, India

Email: farooq.i_g03648@utp.edu.my

Abstract. Additive manufacturing (AM), also known as 3D Printing, is a revolutionary manufacturing technique which has been developing rapidly in the last 30 years. The evolution of this precision manufacturing process from rapid prototyping to ready-to-use parts has significantly alleviated manufacturing constraints and design freedom has been outstandingly widened. AM is a non-conventional manufacturing technique which utilizes a 3D CAD model data to build parts by adding one material layer at a time, rather than removing it and fulfills the demand for manufacturing parts with complex geometric shapes, great dimensional accuracy, and easy to assemble parts. Additive manufacturing of metals has become the area of extensive research, progressing towards the production of final products and replacing conventional manufacturing methods. This paper provides an insight to the available metal additive manufacturing technologies that can be used to produce end user products without using conventional manufacturing methods. The paper also includes the comparison of mechanical and physical properties of parts produced by AM with the parts manufactured using conventional processes.

1. Introduction

Additive manufacturing has been around for decades and it is continuously evolving at an immense rate. This advancement in additive manufacturing techniques has attracted interest from both the academic community and the business world. AM is an umbrella which accommodates wide range of technologies and is known by many nick names such as Freeform Fabrication, Solid Freedom Fabrication, layer-Based Manufacturing, Rapid Prototyping, Additive Layered Manufacturing, Rapid Manufacturing, Additive Fabrication etc. [1-3]. The official definition of Additive Manufacturing (AM) is given by American Society for Testing and Materials (ASTM) F2792 as "process of joining materials to make objects from three-dimensional (3D) model data, usually layer upon layer, as opposed to subtractive manufacturing methodologies" [4]. This approach of AM to create products with increased efficiency, accuracy and reduced wastage makes it a more sustainable manufacturing process. Additive manufacturing stands out the most among other emerging manufacturing techniques and has the potential to replace conventional manufacturing process, allow for new products and change society as whole. However, this process is still in its early existence, but it has been showing signs of great future potential and has been enthusiastically embraced by a small number of early global adopters [5-7].

IOP Conf. Series: Materials Science and Engineering **328** (2018) 012005 doi:10.1088/1757-899X/328/1/012005

2. Metal Additive Manufacturing

Metal additive manufacturing consists of many different technologies, These technologies can be divided into two categories; the "direct" way is where the metal powder completely melts and solidifies to form the final part. In "indirect" way, a binder is used to join the particles of metal powder together and post processing is necessary to achieve desired density. Classification of AM processes for metals is shown in Figure 1. "Direct" methods for manufacturing metal parts using additive manufacturing including Selective Laser Melting (SLM), Laser Metal Deposition (LMD) and Electron Beam Melting (EBM). In all these technologies, metal powder fully melts and then solidifies to form the end product. All these technologies can produce denser parts compared to other AM technologies.

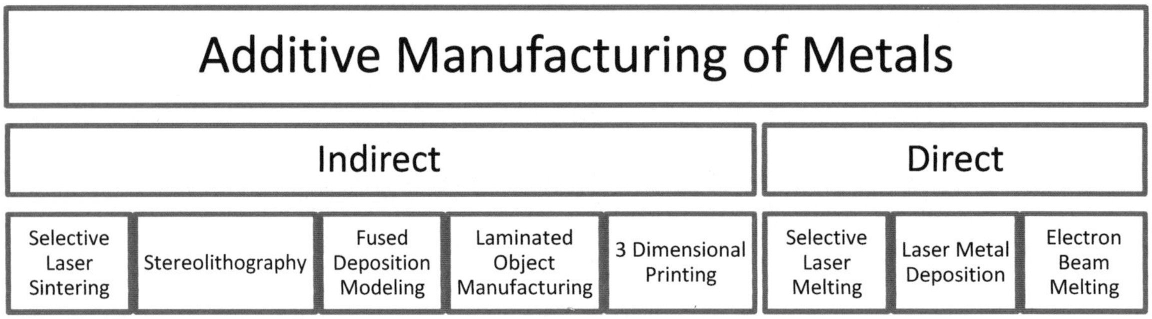

Figure 1. Classification of metal AM processes [8]

2.1. Selective Laser Melting (SLM)

Selective Laser Melting is one of the industry's leading additive manufacturing technologies. It is precise and fast compared to other AM technologies. SLM is similar to SLS, but it goes a step further and rather than just sintering the powder particles to form the 3D part, SLM fully melts the powder particles to form the final part. SLM technology makes it possible to achieve approximately 100% density and a much stronger part, eliminating the need for postprocessing steps like infiltration, which are highly required in SLS or DMLS. SLM provides unlimited freedom of design and can produce highly complex shapes which can be impossible to manufacture using conventional manufacturing processes. [9] in his experimental study fabricated an impression block with conformal cooling channels for die casting using SLM. When compared to conventionally produced block, it was discovered that due to the conformal channels, the cooling rate increased and resulted in reduced need for spray cooling. It was also observed that the surface finish of casted parts improved and the cycle time was reduced [9]. As other AM processes, SLM process starts with a computerized 3D model which after sliced into layers, the information is fed to the SLM machine. A thin metal powder layer is spread across the metal build plate. Then the laser selectively scans the cross section on the material in x-y axes per the 3D data provided earlier. As the laser scans each layer, the powder particles are melted and consolidated into a homogenous part. Material which is not part of the model geometry is left unaffected and acts as a support structure. The build platform then lowers by a single layer thickness and the levelling blade sweeps across the build platform and covers it with another layer of powdered metal. The laser then scans the next layer and the process repeats building layer by layer until the part is completed [10-12]. The process is illustrated in Figure 2.

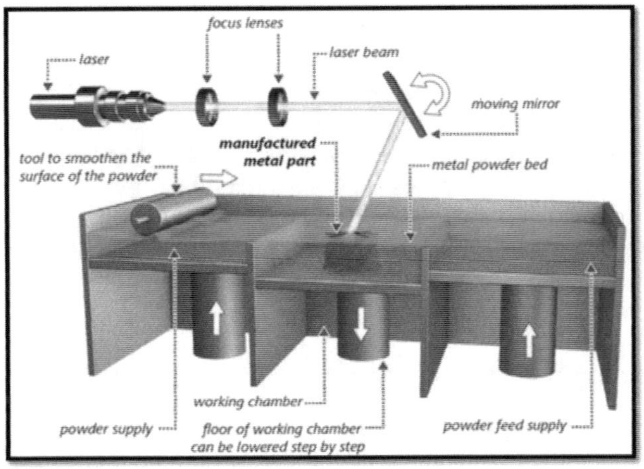

Figure 2. Selective Laser Melting (SLM) mechanism (Source: empa.ch)

2.2. Electron Beam Melting (EBM)

Electron Beam melting is another additive manufacturing technology which forms 3D parts by full melting of powder particles. The key difference between laser based additive manufacturing technologies and EBM is the heat source. EBM, as the name suggests uses an electron beam instead of laser, which requires that the procedure is carried out under vacuum conditions to prevent dissipation of the electron beam. After being heated at a high temperature, the electrons are emitted from a filament. The beam of electrons is accelerated up to half the speed of light, controlled by two magnetic fields. One field acts as a magnetic lens and helps the beam to focus to desired diameter, whereas the second field deflects the focused beam to the build platform at a desired point. In EBM printer, as shown in Figure 3, the machine spreads a layer of powder material on the build platform. The electron beam melts the material powder per the data provided to it. The build platform is lowered and next layer of powder is spread. The beam traces the cross section of the next layer and melts the powder. The process repeats until the 3D part is built [13-14].

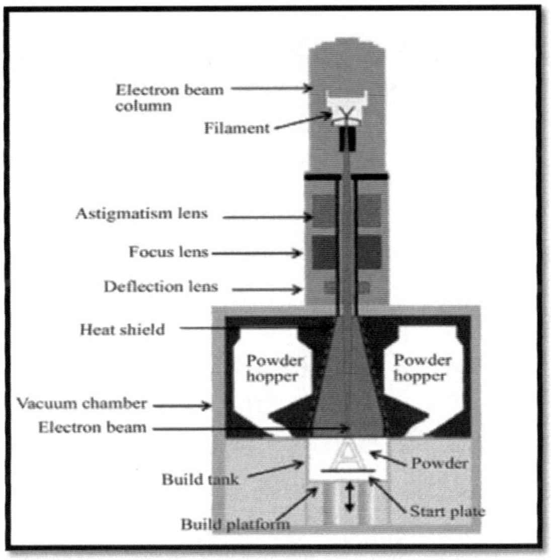

Figure 3. Electron Beam Melting (EBM) mechanism (Source: arcam.com)

IOP Conf. Series: Materials Science and Engineering **328** (2018) 012005 doi:10.1088/1757-899X/328/1/012005

2.3. Laser Metal Deposition (LMD)

Laser metal deposition, also called as direct energy deposition and laser cladding, is a powder based additive manufacturing process, which is used to build 3D parts, repair metal components deemed nonrepairable by conventional methods or add features to existing parts. The process is very simple and it begins with a 3D model like other AM technologies. The model is divided into layers and the information is fed to the LMD machine. Molten pool is created by the laser beam on metallic substrate, then metal in powder form is applied through the nozzle into the melt pool by a meticulousness powder fed system as shown in Figure 4 below. The deposited powder then melts and is metallurgically bonded to the base material creating a weld bead. The process repeats building up layer by layer until the part is complete [15].

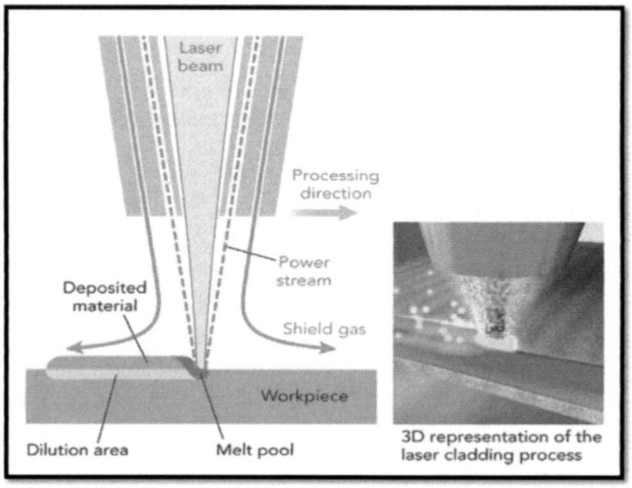

Figure 4. Laser Metal Deposition (LMD) mechanism (Source: industrial-lasers.com)

3. Physical Properties

When printing a metal part, the surface finish, mechanical properties and geometrical accuracies are the main concerns The size of the feedstock and the diameter of the heat source determine the minimum feature size. Smaller feature sizes lead to a smaller layer thickness which, as a result, give better resolution but at the cost of build rate/deposition rate. SLM has the minimum beam diameter and the smallest layer thickness compared to EBM and LMD. Thus, SLM has a higher resolution among the three. SLM has a minimum beam diameter of 50 μm compared to EBM's 140 minimum beam diameter. Compared to LMD's minimum layer thickness of 100 μm, SLM has a minimum layer thickness of 20 μm. Minimum feature size is also advantageous for SLM and EBM for printing metal mesh structures. SLM and EBM can also be utilized to build complex geometries with overhangs whereas with LMD, support structure may be required (5 axes can be utilized to eliminate support structures) [16]. Two factors contribute to surface roughness of metal; Layer roughness and actual roughness [17], as shown in Figure 5..

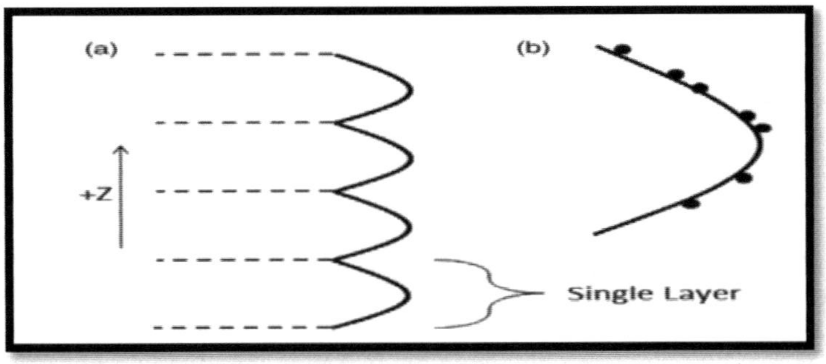

Figure 5. Factors for surface roughness in AM (a) Layer roughness (b) Actual roughness [17]

Layer roughness effect can be reduced by decreasing layer thickness, but doing so would result in longer build time, whereas actual roughness of metal surface depends on the machine used, build direction and process parameters. As 3D parts in AM are built layer by layer, reducing the layer thickness would result in higher number of layers. Thus, EBM and LMD with a larger layer thickness have a higher build rate compared to SLM. This higher build rate comes at the expense of geometrical accuracy. Larger layer thickness comes in the way of achieving near net shape part and machining is required to accomplish final geometry and details.

4. Mechanical Properties

The yield strength, ultimate Tensile Strength, % elongation and hardness from published data have been summarized in Table 1. The data from EBM, SLM and LMD is compared with the values obtained from conventionally cast and wrought test specimens. EBM processed specimens show superior mechanical properties compared to cast specimens, but are almost equivalent when compared to wrought specimens. SLM on the other hand has superior UTS and Ys compared to EBM, wrought and cast parts. However, it should be noted that the ductility is lower compared to them. LMD like SLM has higher UTS and Ys compared to the EBM, wrought and cast samples but a lower ductility. When comparing SLM and LMD, SLM shows either superior or equivalent mechanical properties. [18] in his study demonstrated that SLM specimen of Ti6Al4V ELI exhibited a superior fatigue limit (550 MPa), whereas EBM samples of Ti6Al4V ELI showed an inferior fatigue limit (340 MPa).

Process	Material	Yield Strength (MPa)	Ultimate Tensile Strength (MPa)	Elongation (%)	Hardness (HV)	Reference
Wrought	*Ti6Al4V ELI*	860	931	14	327	[19]
	Ti6Al4V	860	930	10	-	[20]
Cast	*Ti6Al4V ELI*	734	851	4.4	333	[19]
	Ti6Al4V	758	860	8	-	[20]
EBM	*Ti6Al4V ELI*	930	970	16	318	[21]
		869	928	9.9	327	[18]
	Ti6Al4V	950	1020	14	327	[20]
		883.7 - 938.5	993.9 – 1029.1	13.6 – 13.2	-	[22]
SLM	*Ti6Al4V ELI*	1143	1219	4.89	403	[18]
		996 ± 10	1110 ± 13	7 ± 4	399 ± 4	[23]
	Ti6Al4V	990 ±5	1,095 ± 10	8.1 ± 0.3	-	[24]
		1116 ± 61	1286 ± 57	8 ± 2	384 ± 5	[25]
LMD	*Ti6Al4V*	1062	1157	6.2	-	[26]
		976 ± 24	1099 ± 2	4·9 ± 0·1	360 ± 10	[27]
		1025 - 1085	1138 - 1168	3.4 – 3.7	390 ± 2	[28]
		973	1077	11	-	[29]

Table 1. Comparison of Mechanical Properties of Different AM Technologies

5. Conclusion

As the interest of different industries is growing in AM, more and more AM equipments are available in the market. Different AM technologies serve different purposes from rapid prototyping of novel products to rapid manufacturing of end products. Many different AM technologies are available with numerous different parameters which affect the final product.[30] estimated that there are approximately 130 parameters that can impact the SLM process, of which, 13 are critical to the quality characteristics of the final manufactured parts. In SLM, EBM and LMD many different process parameters affect the final product and the mechanical properties of these processes are either equivalent or higher compared to conventional manufacturing techniques (wrought, cast, etc.). Regardless of the processes, what matters more is the suitability of a final product for a specific application. For instance, SLM can be chosen for an application requiring superior part strength, whereas EBM can be selected if the application requirement is a moderate strength ductile part. Therefore, selection of a specific process, may it be LMD, EBM or SLM, largely depends on the requirements for end-use. Thus, proper understanding of the final product application is necessary for choosing the right AM process.

References

[1] V. Petrovic, J. Vicente Haro Gonzalez, O. Jordá Ferrando, J. Delgado Gordillo, J. Ramón Blasco Puchades, and L. Portolés Griñan, "Additive layered manufacturing: sectors of industrial application shown through case studies," Int. J. Prod. Res., vol. 49, no. 4, pp. 1061–1079, 2011.

[2] W. K. Chiu and K. M. Yu, "Direct digital manufacturing of three-dimensional functionally graded material objects," CAD Comput. Aided Des., vol. 40, no. 12, pp. 1080–1093, 2008.

[3] J.-H. Shim et al., "Effect of solid freeform fabrication-based polycaprolactone/(polylactic-co-glycolic acid)/collagen scaffolds on cellular activities of human adipose-derived stem cells and rat primary hepatocytes.," J. Mater. Sci. Mater. Med., vol. 24, no. 4, pp. 1053–65, 2013.

[4] ASTM International, F2792-12a - Standard Terminology for Additive Manufacturing Technologies. 2013.

[5] H. Lipson and M. Kurman, "FACTORY@HOME THE EMERGING ECONOMY OF PERSONAL MANUFACTURING OVERVIEW AND RECOMMENDATIONS A Report Commissioned by the US Office of Science and Technology Policy," 2010.

[6] S. H. Huang, P. Liu, A. Mokasdar, and L. Hou, "Additive manufacturing and its societal impact: A literature review," International Journal of Advanced Manufacturing Technology, vol. 67, no. 5–8. pp. 1191–1203, 2013.

[7] R. A. of Engineering and Engineering Royal Academy of, "Additive manufacturing: opportunities and constraints," R. Acad. Eng., no. May 2013, p. 21, 2013.

[8] N. Guo and M. C. Leu, "Additive manufacturing: Technology, applications and research needs," Front. Mech. Eng., vol. 8, no. 3, pp. 215–243, 2013.

[9] A. Armillotta, R. Baraggi, and S. Fasoli, "SLM tooling for die casting with conformal cooling channels," Int. J. Adv. Manuf. Technol., vol. 71, no. 1–4, pp. 573–583, 2014.

[10] I. Yadroitsev, P. Bertrand, and I. Smurov, "Parametric analysis of the selective laser melting process," Appl. Surf. Sci., vol. 253, no. 19, pp. 8064–8069, 2007.

[11] S. Bremen, W. Meiners, and A. Diatlov, "Selective Laser Melting," Laser Tech. J., vol. 9, no. 2, pp. 33–38, Apr. 2012.

[12] E. Yasa and J. Kruth, "Application of Laser Re-Melting on Selective Laser Melting Parts," Adv. Prod. Eng. Manag., vol. 6, no. 4, pp. 259–270, 2011.

[13] L. Morgan, U. Lindhe, and O. Harrysson, "Rapid Manufacturing with Electron Beam Melting (EBM) – A manufacturing revolution?" Solid Free. Fabr. Symp. Proc., pp. 433–438, 2003.

[14] L. E. Murr et al., "Metal Fabrication by Additive Manufacturing Using Laser and Electron Beam Melting Technologies," Journal of Materials Science and Technology, vol. 28, no. 1. pp. 1–14, 2012.

[15] S. Ocylok, E. Alexeev, S. Mann, A. Weisheit, K. Wissenbach, and I. Kelbassa, "Correlations of melt pool geometry and process parameters during laser metal deposition by coaxial process monitoring," in Physics Procedia, 2014, vol. 56, no. C, pp. 228–238.

[16] H. Gong, K. Rafi, H. Gu, T. Starr, and B. Stucker, "Analysis of defect generation in Ti – 6Al – 4V parts made using powder bed fusion additive manufacturing processes ☆," Addit. Manuf., vol. 1–4, pp. 87–98, 2014.

[17] W. J. Sames, F. A. List, S. Pannala, R. R. Dehoff, and S. S. Babu, "The metallurgy and processing science of metal additive manufacturing," Int. Mater. Rev., vol. 61, no. 5, pp. 315–360, 2016.

[18] H. K. Rafi, N. V. Karthik, H. Gong, T. L. Starr, and B. E. Stucker, "Microstructures and mechanical properties of Ti6Al4V parts fabricated by selective laser melting and electron beam melting," J. Mater. Eng. Perform., vol. 22, no. 12, pp. 3872–3883, 2013.

[19] M. Koike, K. Martinez, L. Guo, G. Chahine, R. Kovacevic, and T. Okabe, "Evaluation of titanium alloy fabricated using electron beam melting system for dental applications," J. Mater. Process. Technol., vol. 211, no. 8, pp. 1400–1408, 2011.

[20] ArcamEBM, "Ti6Al4V Titanium Alloy." [Online]. Available: http://www.arcam.com/wp-

content/uploads/Arcam-Ti6Al4V-Titanium-Alloy.pdf. [Accessed: 18-Dec-2016].

[21] ArcamEBM, "Ti6Al4V ELI Titanium Alloy Arcam EBM system." [Online]. Available: http://www.arcam.com/wp-content/uploads/Arcam-Ti6Al4V-ELI-Titanium-Alloy.pdf. [Accessed: 18-Dec-2016].

[22] S. S. Al-Bermani, M. L. Blackmore, W. Zhang, and I. Todd, "The origin of microstructural diversity, texture, and mechanical properties in electron beam melted Ti-6Al-4V," Metall. Mater. Trans. A Phys. Metall. Mater. Sci., vol. 41, no. 13, pp. 3422–3434, 2010.

[23] Renishaw, "Renishaw | Data sheets - Additive manufacturing." [Online]. Available: http://www.renishaw.com/en/data-sheets-additive-manufacturing--17862. [Accessed: 18-Dec-2016].

[24] L. Facchini, E. Magalini, P. Robotti, A. Molinari, S. Höges, and K. Wissenbach, "Ductility of a Ti-6Al-4V alloy produced by selective laser melting of prealloyed powders," Rapid Prototyp. J., vol. 16, no. 6, pp. 450–459, 2010.

[25] SLM Solutions Group AG, "3D Metals and Services | SLM." [Online]. Available: https://slm-solutions.com/sites/default/files/downloads/w_metals_services_en_druck.pdf. [Accessed: 18-Dec-2016].

[26] R. T. Terms, "NRC Publications Archive Archives des publications du CNRC," Man. - Guid., 1977.

[27] J. Yu, M. Rombouts, G. Maes, and F. Motmans, "Material Properties of Ti6Al4V Parts Produced by Laser Metal Deposition," Phys. Procedia, vol. 39, pp. 416–424, 2012.

[28] E. Amsterdam and G. Kool, "High Cycle Fatigue of Laser Beam Deposited Ti-6Al-4V and Inconel 718," 25th ICAF (International Comm. Aeronaut. Fatigue) Symp. Rotterdam, Netherlands, 27-29 May, no. May, pp. 1261–1274, 2009.

[29] "Metal 3d Printing Materials - Optomec Additive Manufacturing." [Online]. Available: http://www.optomec.com/3d-printed-metals/lens-materials

[30] O. Rehme and C. Emmelmann, "Reproducibility for properties of selective laser melting products.," Proc. Third Int. WLT-Conference Lasers Manuf., pp. 227–232, 2005.

Investigation on Simultaneous Effects of Shot Peen and Austenitizing Time and Temperature on Grain Size and Microstructure of Austenitic Manganese Steel (Hadfield)

M Beheshti [1], M Zabihiazadboni[2], M C Ismail[1], S Kakooei[1*] and S Shahrestani[1]

[1]Centre for Corrosion Research, Mechanical Department, Universiti Teknologi PETRONAS, Malaysia

[2]F.Z.A Application University & Fouladin Zob Amol Foundry Company, Iran

*saeid.kakooei@utp.edu.my

Abstract. Optimal conditions to increase life time of casting parts have been investigated by applying various cycles of heat treatment and shot peening on Hadfield steel surface. Metallographic and SEM microstructure examinations were used to determine the effects of shot peen, austenitizing time and temperature simultaneously. The results showed that with increasing austenitizing time and temperature of casting sample, carbides resolved in austenite phase and by further increase of austenitizing temperature and time, the austenite grain size becomes larger. Metallographic images illustrated that shot peening on Hadfield steel surface; Austenite - Martensite transformation has not occurred, but its matrix hardened through twining formation process.

Keywords: Hadfield Steel, heat treatment, shot peen, austenitizing, grain size, twining

1. Introduction

Austenitic manganese steel (Hadfield steel) is widely used in cement industry, mining, railways and other heavy industry due to its relatively good toughness and wear properties. Hadfield steels contains about 2.1% C and 12% Mn first produced in 1882 by Robert Hadfield [1, 2]. It exhibits high toughness, high ductility, high work hardening ability and excellent wear resistance. Because of the excellent combination of these properties, this steel has been accepted as a very useful engineering material. It is widely used in the earthmoving, mining, quarrying, oil well drilling, steelmaking, railroading, dredging, naval, lumbering excavators, and mineral crushing equipment [2-4].
Despite the desirable properties of the Hadfield steels, the main limitation of their applications is due to non-dissolved carbides in the austenite matrix from improper heat treatment. The wear resistance of these steels can be improved by increasing hardness through addition of limited content of carbon. Their wear resistance could also be increased by adding elements such as titanium and vanadium as inoculations [5 - 7]. There are two significant steps in Hadfield steel production, namely melting and developing of appropriate composition and secondary, proper heat treatment on casting parts [6]. Strength, ductility and wear resistance properties of Hadfield steels increase with controlled heat

treatment at the suitable temperature. If heat treatment temperature was adjusted improperly, tensile strength and ductility would be reduced [8].
The toughness and the high wear resistance are important properties of the Hadfield steel. The hardness of the Manganese austenitic steel is not high and similar to that of medium carbon steels. The impact resistance of heat-treated Hadfield steel is high and generally are bent on the impact rather than the failure of the device. This hardness of the surface layer increases during the operation which caused them to be the more wear resistant [9]. In practice, due to low surface hardness, Hadfield steel is commonly exposed to work hardening process such as shot blasting in order to increase the wear resistance [10,11]. The deformation mechanisms responsible for the high strain hardening rate include interactions between interstitial carbon atoms and dislocations interactions between dislocations and twins and interactions between twinning systems [12-14].
The casting Hadfield steels are not efficient because the grain boundary carbides formed during solidification and will greatly reduce the flexibility. Deposition of carbides in austenite grain boundary that result to reduce strength of the steel to a minimum. In addition to, the perlite Regions in steel microstructure must be transformed into austenite during heat treatment [15]. Austenitizing makes dissolution of the carbides and form of the single-phase austenitic structure. With an increase in the austenitizing temperature, the carbide solubility and grain size are increased, but excessive austenitizing temperature will cause decarburization of the surface. The austenitizing temperature of the Hadfield steel is in 1050 to 1200 °C and it is kept at this temperature for 1 to 2 hours to dissolve of the carbide phase [16]. Due to the opposite influence of heat treatment on austenite grain size and presence of carbide, this paper investigates to determine the optimal heat treatment temperature and time on the austenite grain size.

2. Experimental Method

In order to examine the simultaneous effects of the austenitizing temperature and shot peening time on the Hadfield steel structure, 19 different samples were casted as detailed in Figure. 1. A sample without heat treatment and shot peen operations as controlled sample was kept aside and 18 samples were heat-treated at different quench and shot peen cycles.
The chemical composition of melt for casting and preparation of samples were performed according to ASTM A-128 standard of the manganese austenitic steels. The chemical composition of melt is shown in Table 1.
Samples heat treatment were performed at different temperatures 970 °C, 1060 °C and 1150 °C and maintenance times were 30 or 60 minutes. Finally, they were cooled by choppy water. Then, samples were shot peened for 30 or 60 minutes. Heat treatment cycle of the samples was performed at 25 °C to 650 °C with rate of 100 °C/h. Then, samples were kept at this temperature for 30 minutes for homogenization and stress relieving.
Finally, in order to reach a final heat treatment, the austenitizing temperature (970, 1060 and 1150 ºC), they were rapidly heated to the desired temperature with rate 100 °C/h. Heat treatment furnace, for the resistive ceramic samples, with a capacity of 250 liter is used. Samples for reduction grain size and increasing the hardness and wear resistance, have undergone 30 and 60 minutes shot peening. Then, grinding for samples with P100, P600 and P2500 was done and samples surfaces were etched using a solution of 2% Nital. The metallographic structures of the sample with the optical microscope and photographic were used. The metallographic images were compared with the standard images (DIN50601) and the grain sizes of the samples were calculated and compared. To study samples microscopic structure, the cross-sections of shot peening were taken, and the results of SEM analysis were compared.

IOP Conf. Series: Materials Science and Engineering **328** (2018) 012006 doi:10.1088/1757-899X/328/1/012006

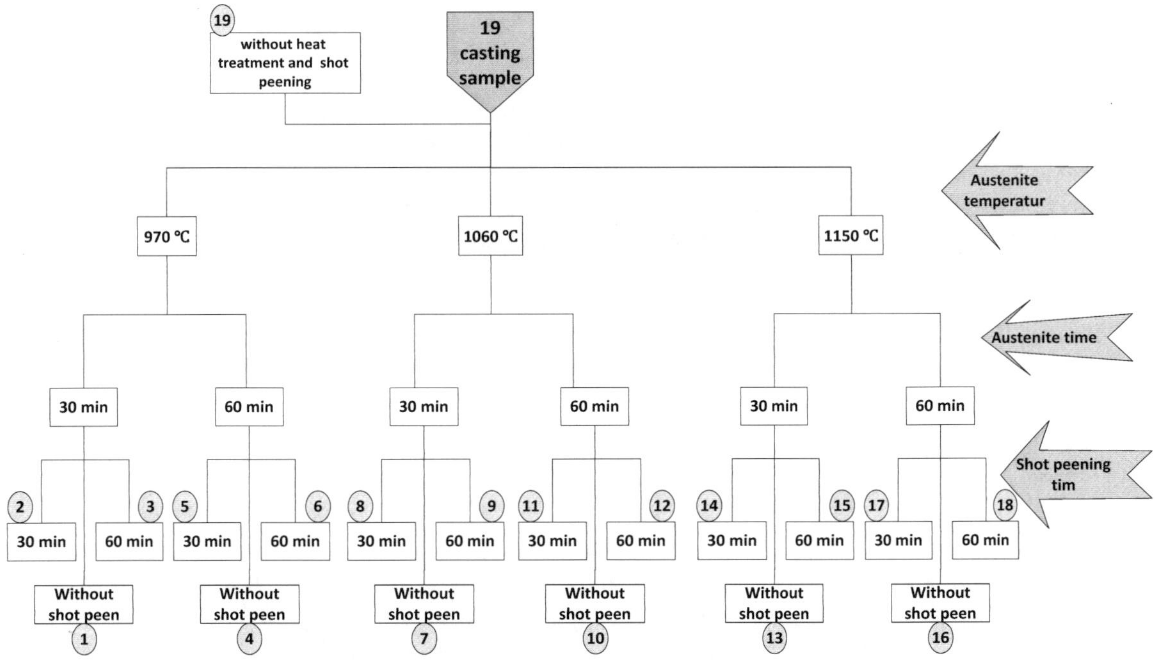

Figure 1 Different heat treatments and shot peening cycles of samples

Table 1 Chemical composition of melt

Element	C	Si	Mn	P	S	Cr
%	1.24	0.64	14	0.044	0.003	2
Element	Fe	Mo	Ni	V	Al	
%	Balance	0.19	0.074	0.031	-	

3. Results and Discussions

Figure 2 shows as-cast steel structure which includes austenitic matrix in bright areas with a high percentage of iron, manganese and chromium carbides which formed a continuous network at grain boundaries. These carbides caused brittleness of austenitic manganese steels.

Figure 3 shows the effect of heat treatment temperature (970 – 1150 °C) at 30 minutes constant maintenance time (before the shot peening operation). The metallographic images show that the samples microstructures include the austenite matrix and non-dissolved carbides. With increasing heat treatment temperature, percentage of non-dissolved carbide decreases, and their formation was removed in grain boundaries. During further increase in temperature (1150 °C), grain boundaries carbides completely disappeared and just a little percentage of carbides remain in the austenite grains.

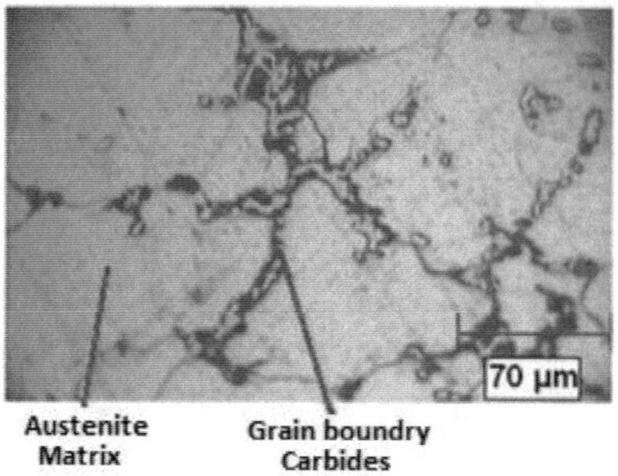

Figure 2 As-cast Hadfield steel microstructure without heat treatment and shot peening.

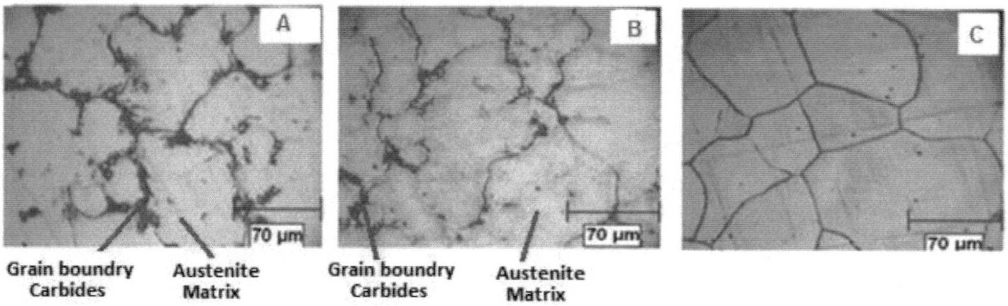

Figure 3 Effect of heat treatment temperature at 30 minutes constant maintenance time without shot peen A: 970, B: 1060 and C: 1150 ° C

Figure 4 shows the effect of heat treatment temperature (970 – 1150 °C) at 30 minutes constant maintenance time (before the shot peen operation). Figures 3 and 4 shows that with increasing temperature, non-dissolved carbide percentage in grain boundary reduced and finally will reach to a very small percentage. To compare Figures 3 and 4, it shows that increasing the maintenance time (30 to 60 minutes) effect on reducing and discontinuity of grain boundary carbide poorly, and austenite grain size was larger. The austenite grain size in various samples as shown in Table 2, where it was obtained by DIN 50 601 standard method. It is clearly evidence that with increasing austenitizing temperature and holding time, the austenite grain size becomes larger and grain size number becomes smaller.

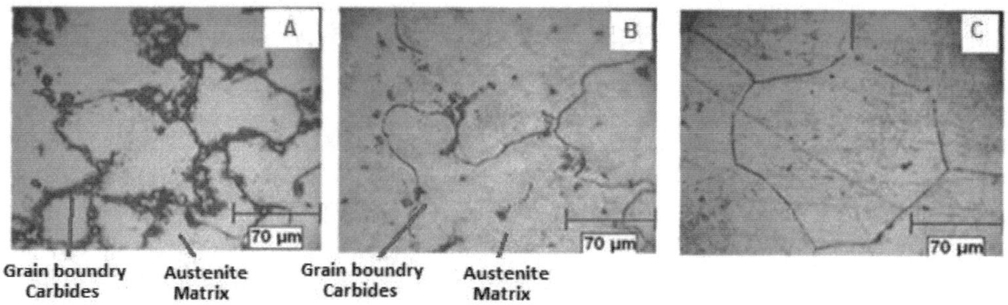

Figure 4 Effect of heat treatment temperature at 60 minutes constant maintenance time without shot peen A: 970,

B: 1060 and C: 1150 ° C

Table 2 Samples Austenite grain size

Grain size number	Austenitic time(min)	Austenitic temperature (°C)	Sample Number
7	30	970	1
7	60	970	4
6	30	1060	7
5	60	1060	10
5	30	1150	13
4	60	1150	16
7	Without Heat Treatment	Without Heat Treatment	19

In totality, the microscopic images of samples with different heat treatment cycles show most of the restructuring and solubility of carbides, and their morphologies are related to the austenitizing temperature. Then, by increasing the austenitizing temperature, the amounts of dissolved carbides in the structure are decreased, and therefore, austenite grain size is increased. It was found that with increasing time of holding of samples at high austenitizing temperature, the austenite grain size would become larger. For the simultaneous effect of austenitizing temperature and the shot peening treatment on the microstructure, the metallographic images of heat treated specimens in 1150 °C for 60 minutes and 60 minutes shot peen are shown in Figure 5.

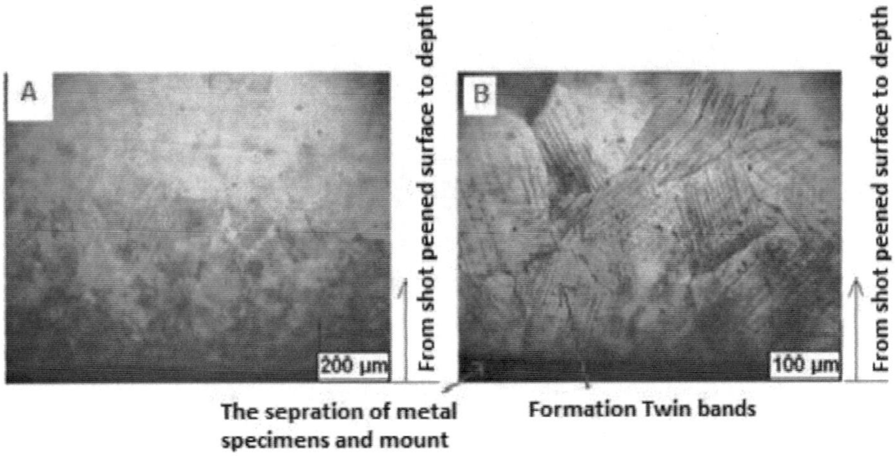

Figure 5 Effects of simultaneous shot peen on metallographic structure of the sample under heat for 60 minutes in 1150°C magnification of (A) 100X, (B) 200X.

As shown in Figure 5, the sample surface has hardened by shot peening operation, but in the centre, the austenite grains remained unchanged. Parallel lines and changes of twin directions can be seen in various grains. Work hardening of austenitic grains in the surface layer, with a thickness about 0.3-0.5mm, are specified in Figure 5 (a). According to Figure 5, it suggests that the surface's shot peening is appropriate, and also, the austenitic to martensitic transformation has not taken place, whereas, the work hardening

IOP Conf. Series: Materials Science and Engineering **328** (2018) 012006 doi:10.1088/1757-899X/328/1/012006

by the twin mechanisms is only done, and also, slip into the surface layer is caused by the shot peening operation. Figure 6 shows the metallographic image of the austenitic grains where both work hardening, and non-work hardening have occurred.

Figure 6 Illustration of the austenite grains after heat treatment in1060 °C for 60 minutes and shot peened for 60 minutes.

To inspect different phases in the electron microscopy images and phase elemental analysis of the EDS, the two heat treatment samples at 1060 °C for 60 min and 1150 ◦C for 30 minutes have been prepared, where both samples were shot peen for 60 min period.

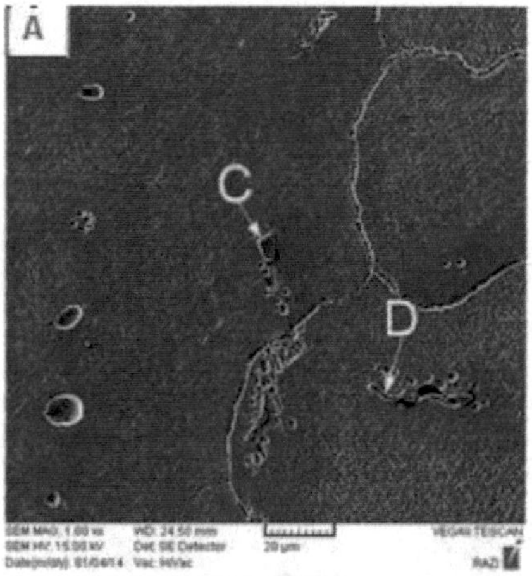

B Spectra: 12-ZONE C

Element	Series	unn. C [wt.-%]	norm. C [wt.-%]	Atom. C [at.-%]
Carbon	K series	3.62	3.72	15.17
Silicon	K series	0.00	0.00	0.00
Chromium	K series	9.19	9.45	8.91
Manganese	K series	21.63	22.23	19.84
Iron	K series	61.18	62.88	55.20
Molybdenum	L series	1.68	1.73	0.88

C Spectra: 12-ZONE D

Element	Series	unn. C [wt.-%]	norm. C [wt.-%]	Atom. C [at.-%]
Carbon	K series	3.47	3.37	13.86
Silicon	K series	0.00	0.00	0.01
Chromium	K series	8.83	8.56	8.14
Manganese	K series	23.34	22.62	20.35
Iron	K series	66.70	64.64	57.22
Molybdenum	L series	0.84	0.82	0.42

Figure 7 A) SEM image B, C) EDS analysis of sample heat treatment in 1060 ° C for 60 min and 60 min shot peening at points C and D (500X magnification)

Figure7 shows SEM image and EDS analysis for heat treatment sample at 1060 ºC for 60 minutes and shot peen for 60 minutes. C and D points indicate a presence of carbide elements of iron, manganese, molybdenum, and chromium. Figure 8 shows twinning treatment in Hadfield steel structure, where the shot peen operation has occurred. The parallel lines of the microstructure was observed. These parallel lines are due to crystalline directions in the austenite grains. Twin lines have formed in different directions. It also shows the effect of the shot peening treatment whereby the twinning treatments have formed, but martensitic transformations do not take place.

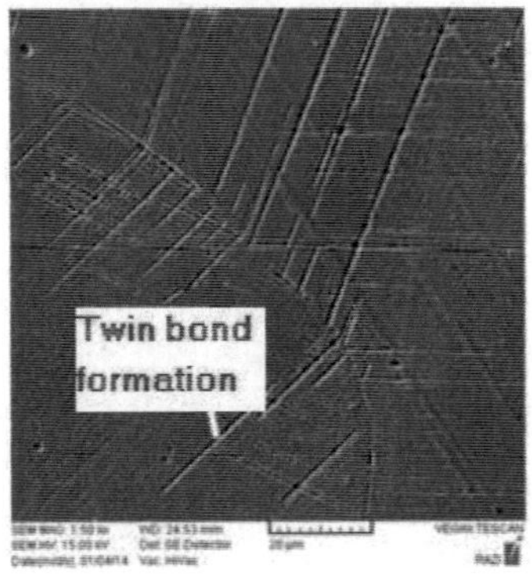

Figure 8: SEM image of the sample heat treatment at a temperature of 1050° C for 30-minute shot peening for 60 min at 1500X magnification

4. Conclusions

A fully austenitic manganese steel with a grain size suitable can be created using a proper heat treatment, such as heat treatment at 1150 ° C for 60 minutes per inch thickness pieces and then quenching in water bubbling on the structure of the Hadfield ASTM128A grade C.

By various cycles of the heat / quench treatment for the manganese steel, according to the absence of the new grain boundaries, the grain size of the Hadfield manganese steel cannot be smaller. The austenite grain size can be increased with the increasing of the temperature and time of the austenite treatment. At the lower temperature of the austenite heat / quench treatment (970°C), the continuous carbides is maintained even with an increase in the heat treatment time.

With the shot peening mechanical operations on the surface, the transformation of austenite to martensitic steel (Hadfield) does not take place. The work hardening signs have been affected by the phenomenon of the twins that is indicated in the visible parts by the metallographic images. To investigate of the metallographic structure of the shot peened surface layer, the twin bands difference direction in the orientation of the crystalline arrangement of the grains have observed.

References

[1] J. R. Davis, A.R. Gardeshzadeh , 2007, translated, "Alloying ", *Nopardazan*, , pp.301-314.

[2] Chijioke Okechukwu, Olurotimi Akintunde Dahunsi, Peter Kayode Oke, Isiaka Oluwole Oladele, Mohammed Dauda, 2017, Prominence of Hadfield Steel in Mining and Minerals Industries: A Review, *Internat. J. of Engineer. Techn.* -IJET, **Vol.3**, No.2, p83.

[3] A. K. Srivastava, K. Das, June, 2008, "Microstructural characterization of Hadfield Austenitic manganese steel ", *J. of Mat. Sci.*, **vol. 43**, issue 16, pp. 5654-5658.

[4] Shankha Pathak, Abhimanyu Maharshi, 2015, Development of a New Etchant for Austenitic Manganese Steel, *J. Adv. Res. Manu. Mat. Sci. Metall. Engi.*; **2**(1): 1- 9.

[5] C. Chen, F.C. Zhanga , F. Wang , H. Liu , B.D. Yu, 2017, Effect of N+Cr alloying on the

microstructures and tensile properties of Hadfield steel, *Mat. Sci. & Engin.*, **A 679**, p 95.
[6] F. Jahanbakhsh, M. A. Golozar, A. Najafizadeh,1996, Optimizing wear and Mechanical properties of hadfield steel for Industrial uses, *Steel Symposium, Metal sci. uni. of Esfahan.*
[7] Agunsoye, J.O.a,*, Talabi, S.I.b , Bello, O.a, 2015, Wear characteristics of heat-treated Hadfield austenitic manganese steel for engineering application, *Adv. in Produc. Eng. & Manag.* **10**(2).
[8] Shabnam Hosseini and Mohammad Bagher Limooei, 2011, "Optimization of Heat Treatment to Obtain Desired Mechanical Properties of High Carbon Hadfield Steels ", *J. World Appl. Sci.*, **vol. 10**, 1421-1424.
[9] C. H. Avner, A. F. and M. Erfanian, 2002, translated, Introduction of Physical metallurgy, *Nashre Daneshgahi Tehran center, First Publish,* pp. 362-363.
[10] M. Marashi, 2003, Functional metallurgy of Steels, *third Publish, Azadeh Publishers*, **vol. 1**, pp.157-188.
[11] A. Ahani, A. R. Aliporjahani, 2002 , Study on Select special Hadfield steel for lining parts of shotblasting mechine ,*Steel Symp. at Zob Ahan Isfahan.* pp. 88-91.
[12] S. *Kuyucak*, 2004, "Austenitic Manganese Steels Castings, Metallography and Microstructures ", *ASM Metals Handbook, ASM international,* **vol.9**, pp.701-708.
[13] V. *Tsakiris*, D. V. Edmonds, 1999, "Martensite and deformation twinning in austenitic steels ", *Mat. Sci. and Eng.* **A273–275**, pp. 430–436.
[14] C.Efstathiou, H.Sehitoglu, 2010, "Strain hardening and heterogeneous deformation during twinning in Hadfield steel", *Act. Material,* **Vol. 58**, pp.1479–1488.
[15] G.F. Liang, and C.j. Song, 2005, "Eutectic Decomposition in Ca-Si Modified Austenitic Medium Mn Steel After Solidification ", *Material Sci.*, **Vol.40,** pp.2081-2084.
[16] C. Chen, X.Y. Feng, B. Lv, , Z.N. Yang , F.C. Zhang, 2016, A study on aging carbide precipitation behavior of hadfield steel by dynamic elastic modulus, *Mat. Sci. & Eng.*, **A 677**, pp 446–452.

Numerical Investigation of Liquid Carryover in T-Junction with Different Diameter Ratios

William Pao*, **Ban Sam, Ahmed Saieed, Cong Minh Tran**

Universiti Teknologi PETRONAS, 32610 Bandar Seri Iskandar,

Perak, Malaysia

*Email: william.pao@UTP.edu.my

Phone: +6053687012; Fax: +6053656461

Abstract. In offshore Malaysia, T-junction is installed at the production header as a compact separator to tap produced gas from reservoir as fuel gas for power generation. However, excessive liquid carryover in T-junction presents a serious operational issue because it trips the whole production platform. The primary objective of present study is to numerically investigate the liquid carryover due to formation of slug, subsequently its liquid carryover at different diameter ratio. The analyses were carried out on a model with 0.0254 m (1 inch) diameter horizontal main arm and a vertically upward side arm using Volume of Fluid Method. Three different sides to main arm diameter ratio of 1.0, 0.5 and 0.3 were investigated with different gas and liquid superficial velocities. The results showed that, while the general trend is true that smaller diameter ratio T-junction has lesser liquid take off capacity, it has a very high frequency of low liquid carryover threshold. In other words, under slug flow, smaller diameter ratio T-junction is constantly transporting liquid even though at a lesser volume in comparison to regular T-junction.

Keywords: T-junction; liquid-gas separation; separation efficiency

1. Introduction

T-junction is an appendage of small diameter pipe, which is attached to the main pipe to tap the fluid source from the main stream. A T-junction consists of three main components, namely main arm, run arm and branch (or side) arm. T-junction configuration is most often seen in the offshore production platform where the T-junction is used to tap gas directly from the production header. The tapped gas is channeled downstream as fuel gas for power generation or source gas for other purposes. When a two-phase mixture flows pass a T-junction, lighter phase will incline to enter the branch arm while the heavier phase tends to continue flowing into the run arm. The higher the density difference between the two-phase, the better the phase separation is. Liquid carryover occurred when gas phase take-off to the branch arm reaching a certain limit, creating sufficient suction pressure to draw heavier liquid phase into the branch arm. Liquid carryover is a frequent problem that occurs in T-junction whereby excessive amount of liquid is taken off into the gas stream and channeled into downstream equipment which is not designed to handle liquid. Consequently, platform trips and production has to halt to drain out the excessive liquid.

The deceivingly simple T-junction with a simple geometry is an unbelievably complicated topic [1]. The geometry alone consists of eight different parameters that could have many combinations. Compounding to the complexity are flow parameters such as flow rate and operating pressure etc., and fluid properties such as density, viscosity and surface tension

variation. Experiments performed for side arm pointing at 6 o'clock position showed a significant pressure loss between the inlet and side-arm but a pressure increase in the run arm [2, 3]. This recovery is attributed to an affect similar to that of Bernoulli for single phase flows, produced as a result of the decrease in the mixture velocity in the run arm [2]. Under the dominance of gravity, the liquid will be accelerated, leading to pressure drop. When the side arm is pointing at 12 o'clock, the liquid take-off in the side arm is flowing against the gravity. The force by which the liquid climbs up the side arm is its own momentum force minus the gravity force. The momentum transfer by the gas phase is important if liquid is to take-off in the side arm as reported by [5]. [6], in their experiments, concluded that a lower pressure gradient in the branch arm provides the momentum change for the fluids moving from the inlet into the branch arm. [7] studied the effect caused by a reduction in side-arm diameter concluded that less liquid take-off in a reduced diameter T-junction. [8] experiment also showed that there is an increase in the fractional liquid take off with a decrease in the branch arm diameter. However, his experiments showed negligible impact of diameter ratio in stratified flow but more liquid take off in the regular T compared to a reduced T for stratified-wavy flow. [9] studied the scale effect of branch arm to main arm diameter ratio on phase split of liquid-gas flow at T-junction for annular and stratified flow. For the small scale T-junction, with 1.5" diameter pipe, data shows that decreasing the diameter ratio of a T-junction leads to lower liquid take off for both stratified and annular flow. For the larger scale T-junction, with 5" diameter pipe, similar conclusion was reached. However, he reported that the effect of diameter ratio is most significant at *lower gas rates* while least at lower liquid and high gas flow rate conditions. [10] experiment also showed reduced liquid take off with reducing T-junction diameter ratio.
Interestingly, there are very limited research which focus on the flow regime behavior before the two-phase flow reach the T-junction and its separation thereafter. The objective of the present paper is to investigate the effect of T-junction's diameter ratio on liquid carryover in slug flow regime.

2. Methodology

Governing Equations

Unlike the approach followed in [11, 12, 13] by the same group of authors, this paper used Volume of Fluid (VOF) Method to simulate the T-junction liquid carryover. The authors found that it was difficult to use Eulerian Multiphase Model to mimic the required flow regime. Even though mass flux values taken directly from Baker's map were taken as inputs, the Eulerian Multiphase Model cannot reproduce the flow regime contour visually. Consequently, an alternative mathematical model for T-junction was sorted after. In the VOF Model, only one momentum equation is required to represent the entire gas-liquid mixture. The mixture mass conservation equation and the momentum for the mixture are given by

$$\frac{\partial \rho}{\partial t} + \nabla \cdot (\rho \mathbf{U}) = 0 \tag{1}$$

$$\frac{\partial}{\partial t}(\rho \mathbf{U}) + \nabla \cdot (\rho \mathbf{U}\mathbf{U}) = -\nabla p + \nabla \cdot \left[\mu\left(\nabla \mathbf{U} + \nabla \mathbf{U}^T\right)\right] + \rho \mathbf{g} + \mathbf{F} \tag{2}$$

respectively, where ρ is the mixture density, $\mathbf{U}$ is the mixture velocity, p is the mixture pressure, μ is the mixture viscosity, $\mathbf{g}$ is the gravity acceleration and $\mathbf{F}$ is the body force vector. Equations (1) and (2) need to be complemented by the α-phase continuity equation from VOF given by

$$\frac{\partial \alpha_g}{\partial t} + \mathbf{U} \cdot \nabla \alpha_g = 0 \tag{3}$$

where α_g is the volume fraction of gas.

3. Results and Discussion

Validation of Slug Flow with Experiment

The validation makes use of [17], who conducted experiment in transparent rectangular conduit of 4 m long, 100 mm height and 30 mm width. The slug flow measurement of 5 s was taken with 5 m/s superficial inlet air velocity and 1 m/s superficial water inlet velocity. In order to achieve the superficial velocities described by the experiments, a cross sectional plane very near to the inlet was cut out and the averaged superficial velocities of each phase was adjusted on that plane to be as close as possible to the experimental values by tuning the mass flow rate. Figure 1 showed the side to side comparison of the present simulation model at 1.92 s, 1.96 s and 2.06 s respectively. There are some discrepancies which could be attributed to the inaccurate boundary conditions, particularly at the downstream of the conduit. Nevertheless, the present model is able to mimic the two-phase slug flow captured by experiment.

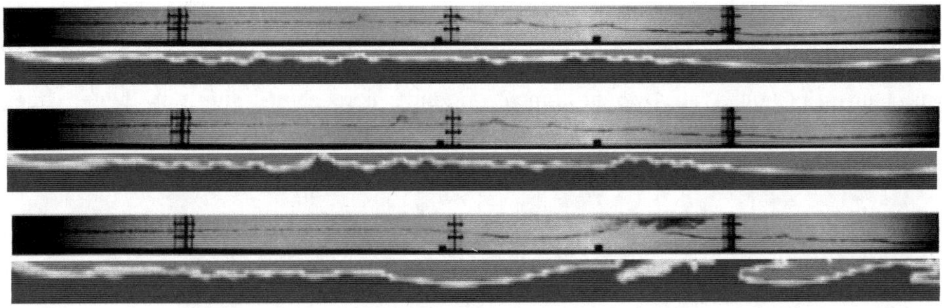

Figure 1. Comparison of experimental image of Christophe et al. [17] and the present simulation model

T-junction Separation with Reduced Branch Arm Diameter and Slug Flow

The schematic model of the T-junction is shown in Figure 2. The main and run arms were 0.0254 m (1 inch) in diameter and was kept constant. Three different branch arm diameters with the same height of 0.3 m were used, resulting in branch to main arm diameter ratio of 0.3, 0.5 and 1.0, respectively.

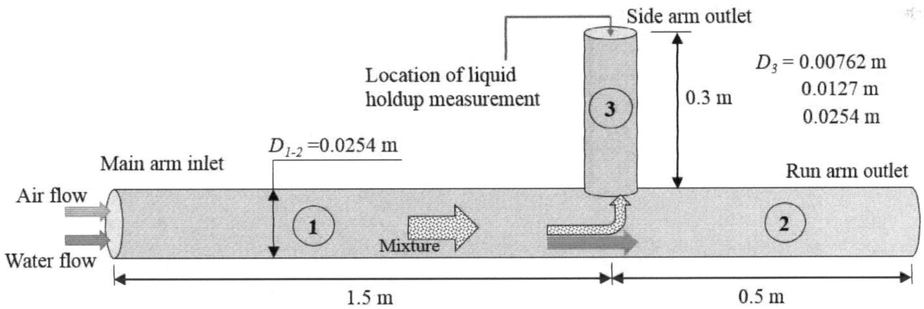

Figure 2. Schematic diagram of the T-junction for parametric study

The air and water mass flow rate (kg/s) were carefully calibrated so that the inlet flow generated the required slug in the main arm. After some tedious trial and error experiments, air and water velocities were set at 1 m/s and 2 m/s for each phase in a combinatorial fashion, making up 4 combination of velocities with identifiable high and low values. Table 1 listed the corresponding inlet air and water mass flow rate and the associated superficial velocities for the parametric study. This resulted in 12 simulations (4 variation of superficial velocities for 1.0, 0.5 and 0.3

diameter ratios). The liquid holdup and its frequency at the outlet of side arm were recorded from 0 to 5 seconds. In the present, the liquid holdup is defined as

$$H_L = \frac{A_L}{A} \tag{4}$$

where A_L is the cross-sectional area occupied by the liquid and A is the total cross-sectional area of the pipe. Liquid carryover is assumed to occur when $H_L \geq 0.75$ at the side arm outlet.

Table 1. Air and water mass flow rate for parametric study

	Superficial velocity	
Mass flow rate	1 m/s	2 m/s
Air flow rate (kg/s)	0.000621	0.001241
Water flow rate (kg/s)	0.506	1.012

The mechanism of liquid carryover in branch arm under approaching slug flow regime in the main arm is shown in Fig. 3 in order to illustrate the physical meaning of this parametric study.

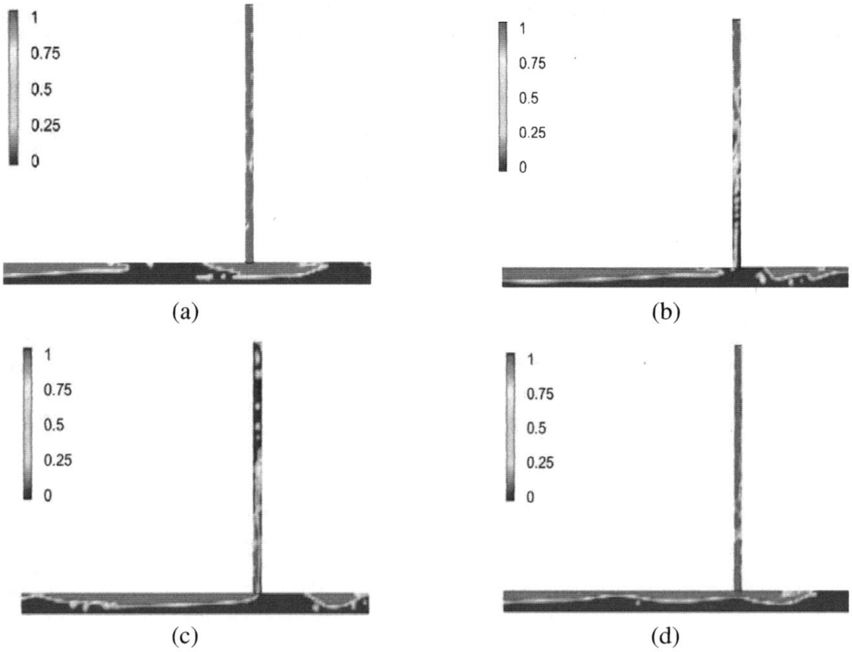

Figure 3. Liquid carryover in T-junction with an approach slug flow

In Figure 3(a), a slug body is approaching the T-junction. The liquid slug body is fully bridging the upper part of the pipe, with clearly identifiable slug nose and tail. In Figure 3(b), the slug body passed through the T-junction and broken up. The liquid jumped into the branch arm to about 0.2 m height, covering almost two third of the height of the branch arm. Liquid carryover occurred in the branch arm. When the liquid filled up 75% of the cross-section of the branch arm outlet, liquid holdup = 1. Figure 3(c) showed that the tail of the slug body is about to pass through the T-junction. The outlet of the branch arm is now completely filled with water. The air pocket is about to pass through the branch arm and portion of air started slipping into the appendage. The bottom of the branch arm is partially filled with air water and partially filled with air. The top section of the branch arm is fully saturated with water. Figure 3(d) showed that

IOP Conf. Series: Materials Science and Engineering **328** (2018) 012007 doi:10.1088/1757-899X/328/1/012007

the slug body completely passed the T-junction. All the holdup water in the branch arm has fallen back into the main arm. The branch arm is now filled up again with air, but traces of water can still be detected. Note that traces of water are detected close to the inner surface of the branch arm. The air velocity has barely enough momentum to drag the water film up into the branch arm.

Figure 4 summarized the normalized percentile versus different range of liquid holdup for four different cases of velocities and for three different diameter ratios. It is clearly evident that smaller diameter ratio T-junction has a better separation efficiency, or less liquid carryover, concurred the conclusion from [1, 3, 7, 9]. A further careful study revealed that while this is overall true for $H_L \geq 0.75$, smaller diameter ratio T-junction showed a very high percentile frequency when $H_L \leq 0.2$ compared to equal diameter T-junction. This is an interesting observation that was never reported in open literature. The numerical results showed that smaller diameter ratio T-junction has very low liquid carryover threshold, which was later confirmed by in-house experiments.

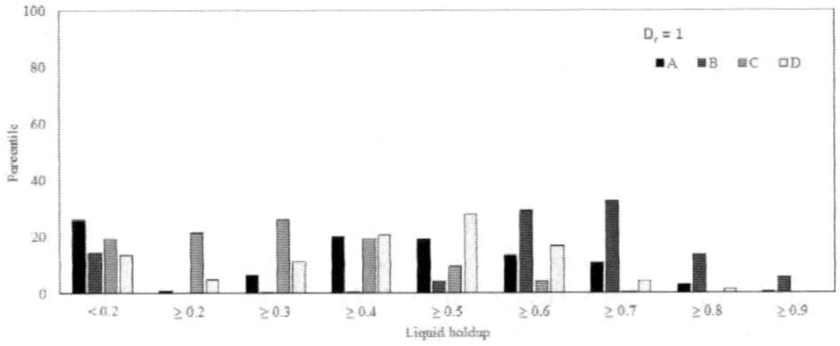

(a)

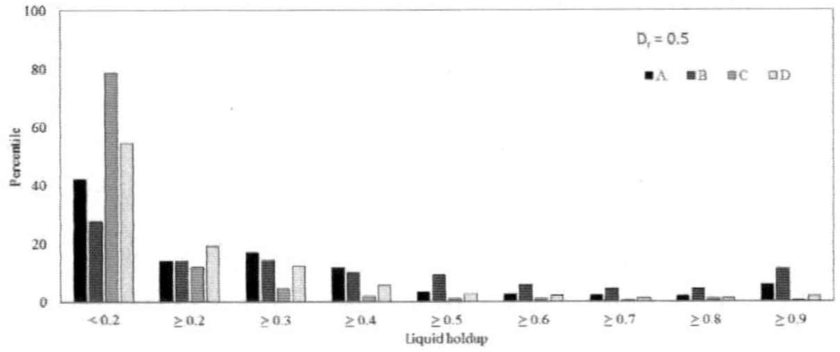

(b)

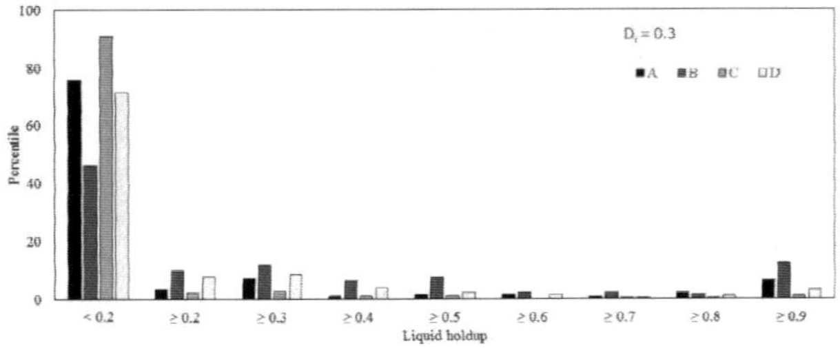

(c)

Figure 4. Liquid holdup distribution versus percentile frequency for diameter ratio (a) 1.0; (b) 0.5 and (c) 0.3

4. Conclusion

A numerical study is carried out using FLUENT software Volume of Fluid Method to investigate the effects of liquid holdup in a T-junction under different branch to main arm diameter ratios under the influence of different superficial phase velocity for slug flow. Numerical experiments found that liquid holdup in the branch arm decreases with decreasing diameter ratios, consistent with experimental conclusion. However, the smaller diameter ratio T-junction has very low liquid carryover threshold.

Acknowledgements

The autors acknowledge Universiti Teknologi PETRONAS for providing a platfrom for conducting this research. Special thanks to MOHE for finanically supporting this project via FRGS 0153AA-L03 and PETRONS Research Foundation via YUTP grant 0153AA-E03.

References

[1] Rea S, Azzopardi BJ. The split of horizontal stratified flow at a large diameter T-junction. Chemical Engineering Research and Design 2001; 79(4): 470-476.

[2] Baker G. *Separation and control of gas-liquid flows at horizontal T-junction.* (Doctoral dissertation, University of Nottingham).

[3] Wren EMK. *Geometric effects on phase split at a large diameter T-junction.* (Doctoral dissertation, University of Nottingham).

[4] Penmatcha VR, Ashton PJ, Shoham O. Two-phase stratified flow splitting at a T-jun. International Journal of Multiphase Flow 1996; 22(6): 1105-1122.

[5] Hart J, Hamersma PJ, Fortuin JMH. Phase distribution during gas liquid flow through horizontal dividing junctions. Nuclear Engineering and Design 1991; 126: 293-312.

[6] Penmatcha VR, Ashton PJ, Shoham O. Two-phase stratified flow splitting at a T-junction with an inclined branch arm. *International Journal of Multiphase Flow* 1996, 22: 1105-1122

[7] Azzopardi BJ. The effect of the side arm diameter on the two-phase flow split at a T junction. *International Journal of Multiphase Flow* 1984, 10: 509-512.

[8] Pandey S, Gupta A, Chakrabarti DP, Das G, Ray S. Liquid– Liquid Two Phase Flow Through a Horizontal T-Junction. *Chemical Engineering Research and Design* 2006, 84(10): 895-904.

[9] Azzopardi BJ. The effect of side arm diameter on phase split at T-junctions. in *SPE Annual Technical Conference and Exhibition* 1999 October 3-6, Houston, Texas, USA, pp. 1-9

[10] Marti S, Shoham O. A unified model for stratified flow splitting at a reduced diameter T-junction with an inclined branch arm. *International Journal of Multiphase Flow* 1997; 23(4): 725-748.

[11] Pao W, Hashim FM, Ming LH. Numerical investigation of gas separation in T-junction. In AIP Conference Proceedings, pp. 070001; 2015.

[12] Saieed A, Ban S, Pao W, Hashim FM. Numerical investigation of side arm gas volume fraction in two phase T-junction. *Journal of Mechanical Engineering and Sciences* 2016; 10(3): 2311-2323

[13] Ban S, Pao W, Nasif MS, Ofei TN. Preliminary numerical simulation of steady-state gas-liquid flow in horizontal T-junction. *ARPN Journal of Engineering and Applied Sciences* 2017; 12(8): 2570-2575.

[14] Baker O. Simultaneous flow of oil and gas. *Oil and Gas Journal* 1954; 53: 185-195.

[15] Fluent Inc, *Fluent 16.1 User's Guide*, Lebanon, USA, 2016.

[16] De Schepper SCK, Heynderickx GJ, Marin GB. CFD modeling of all gas–liquid and vapor–liquid flow regimes predicted by the Baker chart. *Chemical Engineering Journal* 2008; 138: 349-357.

[17] Christophe V, Thomas H, Prasser HM, Tobias S. Experimental investigation and CFD simulation of horizontal stratified two-phase flow phenomena. *Nuclear Engineering and Design* 2008; 238(3): 637-646.

Bit selection using field drilling data and mathematical investigation

M S Momeni[1], S Ridha[1], S J Hosseini[1,*], B Meyghani[2] and S S Emamian[2]

[1]Department of Petroleum Engineering, Universiti Teknologi PETRONAS, 32610 Bandar Seri Iskandar, Perak Darul Ridzuan, Malaysia
[2]Department of Mechanical Engineering, Universiti Teknologi PETRONAS, 32610 Bandar Seri Iskandar, Perak Darul Ridzuan, Malaysia

*Corresponding Author: javad.hosseini@utp.edu.my

Abstract. A drilling process will not be complete without the usage of a drill bit. Therefore, bit selection is considered to be an important task in drilling optimization process. To select a bit is considered as an important issue in planning and designing a well. This is simply because the cost of drilling bit in total cost is quite high. Thus, to perform this task, aback propagation ANN Model is developed. This is done by training the model using several wells and it is done by the usage of drilling bit records from offset wells. In this project, two models are developed by the usage of the ANN. One is to find predicted IADC bit code and one is to find Predicted ROP. Stage 1 was to find the IADC bit code by using all the given filed data. The output is the Targeted IADC bit code. Stage 2 was to find the Predicted ROP values using the gained IADC bit code in Stage 1. Next is Stage 3 where the Predicted ROP value is used back again in the data set to gain Predicted IADC bit code value. The output is the Predicted IADC bit code. Thus, at the end, there are two models that give the Predicted ROP values and Predicted IADC bit code values.

Keywords--Artificial Neural Network (ANN), International Association of Drilling Contractors (IADC), Rate of Penetration (ROP).

1. Introduction

1.1. Background

The basic form of bit selection is normally done based on cost per foot. This method is simply choosing the bit that will provide the lowest cost per foot over the upcoming interval. In addition to that, other factors are taken into consideration as well such as offset, journal angle, and other design aspects. This differentiates one bit to another according to the specific environments. Therefore, understanding bit types is a vital step before moving on to bit design as well as bit selection.

1.2. Bit Types

Rotary drilling bits can be generally classified as either drag bit or rolling cutter bits according to the design features. Drag bits have fixed cutter blades in common. These blades are integrated within the body of the bit. The rotation takes place with the drill string as one unit. On the other hand, the rolling cutter bits normally have 2 or more cones which have basic cutting elements. These cutters rotate about the axis of the cone during the bottomhole rotation [1]. There are two main types of bits which are Drag Bit (PDC Bit) and Rolling Cutter Bit (Tricone Bit).

1.3 Bit Design

1.3.1 Bit Design for Rolling Cutter Rock Bits

Basically, there are 3 basic types of bearings:

- The Non-Sealed Roller Bearings (NSRB)
- The Sealed and Lubricated Roller Bearings (SLRB)
- The Sealed and Lubricated Journal Bearings (SLJB)

1.3.2 Bit Design for PDC Bits

The PDC bit is known to be a solid one piece tool holding polycrystalline diamond cutters. The synthetic diamonds are molded into a thin layer to be attached to a tungsten carbide disc. This is done via a HPHT process. The shock load propagates through entire cutting via the random orientation of the cleavage planes that indirectly allows reduction in breakage. PDC bits are known to shear the rock which helps save energy. Thus, optimized drilling can be gained with the usage of less WOB. Given a favorable formation, PDC bits are acknowledged to perform longer and harder. The effectiveness is about 3 times than conventional rolling cutter bit [2]. Having all these plus points, PDC bits are quite expansive and can be destroyed by gumbo type formations. Therefore, a proper geologic analysis together with PDC bit compatibility must be done before drilling. The detailed field analysis for PDC bit has not been completed and this report is based on the data limitations [3].

1.4 Bit Selection Methods

A bit cost might be relatively small in a well's budget which is approximately 5% of the total expense of a well, but the impact of bit performance on overall well cost might end up being considerably large [4-6]. Bit selection is basically classified into three categories namely:

- Cost Analysis
- Offset Well Log Analysis
- Bit Performance Modeling Using ANN and Genetic Algorithm

2. Methodology

ANN is a computational technique used to solve complex problems. MLP network is one of the most popular neural network architectures for modeling process [7]. It consists of input layer of source nodes, hidden layer of computation nodes (neurons), and output layer. The number of nodes that is being used in the input as well as the output layer is completely dependent on the number of input and output variables being used respectively [8]. The figure shown is a schematic of MLP network. Further theoretical details about MLP networks is presented by [9]. In this project, the MLP network is trained by the usage of LM technique. One of the advantages of using the LM technique is that it produces efficient and faster second order convergence rate and is capable to maintain the stability at the same time [10, 11]. Based on the research papers, it is clearly stated that the single layer is good enough to approximate a complex function [12].Therefore, in this study, a three layered Feed-Forward Network is developed namely input, hidden, and output layers. Among the available data sets which are gathered from different offset wells, 70% is used for training, 15% is applied for validation process and the remaining 15% is used to test gained results from the bit being modeled as well as the ROP functions. Stage 1 was to find the IADC bit code by using all the given filed data. This data includeSize(in), Flow area (in2), Depth out (m MD), Bit meter (m), Rotation hours (hrs), ROP(m/hr), Minimum Rotation (rpm), Maximum Rotation (rpm), Total bit revolution, Minimum Weight (kN), Maximum Weight (kN), Minimum Flow (l/min), Maximum Flow (l/min), Minimum Pump Pressure (bar), and Maximum Pump Pressure (bar). The output is the Targeted IADC bit code.

Table 1. Method in Obtaining ANN Results

Model	Data Being Used	Description
ANN	Drilling Data	10= 512X 20=434X 30=532M
Output IADC Bit Code	**Example** 10,20,30	Ex: ANN predicts the numbers as perhaps 11.5 (10), 23.4 (20), and 38.7 (30)

The results of the three methods discussed above are considered for ROP prediction and improve predicted IADC bit code value.

3. Results and Discussion

For the ANN simulation process, the Matlab software was used. The Neural Network tool which was prebuild in this Matlab software was used to gain all the results and generate all the graphs needed. The type of data used for this project comprises of drilling data. The table below shows an example of data set that was used. The following table represents the range of the data which was used.

Table 2. Data Type Example

Type	MITO	Rotation min(rpm)	60
Size(in)	17,50	Rotation max (rpm)	60
IADC Code	115M	Total bit revolution	86400
Input For ANN	1	Weight min (kN)	20
Flow area (in2)	0.994	Weight max (kN)	60
Depth out (m MD)	68	Flow min (l/min)	2500
Bit meter (m)	55	Flow max (l/min)	2500
Rot hours (hrs)	24,00	Pump min/max (bar)	20
ROP(m/hr)	2,3	Pump max (bar)	22

Table 3. Data Type Range

Type	-------- -------	Rotation min (rpm)	0-271

Size(in)	4.13-36.00	Rotation max (rpm)	0-335
IADC Code	-------- -------	Total bit revolution	0-1829000
Input For ANN	1-150	Weight min (kN)	0-240
Flow area (in2)	0.000-2.227	Weight max (kN)	0-400
Depth out (m MD)	52-4762	Flow min (l/min)	0-4000
Bit meter (m)	0-569	Flow max (l/min)	0-4100
Rot hours (hrs)	0.44-190.78	Pump min (bar)	0-242
ROP(m/hr)	0.00-50.00	Pump max (bar)	0-297

The following was done to gain the results that will be discussed below. The Azar Well data was used in the ANN modeling process. The modeling process was done in three stages. Stage 1 was to find the IADC bit code by using all the given filed data. This data includes Size(in), Flow area (in2), Depth out (m MD), Bit meter (m), Rotation hours (hrs), ROP(m/hr), Minimum Rotation (rpm), Maximum Rotation (rpm), Total bit revolution, Minimum Weight (kN), Maximum Weight (kN), Minimum Flow (l/min), Maximum Flow (l/min), Minimum Pump Pressure (bar), and Maximum Pump Pressure (bar). The output is the Targeted IADC bit code. Stage 2 was to find the Predicted ROP values using the gained IADC bit code in Stage 1. This time, the data used as input in the ANN modeling process include Targeted IADC bit code, Size(in), Flow area (in2), Depth out (m MD), Bit meter (m), Rotation hours (hrs), Minimum Rotation (rpm), Maximum Rotation (rpm), Total bit revolution, Minimum Weight (kN), Maximum Weight (kN), Minimum Flow (l/min), Maximum Flow (l/min), Minimum Pump Pressure (bar), and Maximum Pump Pressure (bar). The output is the Optimized ROP. Next is Stage 3 where the Predicted ROP value is used back again in the data set to gain Predicted IADC bit code value. The input parameters were Size(in), Flow area (in2), Depth out (m MD), Bit meter (m), Rotation hours (hrs), Predicted ROP(m/hr), Minimum Rotation (rpm), Maximum Rotation (rpm), Total bit revolution, Minimum Weight (kN), Maximum Weight (kN), Minimum Flow (l/min), Maximum Flow (l/min), Minimum Pump Pressure (bar), and Maximum Pump Pressure (bar). The output is the Predicted IADC bit code. The results were gained and they were used to plot regression graphs. The graphs are shown below. In all the simulations done under the ANN modeling, the data was used with 70% for training data, 15% for validation data, and 15% for test data.

For the simulation process, the number of hidden neurons can be varied from a range of 1 until 23. The best simulation results were gained when the number of hidden neurons was set to 10. The following table shows the Mean Square Error values and R values for Training Data, Validation Data, and Test Data.

Table 4. MSE And R Value For 10 Number Of Hidden Neurons (Stage 1)

	SAMPLES	MSE	R
TRAINING	71	498.46659	0.82799
VALIDATION	16	1287.67076	0.68085
TESTING	16	2095.87942	0.83949

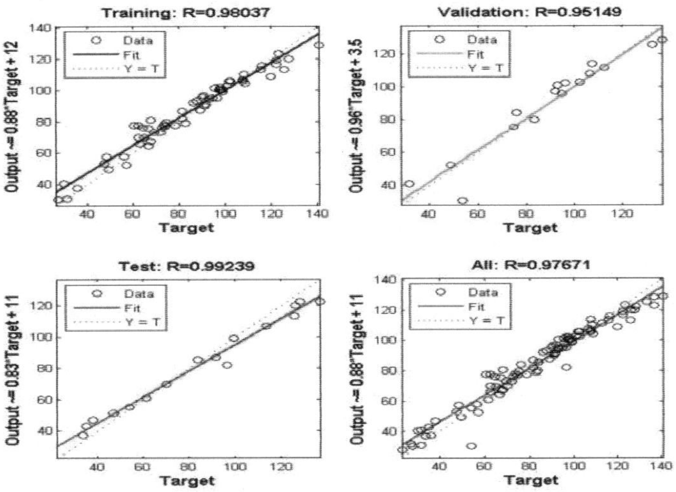

Figure 1. Results Of The Regression Graphs For Output Of Optimized IADC Bit Code (Stage 2)

For the stage 3 simulation process, it is seen that the number of hidden neuron layers of 17 showed a better response compared to number of hidden neuron layers of 10.

Table 5. Predicted IADC Bit Code Value And Predicted ROP Value Examples

IADC Bit Code	ANN Output	Predicted IADC Bit Code Value	ROP (ft/hr)	Predicted ROP Value (ft/hr)
1,1,1	10	8.540809	19.263360	20.03738
1,1,1	10	10.49332	16.2142857	18.74017
1,3,4	70	74.78044	7.77586207	3.43065
2,1,4	90	93.24648	3.37209302	2.866403
2,1,4	90	85.00292	3.09090909	2.379206
1,3,4	70	65.81053	2.80991736	1.642444
1,3,5	80	83.09457	2.80991736	2.324059
2,1,4	90	90.54357	1.39534884	1.711341
2,1,4	90	70.67985	1.93478261	1.633869
2,1,4	90	71.96897	1.90697674	1.499116
2,1,4	90	71.90172	2.0000000	1.35653
2,1,4	90	72.16207	1.60655738	1.455876
2,1,4	90	70.22408	1.5106383	1.02429
2,1,4	90	88.04242	1.41176471	1.29425
2,1,4	90	83.33124	2.63157895	1.329831
1,1,1	10	6.796307	13.5396825	12.9523
1,1,1	10	8.319983	6.4000000	5.959551
1,1,1	10	6.990621	4.30769231	3.70865
1,1,1	10	9.344001	3.3800000	3.596977

These values can be concluded as a good approximation as the predicted values are almost near to target values provided. The graph below shows the comparison of predicted IADC Bit Code versus the targeted IADC Bit Code.

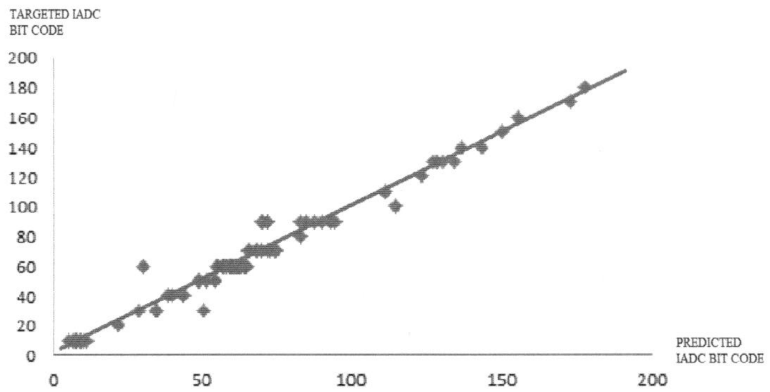

Figure 2. Results Of The Regression Graphs For Output Of Predicted IADC Bit Codes

Based on the second set of data set used to recheck the accuracy of ANN simulation, it can be said that the ANN simulation process is quite accurate to be used as a prediction tool. Thus, a third set of data was taken into consideration to confirm the accuracy of the ANN simulation process whereby the results were predicted to be at least 95% accurate.

4. Conclusion

In conclusion, the objective of the project has been achieved. The ANN simulation modeling process has been used to Predict ROP value and the Predicted IADC Bit Code has been found. Generally, the training is done for about 10 to 15 times for each Number Of Hidden Neurons Layers to gain the best output of Mean Square Error, MSE and Regression value, R. It has been seen during the simulation process that the most suitable hidden neuron number is 10 and 17. For the Azar Well 1 and 2 data sets, the final accuracy levels of the Predicted IADC Bit Code Values were 0.99 and the final accuracy level of the Predicted ROP Values were 0.95. This clearly shows that the usage of ANN simulation as a prediction tool can provide and accuracy level of more than 95 percent which is fairly accurate. In order to reconfirm that these values were accurate enough, a new set of data was also used to see if it gives a proper prediction. Once again, the final accuracy levels of the Predicted IADC Bit Code Values were 0.96 and the final accuracy level of the Predicted ROP Values were 0.95 for the second set of data set. Therefore, it is clearly seen that this project has managed to utilize the ANN simulation to predict the IADC Bit Code and ROP values quite effectively. However there are still some errors in the prediction. This is simply due to certain reasons which are insufficient data, limited time frame to complete the project, and also the absence of Log Data.

REFERENCES

1. Fasheloum M. Investigation of drilling parameters indicators: University of Nottingham; 1997.
2. Nygaard R, Hareland G, editors. How to select PDC bit for optimal drilling performance. Rocky Mountain Oil & Gas Technology Symposium; 2007.
3. Oteri VA. Drilling Optimization- Drill Bit Performance Optimization Using DROPS Simulator. THESIS. 2010 11-30.
4. Lummus J. Acquisition and analysis of data for optimized drilling. Journal of Petroleum Technology. 1971;23(11):1285-93.
5. Yılmaz S, Demircioglu C, Akin S. Application of artificial neural networks to optimum bit selection. Computers & Geosciences. 2002;28(2):261-9.

IOP Conf. Series: Materials Science and Engineering **328** (2018) 012008 doi:10.1088/1757-899X/328/1/012008

6. Mostafavi H, Jamshidi E, editors. Soft Computation Application to Optimize Drilling Bit Selection Utilizing Virtual Inteligence and Genetic Algorithms. International Petroleum Technology Conference; 2013: International Petroleum Technology Conference.
7. Feng S, Li L, Cen L, Huang J. Using MLP networks to design a production scheduling system. Computers & Operations Research. 2003;30(6):821-32.
8. Pinar E, Paydas K, Seckin G, Akilli H, Sahin B, Cobaner M, et al. Artificial neural network approaches for prediction of backwater through arched bridge constrictions. Advances in Engineering Software. 2010;41(4):627-35.
9. Haykin S. Neural networks: a comprehensive foundation: Prentice Hall PTR; 1994.
10. Hagan MT, Menhaj MB. Training feedforward networks with the Marquardt algorithm. Neural Networks, IEEE Transactions on. 1994;5(6):989-93.
11. Cigizoglu H, Kisi O. Flow prediction by three back propagation techniques using k-fold partitioning of neural network training data. Nordic Hydrology. 2005;36:49-64.
12. Wilamowski BM, Iplikci S, Efe MO, editors. An algorithm for fast convergence in training neural networks. Neural Networks, 2001 Proceedings IJCNN'01 International Joint Conference on; 2001: IEEE.

IOP Conf. Series: Materials Science and Engineering **328** (2018) 012009 doi:10.1088/1757-899X/328/1/012009

Modified SPC for short run test and measurement process in multi-stations

C K Koh[1]*, J F Chin[1] and S Kamaruddin[2]

[1]School of Mechanical Engineering, Universiti Sains Malaysia, 14300, Nibong Tebal, Pulau Pinang, Malaysia
[2]Department of Mechanical Engineering, Universiti Teknologi PETRONAS, 32610, Seri Iskandar, Perak, Malaysia

*Email: kck_79@hotmail.com

Abstract. Due to short production runs and measurement error inherent in electronic test and measurement (T&M) processes, continuous quality monitoring through real-time statistical process control (SPC) is challenging. Industry practice allows the installation of guard band using measurement uncertainty to reduce the width of acceptance limit, as an indirect way to compensate the measurement errors. This paper presents a new SPC model combining modified guard band and control charts ($\overline{Z}$ chart and W chart) for short runs in T&M process in multi-stations. The proposed model standardizes the observed value with measurement target (T) and rationed measurement uncertainty (U). S-factor (S_f) is introduced to the control limits to improve the sensitivity in detecting small shifts. The model was embedded in automated quality control system and verified with a case study in real industry.

1. Introduction

The key characteristics of test and measurement (T&M) manufacturing are short production runs of multi-product families and testing at multi-stations. Classical Shewhart control charts, namely $\bar{x}$ chart and R chart have been widely used in statistical process control (SPC) to detect process variable shifts in mean and variance [1]. Short production runs render these charts inefficacious as inherent meager data do not warrant meaningful control limits [2]. Furthermore, Costa and Castagliola [3] underscore the problem of growing risk of false acceptance in SPC due to measurement error. This leads to consequences such as unnecessary process adjustment and loss of confidence in SPC.

In these premises, this research proposes a modified SPC model which combines modified guard band and control charts ($\overline{Z}$ chart and W chart) to address the issues caused by short production runs and measurement errors. The proposed model standardizes the observed value with measurement target (T) and rationed measurement uncertainty (U). The organization of the paper is as follow: A review of the literature of measurement uncertainty and short run techniques and will be presented in the next section. The proposed model and the industrial case study are described in section 3 and section 4 respectively. Section 5 discusses the results obtained from the case study. Finally, section 6 draws conclusion.

2. Literature Review

2.1. Short Run approaches

Several short run charts have been proposed using statistics techniques such as difference chart, DNOM (deviation from nominal) chart and standardized charts [1,4]. Difference chart (also called X-nominal) registers differences between the mean to its nominal/target value [5]. DNOM chart is based on ratios of subgroup means to nominal [6]. Useful in a scenario with shifting variances, standardized charts with short runs techniques was proposed by Bothe [2] to revise the target value proportionally to the amount of dispersion from product to product. It allows the operator to plot different product families on a single $\bar{x}$ chart and R chart.

Wheeler [5] extends the approach using the Z chart (also called Zed chart) and W chart to measure central tendency and dispersion respectively. Consider a process where x_{ij} represents the ith response on the jth part family member, $i = 1, \ldots, m$, and $j = 1, \ldots, p$. The process response is normalized using an estimate of the population standard deviation, denoted by $\hat{\sigma}_{x_j}$ as:

$$\hat{\sigma}_{x_j} = \frac{\bar{R}_j}{d_2} \tag{1}$$

where $\bar{R}_j$ is the average range for the jth part family member; d_2 is a statistic constant for subgroup size (n) displaying a normally distributed quality characteristic. The Z_{ij} values for Z chart are then calculated as:

$$Z_{ij} = \frac{x_{ij} - T_j}{\hat{\sigma}_{x_j}} \tag{2}$$

where T_j is the target value for each part family member that can be determined by using historical data, specification, or prior experience on the similar parts [1]. Wheeler [5] introduced $\bar{Z}$ chart (also called Zed bar chart) for subgroup data when $n > 1$. The mean values for $\bar{Z}$ chart is then calculated as:

$$\bar{Z}_{ij} = \frac{\bar{x}_{ij} - T_j}{\hat{\sigma}_{\bar{x}_j}} \tag{3}$$

where $\hat{\sigma}_{\bar{x}_j}$ is an estimate of the population standard deviation of the subgroup mean as:

$$\hat{\sigma}_{\bar{x}_j} = \frac{\bar{R}_j}{d_2\sqrt{n}} \tag{4}$$

and the ranges, R_{ij} for the W chart are transformed as follows:

$$W_{ij} = \frac{R_{ij}}{\hat{\sigma}_{x_j}} \tag{5}$$

$\bar{Z}$ chart will have a central line at zero with the upper and lower control limits set at +3 and -3, respectively. W chart set its central line at d_2 as R= $\bar{R}$; upper and lower control limits at $d_2 + 3d_3$ and $d_2 - 3d_3$ respectively, where d_3 is a statistic constant for subgroup size (n).

The advantages of $\bar{Z}$ chart and W chart, are that the measured data from multiple product families can be plotted on a single chart and relatively low average run length (ARL) required in comparison to $\bar{x}$ chart and R chart [7]. The disadvantages of $\bar{Z}$ chart and W chart, are that the sigma is often unknown, a separate estimate is usually obtained for each product and this method requires sufficient data to negate the fact that the true parameter values used in the calculation of the control limit estimation [8]. These charts are inefficiently adopted in some T&M processes due to the measured data are affected by measurement error.

2.2. Guard band approaches

The most immediate approach to minimize the measurement errors is by utilizing guard band to reduce the width of acceptance limit. Various economic aspects of guard band have been proposed by several researchers (e.g., [9]) to ensure the acceptable risk decision in the product conformity. Industry practice allows the installation of guard band, e.g., through Guide to the Expression of Uncertainty in

Measurement (GUM), published by ISO in 1993 to reduce the width of acceptance limit [10], as an indirect way to compensate the measurement errors. Incorporation of stringent control as in $\overline{Z}$ chart and W chart into guard band could potentially aid to screen for assignable causes and to detect early quality deterioration in the process.

3. Modified SPC model

3.1. Overview

The model standardizes the observed value in relation to the measurement target (*T*) and measurement uncertainty (*U*). Measurement target is referred as an expected value for a predetermined measurement characteristic. The target value is calculated based on repeated measurement with series observations for each test station separately and to be revised when the measurement characteristic changes. Measurement uncertainty refers to the dispersion of the data that could reasonably be attributed to the measurement result [11]. Pythagoras's theorem is used to estimate the process standard deviation with population standard deviation and measurement uncertainty as inputs. S-factor (S_f) is introduced as separate approaches to the control limits to improve the sensitivity in detecting small shifts. Eight tests are proposed to interpret the charts using Nelson's rules.

3.2. Evaluation of measurement uncertainty

Measurement uncertainty will be evaluated using GUM [10], in four steps as follows:

3.2.1. Step 1: Identifying all sources of uncertainty. The first step is to identify the output result, Y, from N input quantities through a function relation *f* as in equation (6):

$$Y = f(X_1, X_2, ..., X_N) \tag{6}$$

where X_i is an input quantity that can significantly affect the measurement result. An estimate of the output Y, denoted by y, using input estimates $x_1, x_2, ..., x_N$, as shown in equation (7):

$$y = f(x_1, x_2, ..., x_N) \tag{7}$$

Determining the uncertainty of y requires the uncertainties of all the input estimates x_i referred as standard uncertainties $u(x_i)$.

3.2.2. Step 2: Evaluating the standard uncertainty. The value for $u(x_i)$, could be obtained from either Type A or Type B evaluation. Type A evaluation is characterized by a statistically estimated sample standard deviation s_i, and the associated number of degrees of freedom v_i. The sample standard deviation of the mean is computed as:

$$u(x_i) = \frac{S_i}{\sqrt{r}} \tag{8}$$

where r was the number of the independent repeated observations. The r should be large enough to ensure the estimated value is reliable that the probability distribution often is assumed to be normal [12].

Type B evaluates a component of measurement uncertainty that has been excluded in Type A evaluation. In the evaluation, $u(x_i)$ refers to an assumed probability distribution based on previous measurement data, input of experienced personnel, manufacturer's specification, data provided in calibration report [13]. Amongst various distributions exist, (e.g., rectangular, triangular, U-shaped, normal.), rectangular distribution is the most common in the analysis of T&M. The accompanying standard measurement uncertainty is:

$$u(x_i) = \frac{a}{\sqrt{3}} \tag{9}$$

where a is the semi-range (or half-width) constant set between the upper and lower limits.

3.2.3. Step 3: Computing the Combined Standard Uncertainty. Combined standard uncertainty, denoted by $u_c(y)$ is compiled from individual $u(x_i)$, using Summation in Quadrature which is often called the law of propagation of measurement uncertainty, as given below:

$$u_c(y) = [\Sigma_1^n[c_i u(x_i)]^2]^{\frac{1}{2}} \quad (10)$$

where c_i is the sensitivity coefficient associated with x_i. Eq. 10 is valid only if the input quantities X_i are independent or uncorrected. In most cases, input quantity X_i is uncorrelated, and sensitivity coefficient can be assumed to be 1 [14].

3.2.4. Step 4: Computing the Expanded Uncertainty. Originated from GUM, expanded uncertainty, U is calculated by factoring in a coverage factor, k into combined standard uncertainty $u_c(y)$:

$$U = ku_c(y) \quad (11)$$

where k is chosen from the effective degrees of freedom vi of all the uncertainty sources, considering a specified coverage probability, calculated through Student's t-distribution. Most commonly, $k = 2$ is chosen to give a level of confidence of approximately 95% [15]. It is recommended that the expanded uncertainty (U) be rounded normally to two significant figures [12].

The doubt of the measurement can be quantified when the true value is within the margin with U and the confidence level [15]. All measurement uncertainty components regardless of classification are modeled by probability distributions quantified by variances or standard deviations [12]. Therefore, the process standard deviation contributed by measurement error can be estimated using equation (12), which is termed standard deviation of the measurement uncertainty and denoted by $\widehat{\sigma}_u$.

$$\widehat{\sigma}_u = \frac{U}{k} \quad (12)$$

3.3. *Standardized measurement target and measurement uncertainty*

The measurement uncertainty is dependent on the accuracy of the equipment used in the individual test station. An equation is formed by extending the equation (3) with *m*th test station instead of *j*th part family member. $\overline{Z}_{im}$ for modified $\overline{Z}$ chart takes measurement uncertainty into account, as depicted in equation (13):

$$\overline{Z}_{im} = \frac{\sqrt{n}\,(\bar{x}_{im} - T_m)}{\widehat{\sigma}_{p_m}} \quad (13)$$

where T_m is an expected value of the measurement target and $\widehat{\sigma}_{p_m}$ is an estimate of the process standard deviation for each test station. Range for modified W chart are extended from equation (5) as follows:

$$W_{im} = \frac{R_{im}}{\widehat{\sigma}_{p_m}} \quad (14)$$

The $\widehat{\sigma}_{p_m}$ provides the tolerance for the control limits. The model iteratively expands the tolerance limit from the lowest to highest value of $\widehat{\sigma}_{p_m}$. For the lowest values, min $\widehat{\sigma}_{p_m} = \widehat{\sigma}_{x_m}$, where $\widehat{\sigma}_{x_m}$ is an estimate of the population standard deviation that can be calculated by:

$$\widehat{\sigma}_{x_m} = \frac{\overline{R}_m}{d_2} \quad (15)$$

where $\overline{R}_m$ is the average range for the mth test station. For the highest values, $\widehat{\sigma}_u$ from equation (12) is taken into account. Pythagoras's theorem is used to estimate the maximum value of the $\widehat{\sigma}_{p_m}$, as given below:

$$\max \widehat{\sigma}_{p_m} = \sqrt{\widehat{\sigma}_{x_m}{}^2 + \widehat{\sigma}_{u_m}{}^2} \quad (16)$$

The central line and control limits of modified charts are fixed with constant values as applied in the $\overline{Z}$ chart and W chart. The charts are interpreted using Nelson's rules as defined in table 1. Rules 1,

2, 5, and 6 are to be applied to upper and lower halves of the chart separately. Rules 3, 4, 7, and 8 are to be applied to whole chart [16]. With this, all rules apply to $\bar{x}$ chart and $\bar{Z}$ chart. However, only rule 1, 2, 5 and 6 can be applied to dispersion chart (R chart and W chart) without modification when the subgroup size is 5 or more which will lead to the symmetric control limits [17].

Table 1. Nelson Rules

Rule	Description
Rule 1	One point is more than 3 standard deviations (σ) from the mean.
Rule 2	Nine points in a row are on the same side of the mean.
Rule 3	Six points in a row are continually increasing (or decreasing).
Rule 4	Fourteen points in a row alternate in direction, increasing then decreasing.
Rule 5	Two out of three points in a row are more than 2σ from the mean in the same direction.
Rule 6	Four out of five points in a row are more than 1σ from the mean in the same direction.
Rule 7	Fifteen points in a row are all within 1σ of the mean on either side of the mean.
Rule 8	Eight points in a row exist with none within 1σ of the mean and the points are in both directions from the mean.

3.4. *Determining $\widehat{\sigma}_{p_m}$ through S-factor*

S-factor is an approach to estimate $\widehat{\sigma}_{p_m}$ is through the expanded uncertainty (U_m) and S-factor (S_{f_m}), with the formula set as below:

$$\widehat{\sigma}_{p_m} = \frac{U_m}{S_{f_m}} \tag{17}$$

S-factor is then delineated as the ratio between the highest value of $\widehat{\sigma}_{p_m}$ and the lowest value of $\widehat{\sigma}_{p_m}$, is given by:

$$S_{f_m} = \frac{\max \widehat{\sigma}_{p_m}}{\min \widehat{\sigma}_{p_m}} \tag{18}$$

and the $\widehat{\sigma}_{p_m}$ is set equal to min $\widehat{\sigma}_{p_m}$ when the estimated value of the $\widehat{\sigma}_{p_m}$ is equal or less than $\widehat{\sigma}_{x_m}$, where:

$$\min S_{f_m} = \frac{U_m}{\widehat{\sigma}_{x_m}} \tag{19}$$

As shown in figure 1, S-factor is useful to rescales the process standard deviation, and it is dependent on the ratio between the measurement uncertainty and the population standard deviation, which is termed as $U/k\sigma_x$. For example, the $\widehat{\sigma}_{p_m}$ estimated by S-factor is equal to max $\widehat{\sigma}_{p_m}$ when $U/k\sigma_x$ ratio is equal to 1, and it is equal to min $\widehat{\sigma}_{p_m}$ when U/kσx ratio is equal or less than 0.577.

Finally, the $\bar{Z}_{im}$ values for $\bar{Z}$ chart are given by:

$$\bar{Z}_{im} = \frac{\sqrt{n}\, S_{f_m}(x_{im} - T_m)}{U_m} \tag{20}$$

and range values for W chart are transformed using the following formula:

$$W_{im} = \frac{S_{f_m} R_{im}}{U_m} \tag{21}$$

3.5. *Model implementation*

In implementation, the model was embedded in an automated SPC system. The SPC application is developed using Visual Studio to gather data from test record database and to perform data analysis using analysis tool developed by Scrucca [18]. The tool uses R package (open-source programming

IOP Conf. Series: Materials Science and Engineering **328** (2018) 012009 doi:10.1088/1757-899X/328/1/012009

language) for quality control charting and violating rules checking. Automated SPC includes four steps: extracting data, creating analysis files, configuring setting and executing the SPC application. Variables used in the model (e.g., T_m, U_m, S_{f_m}) are defined and to be configured at setting stage. Two set of charts are plots, as below:

1. Classical Shewhart control charts ($\bar{x}$ chart and R chart)
2. $\overline{Z}$ chart and W chart with $\hat{\sigma}_{p_m}$ determined through S-factor

Eight tests based on Nelson's rules interpret the charts. False alarm rate is used to compare the performance of the proposed models with classical Shewhart control charts. A false alarm will be counted if the rule test detected when the process behaves as per this normal or in-control behavior.

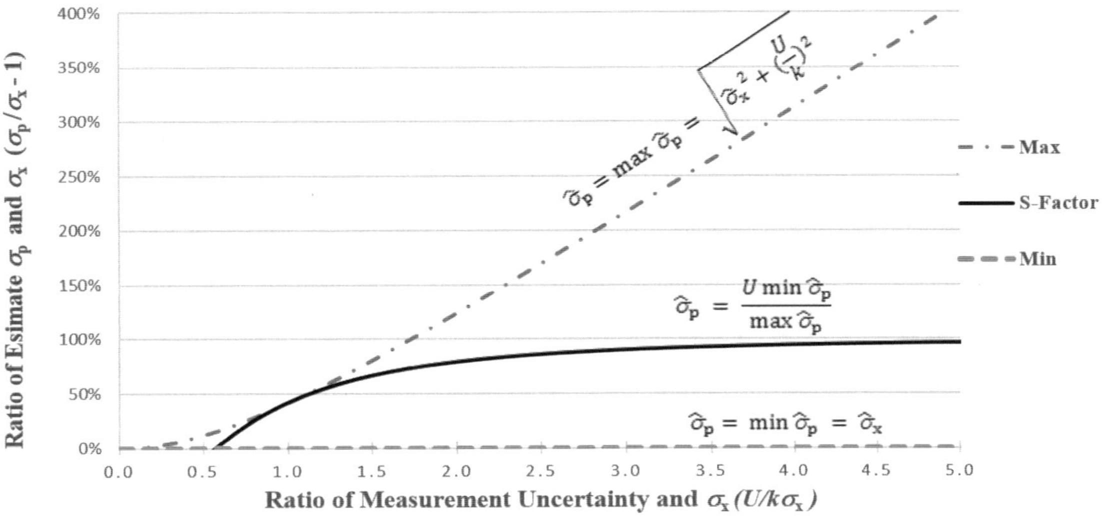

Figure 1. Estimated process standard deviation with S-factor

4. Verification through case study in real industry

A complete year's data samples were collected from a product tested at multi-stations in a T&M manufacturing facility at Bayan Lepas, Penang. The DC vertical gain accuracy (VGA) for oscilloscope test is selected where it consists of 700 measurement results tested at three stations (WH05, WH06, and WH07). Two sets of charts are constructed with usually recommended subgroup size of 5 [19]. First six months' data are used in phase I to construct trail control limits. To produce good results in practice, three sigma control limits were used so that the probability of type I error is kept at 0.0027 [1]. These control limits are used in Phase II to monitor process outcomes with second half six months' data. The implementation and analysis for all test stations are performed using same procedure. In this section, station WH05 is used to demonstrate the implementation with the proposed model.

4.1. Classical Shewhart control charts ($\bar{x}$ chart and R chart)

In figure 2, 5 points and 1 point violate the rules in $\bar{x}$ chart and R chart respectively. However, in actuality, the process is said in compliance to industry standard as all reading and its tolerance coincidentally falls within the acceptance limits with measurement uncertainty guard band as shown in figure 3. With this, the violated points are considered false alarm, except for the points circled in red are considered true signal where there is large shift (rule 1) and small sustained shift (rule 6) on July 2015. To solve the false alarm problem, operator may compensate the error through process adjustment or the control engineer will perform the retrospective analysis to construct a new control limits.

IOP Conf. Series: Materials Science and Engineering **328** (2018) 012009 doi:10.1088/1757-899X/328/1/012009

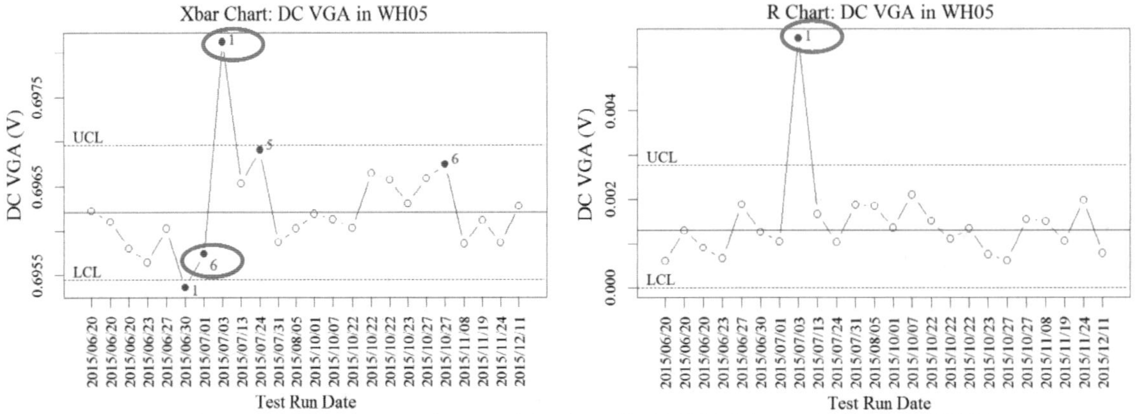

Figure 2. Shewhart x̄ chart and R chart for DC VGA test in station WH05

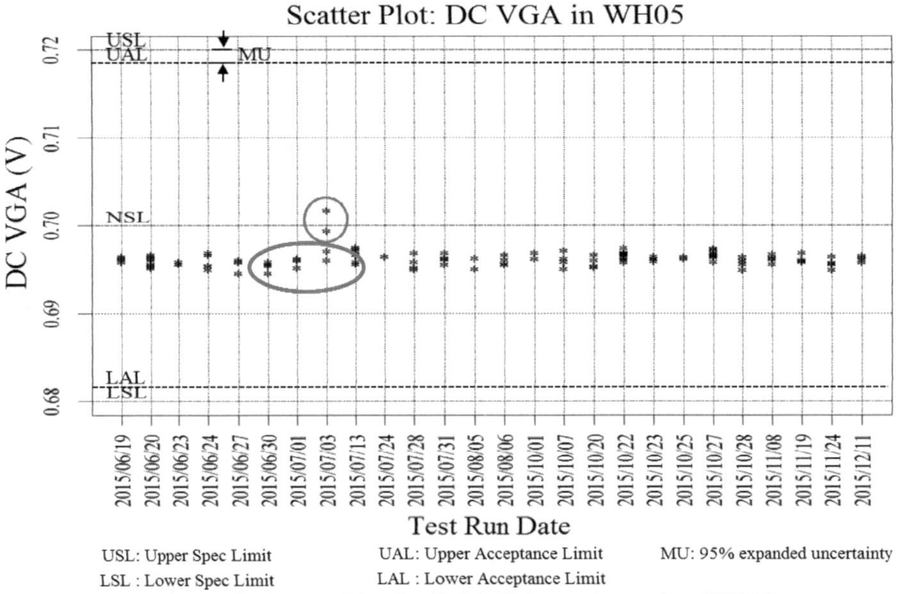

Figure 3. Scatter Plot for DC VGA test in station WH05

4.2. $\overline{Z}$ chart and W chart with $\widehat{\sigma}_{p_m}$ determined through through S-factor

First, measurement target (T) needs to be determined. DC VGA test is measured by comparing the voltage reading between oscilloscope (Scope) and digital multimeter (DMM). T_m value is calculated from repeated measurement for each station and recompiled in the event of DMM readjustment during the calibration or DMM replacement with another unit. Five input quantities (X_i) affecting the measurement result (Y) are identified, as shown in table 2. They undergo either Type A or Type B evaluation to obtain the standard uncertainties $u(x_i)$. X_1 is obtained from 15 repeated observations, the input estimates x_1 is evaluated using Type A. Other input quantities (X_2 to X_5) are the equipment specification and the resolution of the reported measured value, the input estimates x_2 to x_5 are evaluated using Type B with rectangular distribution.

The combined standard uncertainty $u_c(y)$ is computed from individual $u(x_i)$ using equation (10):

$$u_c(y) = \left[[c_1 u(x_1)]^2 + [c_2 u(x_2)]^2 + [c_3 u(x_3)]^2 + [c_4 u(x_4)]^2 + [c_5 u(x_5)]^2\right]^{\frac{1}{2}}$$

$$u_c(y) = 0.638 \text{ mV}$$

The expanded uncertainty (U_m) is calculated with the coverage factor k_m = 2 and rounded to two significant figures.

$$U_m = 2u_c(y) = 1.3 \text{ mV}$$

Table 2. Evaluation of standard uncertainties

X_i	Contributors	Type	v_i	± Limits	Units	Distribution	C_i	$u(x_i)$
X_1	Repeatability tests	A	14	3.21E-04	mV	Sigma	1	3.21E-04
X_2	Voltage accuracy from DMM	B	∞	3.15E-05	mV	Rectangular	1	1.82E-05
X_3	Resolution error of DMM	B	∞	5.00E-03	mV	Rectangular	1	2.89E-03
X_4	Vertical resolution error of Scope with input signal	B	∞	7.81E-01	mV	Rectangular	1	4.51E-01
X_5	Vertical resolution error of Scope without input signal	B	∞	7.81E-01	mV	Rectangular	1	4.51E-01

With the estimated values of $\widehat{\sigma}_{x_m}$ and $\widehat{\sigma}_{u_m}$, the max $\widehat{\sigma}_{p_m}$ is computed using Pythagoras's theorem in equation (16). S-factor (S_{f_m}) is calculated with a ratio between max $\widehat{\sigma}_{p_m}$ and min $\widehat{\sigma}_{p_m}$, as shown in table 3.

Table 3. Estimation of $\widehat{\sigma}_{p_m}$ with S–factor

Station	U_m	$\widehat{\sigma}_{x_m}$	$\widehat{\sigma}_{x_u}$	Min $\widehat{\sigma}_{p_m}$	Max $\widehat{\sigma}_{p_m}$	S_{f_m}
WH05	1.30E-03	5.63E-04	6.50E-04	5.63E-04	8.60E-04	1.527
WH06	1.30E-03	5.61E-04	6.50E-04	5.61E-04	8.58E-04	1.531
WH07	1.30E-03	6.06E-04	6.50E-04	6.06E-04	8.89E-04	1.466

The transformation expression in equation (20) and equation (21) are applied to plot $\overline{Z}$ chart and W chart respectively as shown in figure 4. The results showed that there were no false alarm points found and capable to detect the large shift from mean. As of now, the implementation is completed for station WH05; similar process will be repeated for other stations WH06 and WH07.

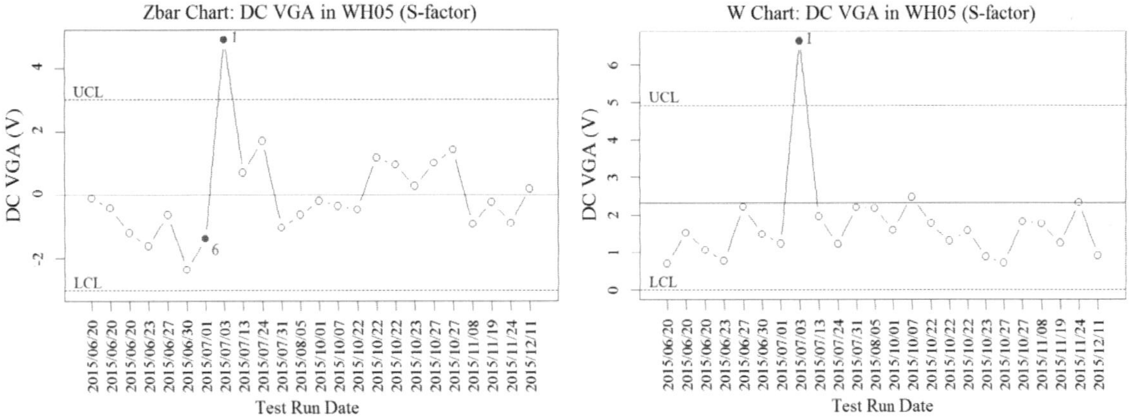

Figure 4. Modified $\overline{Z}$ chart and W chart for DC VGA test in station WH05 with S-factor

5. Result and Discussion

The summarized results of the case study are recorded in table 4. At this case, the $U/k\sigma_x$ ratio is at 1.1, the measurement error marginally affects the process standard deviation. From the analysis results from Shewhart $\bar{x}$ chart, all test stations had many points exceeding the control limits and some abnormal patterns are falsely detected. These false alarms caused many unnecessary process adjustment and loss of confidence in SPC. With the S-factor approach, the analysis results showed that the model performed very well in both $\overline{Z}$ chart and W chart with zero false alarms in all stations. The model reduced false alarms rate by up to 13% in comparison to the classical Shewhart $\bar{x}$ chart. Based on the results of observations, all test stations are in normal and controlled behavior, except for station WH05 where the shift occurs in the process. In this case study, all methods were able to detect these assignable causes. Results of the case study proved that the model with S-factor is efficient and effective for T&M process.

Table 4. Summary result of the case study

Station	Chart Type	Est Std Dev Method	Total Points	False Detected in Nelson Rules Test								False Alarm		Assignable causes detected
				R1	R2	R3	R4	R5	R6	R7	R8	FA	FA%	
DC VGA WH05	$\bar{x}$	Shewhart	24	1	0	0	0	1	1	0	0	3	13%	2
	$\overline{Z}$	S-Factor	24	0	0	0	0	0	0	0	0	0	0%	2
	R	Shewhart	24	0	0	-	-	0	0	-	-	0	0%	1
	W	S-Factor	24	0	0	-	-	0	0	-	-	0	0%	1
DC VGA WH06	$\bar{x}$	Shewhart	26	1	0	0	0	0	0	0	0	1	4%	0
	$\overline{Z}$	S-Factor	26	0	0	0	0	0	0	0	0	0	0%	0
	R	Shewhart	26	0	0	-	-	0	0	-	-	0	0%	0
	W	S-Factor	26	0	0	-	-	0	0	-	-	0	0%	0
DC VGA WH07	$\bar{x}$	Shewhart	23	1	0	0	0	0	0	0	1	2	9%	0
	$\overline{Z}$	S-Factor	23	0	0	0	0	0	0	0	0	0	0%	0
	R	Shewhart	23	0	0	-	-	0	0	-	-	0	0%	0
	W	S-Factor	23	0	0	-	-	0	0	-	-	0	0%	0

6. Conclusions

This paper proposed a modified SPC model combining modified guard band and control charts ($\overline{Z}$ chart and W chart) to address issues caused by short production runs and measurement errors. The model utilized the measurement target and measurement uncertainty to standardize the observed value. S-factor determines the estimation of the process standard deviation for control limits tolerance. The effectiveness of proposed model was demonstrated by a case study in real industry and it shows that the measurement error marginally affected the process standard deviation. The results revealed that the model significantly reduced false alarm rate compared to the one using classical Shewhart control charts. Moreover, the results proved that the model using the S-factor can rescale the process standard deviation in an efficient and effective manner. The proposed model using the modified control charts is practical for T&M manufacturing to eliminate false alarm, without sacrificing sensitivity to level shifts.

References

[1] Montgomery D C 2013 *Statistical quality control: A modern introduction 7th ed.* (Singapore: John Wiley & Sons)
[2] Bothe D R 1989 A powerful new control chart for job shops *43rd Annual Quality Congress Transactions May 1989 (Toronto)* (ASQC) pp 265-270
[3] Costa A F B and Castagliola P 2011 Effect of measurement error and autocorrelation on the $\bar{X}$ chart *Journal of Applied Statistics* **38**(4) 611-673
[4] Barrera J 2012 Analysis of some statistics used in short runs charts *Proc. of 15th International Conference on Experimental Mechanics 22-27 July 2012 (Porto/Portugal)* (ICEM15) pp 1-2

[5] Wheeler D 1991 Short Run SPC *SPC press*
[6] Farnum N 1992 Control charts for short runs: nonconstant process and measurement error *Journal of Quality Technology* **24**(3) 138-144
[7] Moore S S and Murphy E 2013 Process visualization in medical device manufacture: an adaptation of short run SPC techniques *Quality Engineering* **25**(3) 247-265
[8] Del Castillo E, Grayson J M, Montgomery D C and Runger G C 1996 A review of statistical process control for short run manufacturing systems *Communications in Statistics - Theory and Methods* **25**(11) 2723–2737
[9] Pou J M and Leblond L 2015 Control of customer and supplier risks by the guardband method International *Journal of Metrology and Quality Engineering* **6** 205
[10] JCGM 100: 2008 *Evaluation of measurement data– Guide to the expression of uncertainty in measurement (GUM 1995 with minor corrections)* (Joint Committee for Guides in Metrology)
[11] JCGM 200: 2012 *International Vocabulary of Metrology – Basic and General Concepts and Associated Terms (VIM 3rd ed.)* (Joint Committee for Guides in Metrology)
[12] UKAS 2007 *The Expression of Uncertainty and Confidence in Measurement M3003 2nd ed.* (UK: United Kingdom Accreditation Service)
[13] Taylor B N and Kuyatt C E 1994 *Guidelines for Evaluating and Expressing the Uncertainty of NIST Measurement Results* (MD: National Institute of Standards and Technology, USA)
[14] Instone I 1996 Simplified method for assessing uncertainties in a commercial production environment Teste and measurement *Metrology Forum (London)* pp 1–7
[15] Bell S 2001 *Measurement Good Practice Guide No 11 (Issue 2): A Beginner's Guide to Uncertainty of Measurement* (UK: National Physical Laboratory)
[16] Nelson L S 1984 The Shewhart control chart-tests for special causes *Journal of Quality Technology* **16**(4) 237–239
[17] Noskievicova D 2013 Complex control chart interpretation *International Journal of Engineering Business Management* **5** (13:2013) 1-7
[18] Scrucca L 2004 qcc: An R package for quality control charting and statistical process control *R News* **4**(1) 11-17
[19] Reynolds M R and Stoumbos Z G 2004 Should observations be grouped for effective process monitoring *Journal of Quality Technology* **36** 343–366

IOP Conf. Series: Materials Science and Engineering **328** (2018) 012010 doi:10.1088/1757-899X/328/1/012010

The impact of manufacturing complexity drivers on performance-a preliminary study

Suh Huah Leang [1,3], Wan Hasrulnizzam Wan Mahmood [1,2,4], Muhamad Arfauz A Rahman [1,5]

[1]Sustainable and Responsive Manufacturing Research Group, Faculty of Manufacturing Engineering, Universiti Teknikal Malaysia Melaka, 76100 Hang Tuah Jaya, Melaka, Malaysia.
[2]Faculty of Engineering Technology, Universiti Teknikal Malaysia Melaka, 76100 Hang Tuah Jaya, Melaka, Malaysia.
[3]m0510049@student.utem.edu.my
[4] hasrulnizzam@utem.edu.my
[5]arfauz@utem.edu.my

Abstract. Manufacturing systems, in pursuit of cost, time and flexibility optimisation are becoming more and more complex, exhibiting a dynamic and nonlinear behaviour. Unpredictability is a distinct characteristic of such behaviour and effects production planning significantly. Therefore, this study was undertaken to investigate the priority level and current achievement of manufacturing performance in Malaysia's manufacturing industry and the complexity drivers on manufacturing productivity performance. The results showed that Malaysia's manufacturing industry prioritised product quality and they managed to achieve a good on time delivery performance. However, for other manufacturing performance, there was a difference where the current achievement of manufacturing performances in Malaysia's manufacturing industry is slightly lower than the priority given to them. The strong correlation of significant value for priority status was observed between efficient production levelling (finished goods) and finish product management while the strong correlation of significant value for current achievement was minimised the number of workstation and factory transportation system. This indicates that complexity drivers have an impact towards manufacturing performance. Consequently, it is necessary to identify complexity drivers to achieve well manufacturing performance.

1. Introduction

Nowadays, many industries are looking for the best ways to improve productivity. However, there are many factors that influence the productivity reduction. One of the factors is complexity [1]. In the course of this progress, global manufacturing networks have become more and more complex inter alia due to the geographical distribution of sites, a growing product diversity, sophisticated manufacturing processes and many other aspects [2]. Complexity is still one of the biggest challenges in manufacturing due to increased variety, market volatility and distributed global manufacturing [3,4]. This is especially so for identification of the most relevant network segments from a complexity perspective is a critical success factor for a company's competitiveness, since it provides information to management about what kind of improvement measures might be most urgent and might deliver the most favourable impact in a certain economic situation [5].

In general, the drivers can be distinguished between internal and external factors and so-called complexity drivers [6]. To gain an understanding of the complexity, these external and internal drivers need to be identified first and their dependency on products and processes needs to be described subsequently [7]. Aelker et al., [8] stated that the most accepted differentiation is between the complexity arising from within a company (internal complexity), and the complexity caused by factors external to the organisation (external complexity). Yet, reviews the manufacturing complexity drivers, which provides the necessary knowledge to complete the first step of a complexity management initiative. Understanding and analysing the complexity drivers first will be an effective way to proceed to develop a clear strategy to deal with complexity [9].

Therefore, this initial study was conducted to explore the effect of complexity drivers in manufacturing productivity performance, based on two components namely Current Manufacturing Performance (CMP) and Manufacturing Complexity Driver (MCD) with the priority and the current achievement's practice in Malaysia's manufacturing industry. The findings can be used in shaping a more inclusive action to attain higher productivity performance. In addition, the results from this study can be used by the manufacturer, particularly to identify the main of MCD in order to achieve high CMP in production operations. Moreover, academicians can also use the evidence from this study to develop a new manufacturing framework to strengthen the current practice in manufacturing operations with the existing complexity drivers. The following sections would explain the research method underpinning the survey information, the results of statistical analysis from the survey, the measures of MCD and CMP, the relation between MCD and CMP and conclusions.

2. Research Method

The survey questionnaire is one data collection method to obtain quantitative data for statistical testing of the hypothesis from the literature reviews [10, 11]. The analysis was based on data obtained from the quantitative method by using statistical package SPSS. A questionnaire is conducted which that is comprised of 28 items on CMP and 15 items on MCD. For each item, the respondents were asked to rate each item with priority and achievement based on a five points Likert's scale (e.g.: 1=unimportant to 5=critical for priority level and 1=very poor to 5=very good for current achievement). MCD is the dependent variable in the preliminary study to determine the significant driver through the mean score. CMP acts as the dependent variable to define the critical gap between the priority and the current achievement of each item in order to achieve better productivity. The internal consistency and reliability of the survey were tested using the Cronbach's α reliability test [12]. Abolhassani et al., [13] stated that Spearman's correlation evaluates a relationship between two variables, without making any assumptions about the frequency distribution of the variables and it was applied to measure the correlation of the relation of CMP and MCD. Out of 150 questionnaires sent, a total 61 were completed and returned. After a screening process, 54 of those completed with useful information went on for further analysis. With a response rate of 40.7 percent, it was believed that the answers would be reasonable and sampling technique is not used because of the survey not representing any population. The other intention is that this study is only an initial study in manufacturing complexity drivers on performance in Malaysia's industry.

3. Result and Discussion

Descriptive statistics such as frequency, percentage and mean are the main statistical analysis employed. The survey questionnaire has been addressed to the appropriate person with at least two years' experience in the company. The job title of the respondents is categorised into three types which are chief executive officer (CEO)/ director, manager/executive, engineer and others. Most of the respondents are in engineer position with 50.9%. Besides, the highest number of company ownership was from foreign countries (74.5%) typically from Asian company, followed by a local company (25.5%). 72.8% of the companies have more than 50% of the main business operation in manufacturing operation. 72.5% of the companies have more than 50% of original equipment manufacturer (OEM) product.

Most of the respondents were from the electric/electrical products group (36.4%); followed by other products group and automotive products group (16.4%), other product group (14.5%), chemical/scientific products group (12.7%), 10.9% of the company has various product groups such as automotive, electric/electrical and mechanical and lastly, from mechanical engineering products group (9.1%). The majority of the respondents were foreign companies with more than 500 employees (74.5%), while the lowest respondent was from the local companies that 50 to 100 employees (5.5%). Meanwhile, 87.3% of the respondents have experienced in ISO 9001. Furthermore, the other management certification obtained by the company are TS 16949 (30.9%), ISO14001 (67.3%), OHSAS 18001 (52.7%) while 12.7 percent possess other management certification such as ISO 13485 and ISO 9002.

3.1 Analysis of Manufacturing Complexity Driver

In terms of MCD, the Cronbach's alpha analysis shows that all are regularly reliable at a value of 0.874 for priority level and 0.871 for the current achievement. The dispersal of mean score for priority and current achievement of MCD are illustrated in Figure 1. The respondents agreed that all 15 items have fulfilled the priority where most of the manufacturers that prioritise the increase of customers' demands (MCD13) have the highest mean score of 4.27. This is followed by the machine efficiency (MCD7), machine utilisation (MCD1), and capacity planning (MCD2) at a mean score of 4.18, 4.16 and 4.13 respectively. However, process selection (MCD4) is still less popular as this item ranked the lowest score with a mean score of 3.40. Increase of customer demands (MCD13) has the highest mean score too for MCD current achievement value at 3.85, followed by machine utilisation (MCD1) and finish product management (MCD12) at a mean score value of 3.82 and 3.80 respectively. Respondents also agreed that capacity planning (MCD2) and this reported a mean score value of 3.72. In contrast, process selection (MCD4) has less influence as the respondents appraised this with the lowest mean score value of 3.26. Based on Figure 1, increases of customer demands (MCD13) is the main MCD item that prioritised and achieved by most of the company because with customer demands the company have to grow their products unceasingly and adapt their production program and quantity [14].

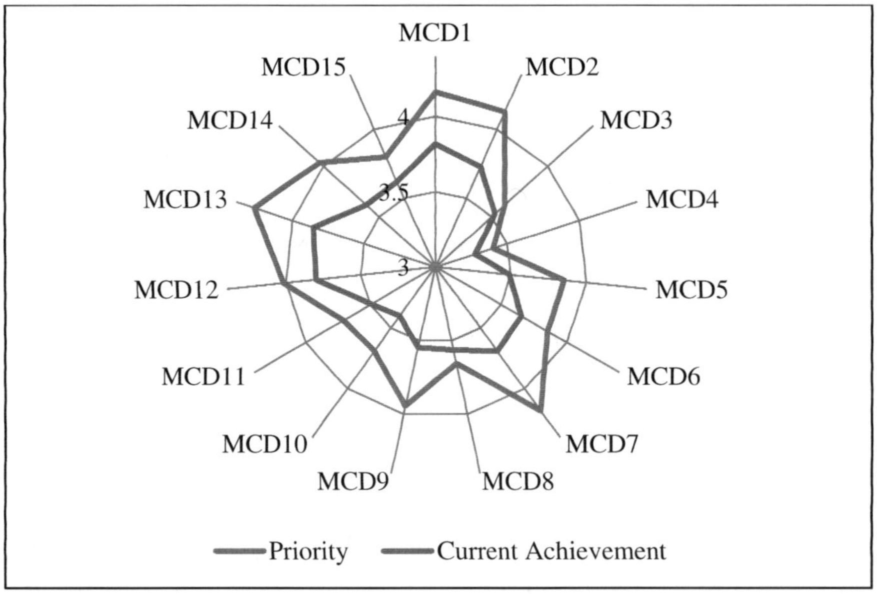

Figure 1. The performance of MCD in priority level and current achievement

3.2 Measures of Current Manufacturing Performance

The Cronbach's alpha analysis shows that all CMP items in the questionnaires were consistently reliable at a value of 0.907 for priority level and 0.931 for the current achievement of CMP. The respondents agreed that all 28 items have fulfilled the priority where most of the manufacturers that prioritised product quality (CMP27) have the highest mean score of 4.64 illustrated in Figure 2. This is followed by the on time delivery (CMP7), minimum cost (CMP26), and efficient space utilisation (CMP24) at a mean score of 4.54, 4.22 and 4.02 respectively. However, zero WIP (one piece flow) (CMP21) is still less popular as this item ranked the lowest score with a mean score of 3.12. As for CMP current achievement, on time delivery (CMP7) has the highest mean score value at 4.10, followed by product quality (CMP27) and minimum cost (CMP26) at a mean score value of 3.96 and 3.78 respectively. The adaptation of new technology (CMP1) had been recognised by the respondents which it is reported a mean score value of 3.68. In contrast, zero WIP (one piece flow) (CMP21) has less influence as the respondents appraised this with the lowest mean score value of 2.98.

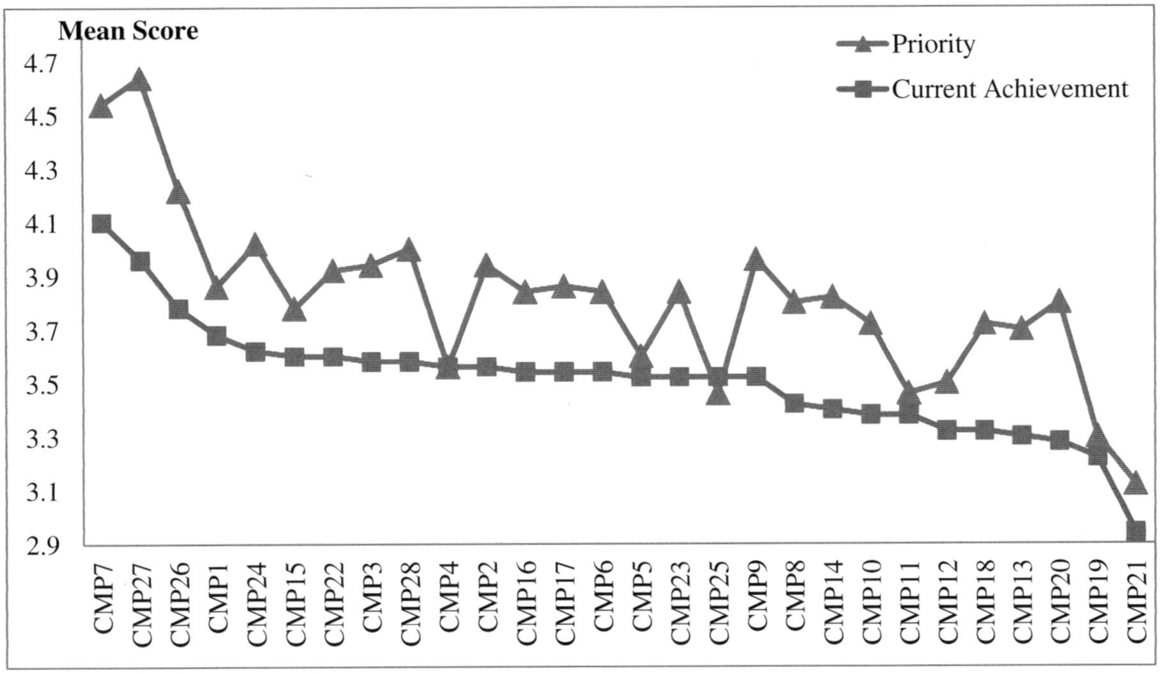

Figure 2. Comparison of mean score differences between priorities and the current achievement of manufacturing performance at manufacturing industry in Malaysia

As shown in Figure 3, the largest gap formed between priority and current achievement is CMP 27 followed by CMP20 and CMP 26 which indicated that the current achievement of CMP is lower than the priority status. This result shows that most of Malaysia's manufacturing industry has a high focus on product quality, on time delivery and minimum cost to achieve better manufacturing performance. This is not surprising because the focus on dealing with product quality and on time delivery issues is crucial to remain competitive in a modern manufacturing era in order to minimise cost. This is evidently shown in Figure 2 where these three components (product quality, on time delivery and minimum cost) have the higher mean score compared to the others component measured. Manufacturing industries are in pursuit of cost and time reduction without decreasing product quality [15, 16]. Improved product quality and on time delivery result in improved performance [17]. Antony et al., [18] implement Six Sigma to ensure there was an improvement in the on time delivery which was able to cost reduction. Most of the manufacturing industries aim the key performance measures are to increase yield, to maintain the quality of products, and to reduce the cost of operations [19].

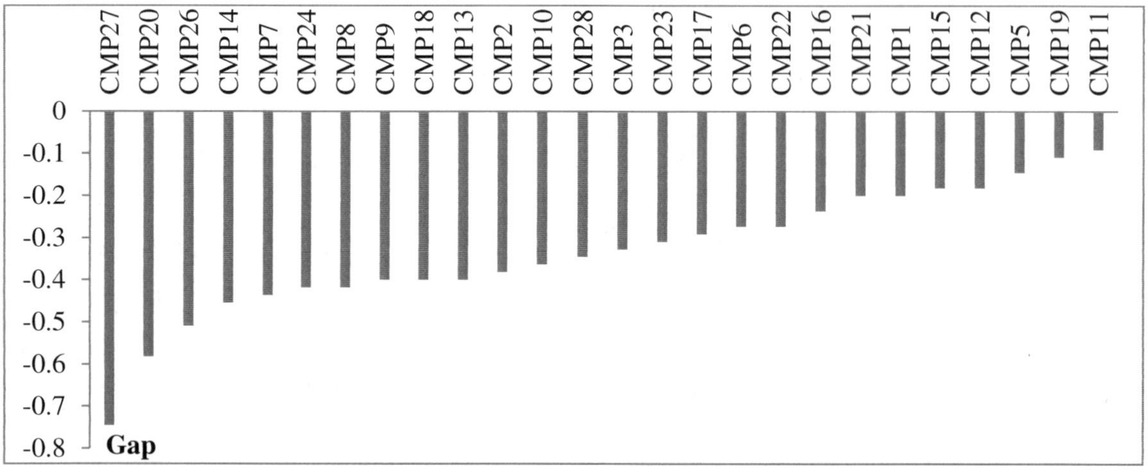

Figure 3. The difference gap between priority level and current achievement status for CMP

Efthymiou et al., [15] stated that demand fluctuation along with the requirements of high product quality, low-cost, short lead time and high customisation may lead to an increase in manufacturing complexity in the globalised and interconnected market. This fact indicates that Malaysia's manufacturing industries nowadays aspire to attain good manufacturing performance particularly to remain competitive in a new manufacturing environment and the complexity arises in the manufacturing. Figure 1 illustrated the lowest mean score for both priority status and current achievement component is zero WIP (one-piece flow). Zero WIP (one piece flow) is one of the lean tools where the parts are moved through operations from step to step ideally without any work-in-progress (WIP) [20]. It showed this component was not preferable by the manufacturing industry because of it is difficult to deploy [13].

3.3 The Relation Between Manufacturing Complexity Driver and Current Manufacturing Performance

In this study, Spearman's correlation analysis between CMP with the MCD is created to form a total of 420 matrices of the relationship. The proposed hypothesis is addressed in this section which is:

H_{0A}: All MCD is not significant correlation with CMP in priority level.

H_{1A}: All MCD is a significant correlation with CMP in priority level.

H_{0B}: All MCD is not significant correlation with CMP in current achievement.

H_{1B}: All MCD is a significant correlation with CMP in current achievement.

Table 1 shows the test results between priority level and current achievement related to CMP and MCD. The Spearman ρ correlation coefficient ranged from 0.021 to 0.716 for priority status and ranged from 0.031 to 0.699 for current achievement, at a significant level of 0.01, which indicated a moderate to the high-linear association between these variables. The analysis also shows that the item in the CMP has a significant relationship with MCD, where a strong correlation relationship existed between the efficient production levelling (finished goods) (CMP17) with finished product management (MCD12) at a value of 0.741 for priority status, the null hypothesis, H_{0A} is rejected for these items. Production levelling is the lean concept for production smoothing [21] and [22] concluded that with production levelling able to manage product flow operation efficiently.

The correlation analysis between CMP and MCD for current achievement showed a significant value, 0.689 within minimising the number of the workstation (CMP25) and factory transportation system (MCD8). The null hypothesis, H_{0B} is rejected for these items since the items are significant correlate with MCD and CMP. Pourvaziri and Pierreval [23] identified that the location of machines is one of the key issues regarding manufacturing systems' performance. Indeed, if the material flow between two machines becomes too high, then it may be necessary to add new transporters to the system, which adds new costs, machines or workstations with high flow rates are allocated close to each other to minimize the total distance of the material flow.

Table 1. Spearman correlation between MCD against Current Manufacturing Performance

Item	Current Manufacturing Performance	
	Priority	Current Achievement
MCD2	CMP20	CMP7,CMP11
MCD5	-	CMP23
MCD6	CMP24	CMP12, CMP25
MCD7	CMP28	CMP18, CMP22
MCD8	-	CMP24,CMP25
MCD9	CMP17	-
MCD10	CMP19	CMP7,CMP18
MCD11	CMP14	-
MCD12	CMP14, CMP17	-

4. Conclusion

As a conclusion, this initial study has shown that the MCD has a strong influence on the manufacturing performance. Coupled with correlation analysis, this showed that each of the practices in MCD has a significant positive correlation with each item in the manufacturing sustainability components. The study is still relatively exploratory. Future studies could investigate the management of complexity with the existing complexity drivers at Malaysia's manufacturing industry. The findings in this study can be used as a basis for the next stage of a study in developing a more widespread approach to manufacturing complexity management, primarily in Malaysia's manufacturing industry.

Acknowledgement

This research was co-funded under Ministry of Higher Education (MOHE), FRGS Grant (FRGS/1/2016/TK03/FTK-AMC/F00324) and Zamalah Scholarship provided by Universiti Teknikal Malaysia Melaka (UTeM).

References

[1] Wan Hasrulnizzam, W. M. (2008). Industrial Types of Manufacturing Complexity. *In Proceedings of the National Seminar on Engineering & Technopreneurship* (pp. 209–213).
[2] Reuter, C., Prote, J.-P., & Stöwer, M. (2016). Evaluation of performance differences between manufacturing sites. *Procedia CIRP*, ***50***, 342–347. https://doi.org/10.1016/j.procir.2016.04.191
[3] Alkan, B., Vera, D., Ahmad, M., Ahmad, B., & Harrison, R. (2016). Design Evaluation of Automated Manufacturing Processes Based on Complexity of Control Logic. *26th CIRP Design Conference*, ***50***, 141–146. https://doi.org/10.1016/j.procir.2016.05.031
[4] Elmaraghy, W. H., Elmaraghy, H. A., Tomiyama, T., & Monostori, L. (2012). Complexity in Engineering Design and Manufacturing. *CIRP Annals - Manufacturing Technology*, ***61***(January), 793–814.
[5] Gerschberger, M., Engelhardt-Nowitzki, C., Kummer, S., & Staberhofer, F. (2012). A model to determine complexity in supply networks. *Journal of Manufacturing Technology Management*, ***23***(8), 1015–1037. https://doi.org/10.1108/17410381211276853
[6] Vogel, W., & Lasch, R. (2016). Complexity Drivers in Manufacturing Companies: a Literature Review. *Logistics Research*, ***9***(1), 25. https://doi.org/10.1007/s12159-016-0152-9
[7] Kohr, D., Budde, L., & Friedli, T. (2017). Identifying Complexity Drivers in Discrete Manufacturing and Process Industry. *Procedia CIRP*, ***0***, 1–6. https://doi.org/10.1016/j.procir.2017.03.290
[8] Aelker, J., Bauernhansl, T., & Ehm, H. (2013). Managing Complexity in Supply Chains: A

Discussion of Current Approaches on The Example of The Semiconductor Industry. *Procedia CIRP*, **7**(14), 79–84. https://doi.org/10.1016/j.procir.2013.05.014
[9] Serdarasan, S. (2013). A Review of Supply Chain Complexity Drivers. *Computers and Industrial Engineering*, **66**(3), 533–540. https://doi.org/10.1016/j.cie.2012.12.008
[10] Özşahin, M., Sezen, B., & Çankaya, S. Y. (2013). Effects of Green Manufacturing and Eco-innovation on Sustainability Performance. *Procedia - Social and Behavioral Sciences*, ***99***, 154–163. https://doi.org/10.1016/j.sbspro.2013.10.481
[11] Zhang, L., & Chen, X. (2016). Role of Lean Tools in Supporting Knowledge Creation and Performance in Lean Construction. *Procedia Engineering*, ***145***, 1267–1274. https://doi.org/10.1016/j.proeng.2016.04.163
[12] Jasti, N. V. K., & Kodali, R. (2014). Validity and reliability of lean manufacturing frameworks. *International Journal of Lean Six Sigma* (**Vol. 5**). https://doi.org/10.1108/IJLSS-12-2013-0057
[13] Abolhassani, A., Ky, L., & Gopalakrishnan, B. (2016). Lean and US manufacturing industry : popularity of practices and implementation barriers. *International Journal of Productivity and Performance Management*, ***65***(7), 875–897. https://doi.org/http://dx.doi.org/10.1108/MRR-09-2015-0216
[14] Cichos, D., & Aurich, J. C. (2016). Support of Engineering Changes in Manufacturing Systems by Production Planning and Control Methods. *Procedia CIRP*, ***41***, 165–170. https://doi.org/10.1016/j.procir.2015.12.049
[15] Efthymiou, K., Pagoropoulos, A., Papakostas, N., Mourtzis, D., & Chryssolouris, G. (2012). Manufacturing Systems Complexity Review: Challenges and Outlook. In *Procedia CIRP* (**Vol. 3**, pp. 644–649). https://doi.org/10.1016/j.procir.2012.07.110
[16] Ringena, G., Aschehouga, S., Holtskogb, H., & Ingvaldsena, J. (2014). Integrating Quality and Lean into a Holistic Production System. *Procedia CIRP*, **17**, 242–247. https://doi.org/10.1016/j.procir.2014.01.139
[17] Psomas, E., & Kafetzopoulos, D. (2014). Performance measures of ISO 9001 certified and non-certified manufacturing companies. *Benchmarking*, **21**(5), 756. https://doi.org/10.1108/BIJ-04-2012-0028
[18] Antony, J., Gijo, E. V, Kumar, V., & Ghadge, A. (2016). A multiple case study analysis of Six Sigma practices in Indian manufacturing companies. *International Journal of Quality & Reliability Management International Journal of Quality & Reliability Management International Journal of Quality & Reliability Management Iss International Journal of Quality & Reliability Management*, **338**(8), 1138–1149. https://doi.org/10.1108/IJQRM-10-2014-0157
[19] Chongwatpol, J. (2015). Prognostic analysis of defects in manufacturing. *Industrial Management and Data Systems*, ***115***(1), 64–87. https://doi.org/10.1108/IMDS-05-2014-0158
[20] Ali, R. M., & Deif, A. M. (2014). "Dynamic lean assessment for Takt time implementation." *Procedia CIRP*, ***17***(128), 577–581. https://doi.org/10.1016/i.procir.2014.01.128
[21] Clotet, J. F. (2015). *Lean Production Planning and Control in Semi-Process Industry*. Norwegian University of Science and Technology.
[22] Papadopoulou, T. C. (2013). Application of Lean Scheduling and Production Control in Non-Repetitive Manufacturing Systems using Intelligent Agent Decision Support.
[23] Pourvaziri, H., & Pierreval, H. (2017). Dynamic facility layout problem based on open queuing network theory. *European Journal of Operational Research*, ***259***(2), 538–553. https://doi.org/10.1016/j.ejor.2016.11.011

Appendices

1. Information on Current Manufacturing Performance (CMP)

What are your current manufacturing performances in your company from the past two (2) years? (On a five point Likert-scale)

Item	Manufacturing Performance
CMP1	Adaptation of new technology
CMP2	Decrease customer lead time
CMP3	Decrease throughput time
CMP4	Enhanced product variety
CMP5	High capacity of Innovation
CMP6	Minimise setup time
CMP7	On time delivery
CMP8	Speed up changeover time
CMP9	Effective machine optimisation
CMP10	Increasing Kaizen activities
CMP11	Minimise machine configuration
CMP12	Operator flexibility and innovativeness
CMP13	Optimise Poka-Yoke
CMP14	Proactive Total Preventive Maintenance (TPM) Practice
CMP15	Process line balancing
CMP16	Effective pace of production (Takt time)
CMP17	Efficient production levelling (finished goods)
CMP18	Organised Just-in-Time (JIT) for raw material
CMP19	Organised Kanban system for Work-in-Progress (WIP)
CMP20	Zero missing/misplace material
CMP21	Zero WIP (one piece flow)
CMP22	Effective layout configuration
CMP23	Effective material flow
CMP24	Efficient space utilisation
CMP25	Minimise number of workstation
CMP26	Minimum cost
CMP27	Product quality
CMP28	Simplified operation procedure

2. Information on Manufacturing Complexity (MC)

What are your complexity drivers in your company from the past two (2) years? (On a five point Likert-scale from unimportant to critical for priority and very poor to very good for current achievement)

Item	Manufacturing Complexity Drivers
MCD1	Machine utilisation
MCD2	Capacity planning
MCD3	High product mix
MCD4	Process Selection
MCD5	Material handling management
MCD6	Space utilisation
MCD7	Machine efficiency
MCD8	Factory transportation system
MCD9	Inventory forecast
MCD10	WIP product handling
MCD11	Raw material handling
MCD12	Finish product management
MCD13	Increases of customer demands
MCD14	Increases of customer numbers
MCD15	Specifications of products

IOP Conf. Series: Materials Science and Engineering 328 (2018) 012011 doi:10.1088/1757-899X/328/1/012011

Wear Detection of Drill Bit by Image-based Technique

Maziyah Sukeri[1*], Mohd Zulhilmi Paiz Ismadi[1], Abdul Rahim Othman[1], Shahrul Kamaruddin[1]
[1]Mechanical Engineering Department, Universiti Teknologi PETRONAS, 32610 Seri Iskandar,Perak.
*maziyahsukeri91@gmail.com

Abstract. Image processing for computer vision function plays an essential aspect in the manufacturing industries for the tool condition monitoring. This study proposes a dependable direct measurement method to measure the tool wear using image-based analysis. Segmentation and thresholding technique were used as the means to filter and convert the colour image to binary datasets. Then, the edge detection method was applied to characterize the edge of the drill bit. By using cross-correlation method, the edges of original and worn drill bits were correlated to each other. Cross-correlation graphs were able to detect the difference of the worn edge despite small difference between the graphs. Future development will focus on quantifying the worn profile as well as enhancing the sensitivity of the technique.

1. Introduction

Drilling is one of many basic machining operations in the industry. This operation is widely used in manufacturing since the beginning of industrial innovation. According to [1], nearly 40% of all drilling operations are in the manufacturing industries. Despite its simplicity, in terms of maintenance aspects, its failure, due to tool wear and breakage, accounts to about 7-20% of the average machine tool downtime [2]. The tool failure could also contribute to machine failure. On top of the machine repair cost, the increasing cumulative non-productive time further adds to the loss of opportunity and profit by the company [3].

Drill bit wear is considered critical, as the wear and tear can give the negative effects to the drill bit itself. High quality products have a relatively narrow tolerance margin and require optimal condition of the cutting tool, to ensure the products produced are well within the quality control. Wear drill bit can cause the drilling process to become longer than usual as the bit loss its sharpness or certain tolerance on the surface. Such problems can affect surface quality of finished workpiece in manufacturing industry.

In recent years, wear monitoring was controlled by computers and sensor. The sensor and signal processing method used for direct or indirect techniques in the chatter, tool wear or breakage and surface roughness [4]. Indirect method of tool condition monitoring (TCM) typically utilizes single or multiple sensors to capture process information that correlates to the tool condition [5]. Indirect method measurement such as the acoustic emission (AE), cutting force, vibration, power or current use of spindle and surface roughness of the machine generally involves gathering signals that can be harvested continuously [6]. Despite the indirect nature of the measurement, the technique is widely used in the industry due to its ease of installation in existing or new machines without directly impacting the machine functions.

Direct method is a common way to measure the actual wear. The method measures the tool wear directly by means of tool images and tactile sensors. Direct method is employed to monitor tool wear images for drilling operation. It is based on observations of the actual tool wear and may require removal of the tool [7]. By using direct measurement, it will give an immediate appraisal of the tool wear such as optical measurement.

The optical method is typically applied in the inspection of geometry using a camera to capture the image of the tool tip. In separate study by [9], they utilized a suitable thresholding technology to convert and quantify the amount of flank wear. Introduced a flank wear measurement method of multilayer coated twist drill [10]. The study discovered the edges of the wear profile on the cutting plane. Detection was done by utilizing the smoothed edges using B-spline technique and spatial moment edge detector with sub-pixel accuracy. The Gaussian was applied at low-pass filtering technique to smoothest the curvature curve and finally, to select the accurate threshold valued for extracting the accurate wear profile for precise measurement of maximum flank wear width by applying a statistical process control measurement.

A study [11] proposed a new technique for precise edge detection algorithm utilizing neural network technique for tool wear detection. The study utilized Scanning Electron Microscopic (SEM) images of the flank wear images of a 4-fluted high-speed steel milling cutter. In [12] study, the error calculated using image processing counted 4% in classified the flank wear and breakage of the cutting tool. The image displayed the flank wear featured texture to be sharper and brighter in the machine vision system. Meanwhile, smooth and rough textures were included in the breakage of the tool. Based on this phenomenon, they measured flank wear area and sorted the types of wear.

In 2006, another research in a classified of wear, resulted in the machine vision system showing the flank wear feature. Developed a new technique measurement namely DEFROL (Deviation from Linearity) to classify the sharp and dull drilling tool by their images [13]. However, due to the wearing effect, the emphasis was done on the change point of angle and linearity deviation of the cutting edges.

This proposed study is based on the image processing technique to characterize the wear. This research is aimed at characterizing the bit's edge profile wear by using cross-correlation technique, applied to the original and worn bits.

2. Methodology

The experiment was conducted using 40 AUT Scantool machine. It started with minimum spindle speed of 120 rpm. In the experiment, High Speed Steel (HSS) drill bit with diameter 3mm was used. The investigation was operated 6mm of the depth of cut. All drilling test were carried out under dry conditions and drilled continuously until wear appeared, which tool approximately made 1000 drills.

Figure 1 shows the schematic diagram of a tool wear measurement. The machine vision system for the tool wear monitoring consists of a camera with 266μm per pixel, light source, a PC for image processing, and MATLAB software. Images of original and used bits were captured at exactly similar location and orientation for further analysis. In the image processing, the image was analysed by several steps as shown in Figure 2.

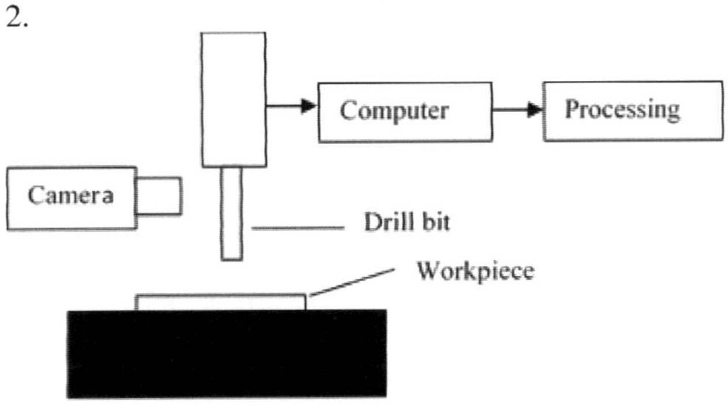

Figure 1. Schematic diagram of tool wear measurement

IOP Conf. Series: Materials Science and Engineering **328** (2018) 012011 doi:10.1088/1757-899X/328/1/012011

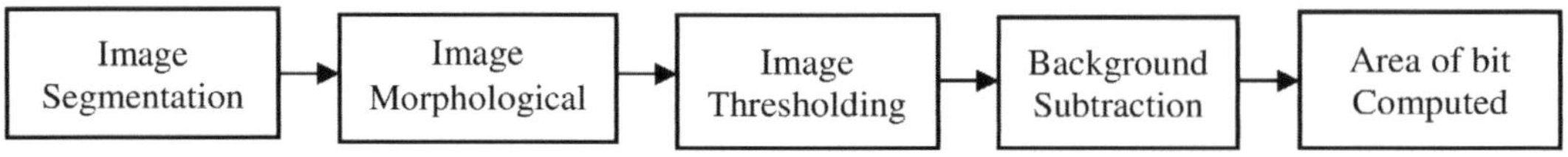

Figure 2. Flowchart of edge detection in image processing

2.1. Image Segmentation

By using image segmentation method, the images were divided into several parts. The function of image segmentation is to identify objects or other relevant information in digital images. Figure 3 shows the segmentation of the image. The texture image described the character of the image regions that are critical for accomplishing robust segmentations.

Figure 3 (a) and (b) shows the original and grayscale image respectively which has a size of 225x535 pixels each. The histogram graph displays distribution of pixel number of grey-scale image (Figure 3(b)) as illustrated in Figure 4. For an 8-bit grayscale image there are 250 different levels between black and white, with 0 being black and 250 represents white colour. Then thresholding was applied to the images that converted grayscale input images to binary images. The binarized image has only 0 (black) and 1 (white). Next, the bit edge, represented by the change of white and black pixel, was characterized and plotted in graphs.

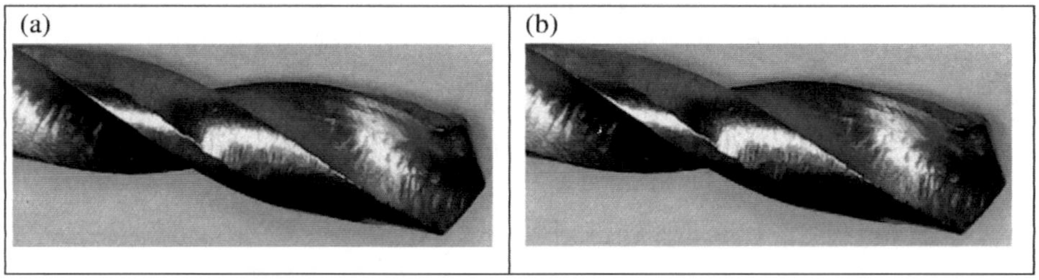

Figure 3. (a) Original and (b) Grayscale image

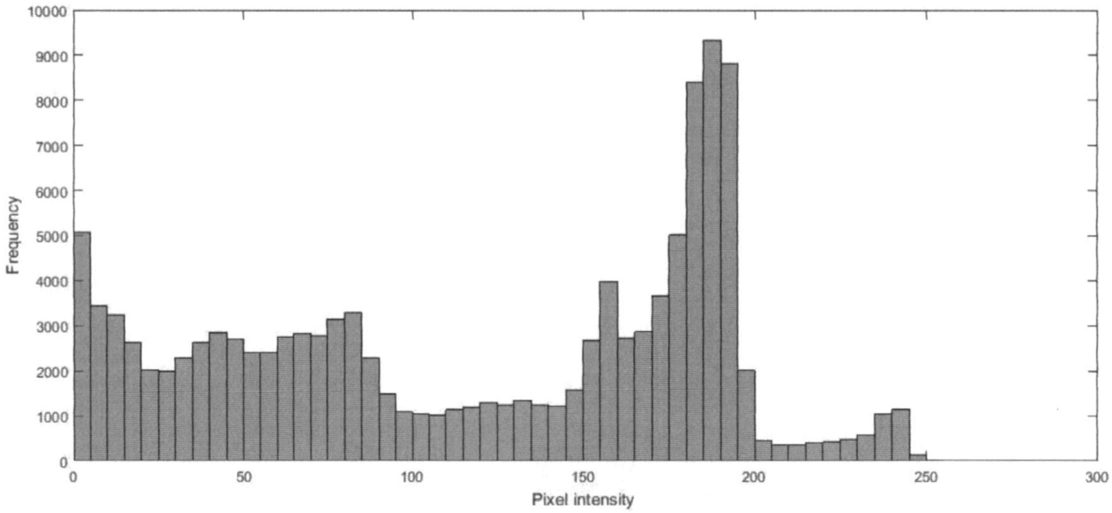

Figure 4. Histogram of grayscale image

IOP Conf. Series: Materials Science and Engineering **328** (2018) 012011 doi:10.1088/1757-899X/328/1/012011

2.2. Cross-correlation technique

Cross-correlation measures the similarity between x and shifted copies of y as a function of the lag. If x and y have different lengths, the function appends zero at the end of the shorter vector. Hence, it has the same value as the other. Equation 1 represents a formula for the cross-correlation technique. The equation is applied to determine the difference between the edge profile of original and worn bits, where f is the image, $\bar{t}$ is the mean of the template, and $\bar{f}_{u,v}$ is the mean of $f(x,y)$. The graphs that represents original and worn bits were correlated to each other and comparison between them were made.

$$\frac{\sum x,y\left[f\left(x,y-\bar{f}_{u,v}\right)\right]\left[t(x-u,y-v)-\bar{t}\right]}{\left\{\sum x,y\left[f(x,y)-\bar{f}_{u,v}\right]^{2}\sum x,y\left[t(x-u,y-v-\bar{t})\right]^{2}\right\}^{0.5}} \qquad (1)$$

3. Result and Discussion

3.1. Pre-processing

To further analyse the image and ensure optimized computational time, only one side of the edge profile is considered by cropping the image. Binarization of the grayscale image through thresholding was performed to the grayscale image and illustrated in Figure 5(a) and 5(b). The edge of the drill bit was clearly defined. The edge of the drill bit was then characterised by determining the change of pixel number from 0 to 1 and vice versa.

The edge of the original and worn bits were plotted and showed in Figure 5(c) and 5(d). It can be noticed that the edge profiles of both worn and new bits are mostly like each other except near the cutting edge. For comparison purposed, the edge profiles were plotted in similar graphs as shown in Figure 6. Despite having similar geometry at other region, near the cutting edge, a small geometric discrepancy was noticed due to some wear or damage to the drill bits, caused by the imposed extensive drilling. Significant wear of approximately 10 pixels in size can be noticed at the sharp edge of the bit. With the small pixel size of 266μm, this method provides high resolution detection and able to quantify slight discrepancy between the profiles.

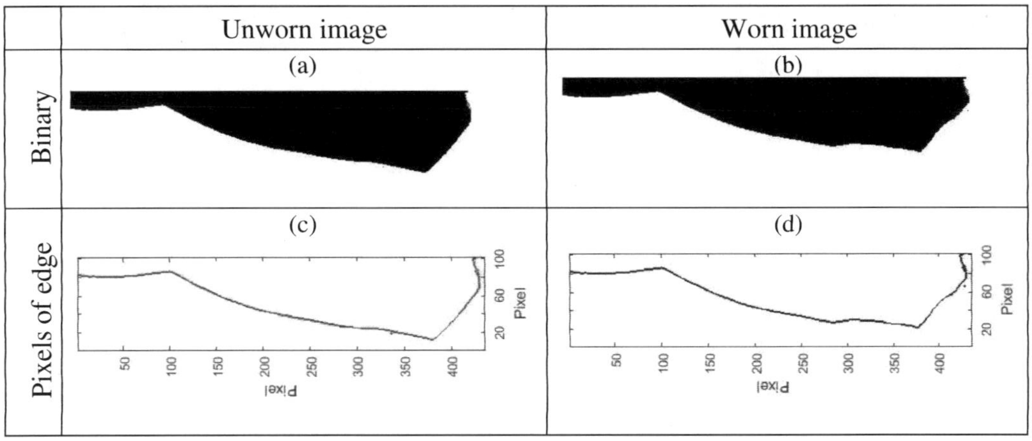

Figure 5. (a) Image crop of unworn bit; (b) Image crop of worn bit; (c) The edge of unworn bit; (d) The edge of worn bit

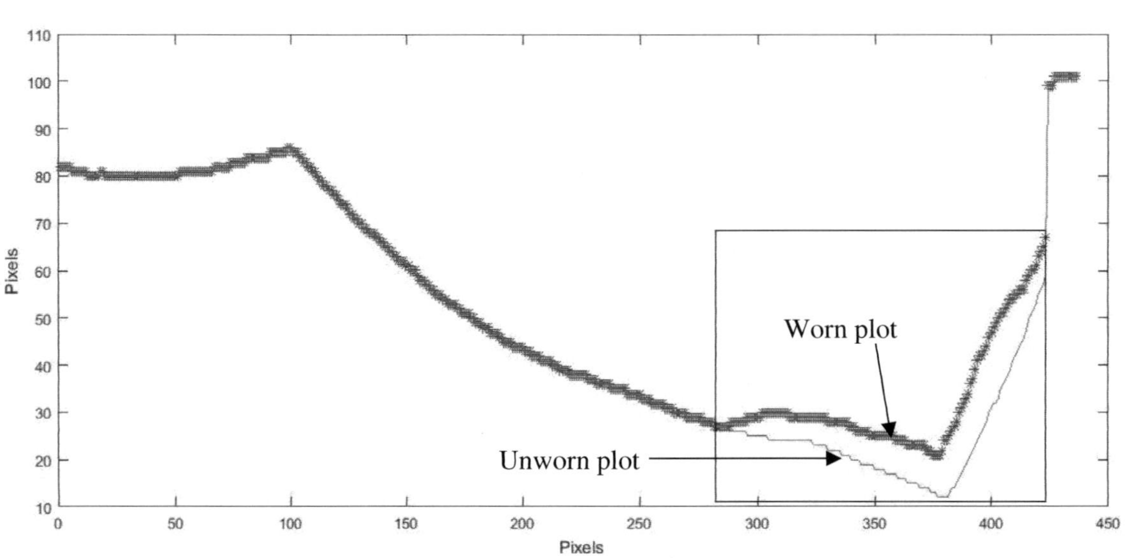

Figure 6. Plot of worn and unworn bit

3.2. Cross-correlation

The original and worn profiles were cross correlated to each other and the original profile was also cross correlated with itself (auto-correlation) and comparison between both graphs are made and shown in Figure 7. Figure 7 shows the cross-correlation graphs for original bit and difference between original and worn bits. Small discrepancy can be seen based on the graphs. Despite the plots being almost similar on the right-hand side of the peak, a different shape can be seen on the left-hand side of the peak. The auto correlation of the new bit showed a minimum value at around -250 whereas the correlation between worn and new bits had minimum value at around -240. The negative values at -240 and -250 represent perfect negative correlation of minimum values for unworn and worn plot. The difference of about 10 pixels between both minimums value of auto correlation shows some pattern to the wear size which has about similar value. Despite being able to detect such small wear size, it is hypothesized that the detection can be further improved by applying smaller interrogation window to the image.

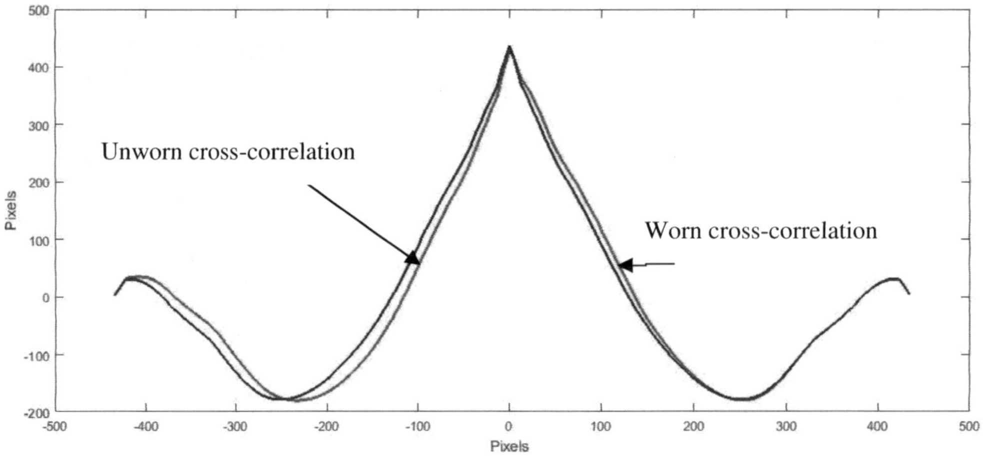

Figure 7. Cross-correlation of unworn with worn plot

4. Conclusion

In this study, an image-based analysis for the tool wear quantification is proposed. Through common image analyses such as filtering and thresholding, the edge of the drill bits was characterized through the change of pixel number of a binary image. The plots, which represent the profile of drill bit edge, were then cross correlated to each other. It was found that the difference of the minimum values between the correlation graphs reflect the physical size. In the experiment performed, the difference of about 10 pixels in the correlation graphs conforms with the wear imposed to the drill bits. With a small pixel size, this technique offers high resolution and sensitivity to the wear detection of cutting bits.

Future study involves the implementation of interrogation window that divides the image into smaller sub-regions. With the implementation of cross-correlation technique to smaller window, it is expected that the cross-correlation graph will show prominent difference between worn and new bit and offer higher resolution and sensitivity to the technique.

Acknowledgement

The authors would like to thank the Ministry of Higher Education, Malaysia, and Universiti Teknologi PETRONAS for the financial support through the Fundamental Research Grant Scheme (FRGS) of FRGS/1/2015/TK03/UTP/03/1 for this research work.

References

[1] E. Brinksmeier, *Prediction of Tool Fracture in Drilling*, CIRP Ann. - Manuf. Technol., vol. 39, no. 1, pp. 97–100, 1990.

[2] A. G. Rehorn, J. Jiang, and P. E. Orban, *State-of-the-art methods and results in tool condition monitoring: A review,* Int. J. Adv. Manuf. Technol., vol. 26, no. 7–8, pp. 693–710, 2005.

[3] J. Chen, Q. Fang, and P. Li, *International Journal of Machine Tools & Manufacture Effect of grinding wheel spindle vibration on surface roughness and subsurface damage in brittle material grinding,* Int. J. Mach. Tools Manuf., vol. 91, pp. 12–23, 2015.

[4] G. Quintana, T. Rudolf, J. Ciurana, and C. Brecher, *Robotics and Computer-Integrated Manufacturing Using kernel data in machine tools for the indirect evaluation of surface roughness in vertical milling operations,* Robot. Comput. Integr. Manuf., vol. 27, no. 6, pp. 1011–1018, 2011.

[5] M. Rizal, J. A. Ghani, M. Zaki, C. Hassan, and C. Haron, *Online tool wear prediction system in the turning process using an adaptive neuro-fuzzy inference system,* Appl. Soft Comput. J., vol. 13, no. 4, pp. 1960–1968, 2013.

[6] Q. Ren and P. Bigras, *An Enhanced Adaptive Neural Fuzzy Tool Condition Monitoring for Turning Process,* 2016.

[7] M. Sortino, *Application of statistical filtering for optical detection of tool wear,* Int. J. Mach. Tools Manuf., vol. 43, no. 5, pp. 493–497, 2003.

[8] K. B. Pedersen, *Wear measurement of cutting tools by computer vision,* Int. J. Mach. Tools Manuf., vol. 30, no. 1, pp. 131–139, 1990.

[9] J. U. Jeon and S. W. Kim, *Optical flank wear monitoring of cutting tools by image processing,* Wear, vol. 127, no. 2, pp. 207–217, 1988.

[10] Y. Liang and Y. Chiou, *An Effective Drilling Wear Measurement based on Visual Inspection Technique,* IAPR Conf. Mach. Vis. Appl. (Tsukuba Sci. City, Japan), pp. 2–5, 2005.

[11] M. J. Jackson, G. M. Robinson, L. J. Hyde, and R. Rhodes, *Neural Image Processing of the Wear of Cutting Tools Coated with Thin Films,* J. Mater. Eng. Perform., vol. 15, no. April, pp. 223–229, 2006.

[12] M. Stemmer, a. Pavim, M. Adur, F. Deschamps, R. Schmitt, and R. Hermes, *Machine Vision and Neural Networks applied to Wear Classification on Cutting Tools,* Proc. EOS Conf. Ind. Imaging Mach. Vis., pp. 139–142, 2005.

IOP Conf. Series: Materials Science and Engineering **328** (2018) 012011 doi:10.1088/1757-899X/328/1/012011

[13] A.V. Atli, O. Urhan, S. Ertürk, and M. Sönmez, *A computer vision-based fast approach to drilling tool condition monitoring,* Proc. Inst. Mech. Eng. Part B J. Eng. Manuf., vol. 220, no. 9, pp. 1409–1415, 2006.

IOP Conf. Series: Materials Science and Engineering **328** (2018) 012012 doi:10.1088/1757-899X/328/1/012012

Investigating Resulting Residual Stresses during Mechanical Forming Process

Stephen A Akinlabi[1], Olawale S Fatoba[2], Peter M Mashinini[3] and Esther T Akinlabi[4]

[1&3]Department of Mechanical & Industrial Engineering Technology, University of Johannesburg, South Africa,
[2&4]Department of Mechanical Engineering Science, University of Johannesburg, South Africa

stephenakinlabi@gmail.com , drfatobasamemi@gmail.com , mmashinini@uj.ac.za , etakinlabi@uj.ac.za

Abstract. Most manufacturing processes such as machining, welding, heat treatment, laser forming, laser cladding and, laser metal deposition, etc. are subjected to a form of heat or energy to change the geometrical shape thus changing the inherent engineering and structural properties of the material. These changes often cause the development of locked up stresses referred to as residual stresses as a result of these activities. This study reports on the residual stresses developed due to the mechanical forming process to maintain a suitable structural integrity for the formed components. The result of the analysis through the X-ray diffraction confirmed that residual stresses were induced in the manufactured parts and further revealed that residual stresses were compressive in nature as found in the parent material but with values less than the parent material.

1. Introduction

The manufacturing industry is observed to be evolving in this modern world of technological advancement. Several manufacturing processes are being employed in the fabrication of different engineering components thereby solving some of the basic societal problems. Among the manufacturing processes commonly employed are machining, welding, grinding, forging, casting, moulding and forming etc. [1], [2]. Fabrication of components through the traditional mechanical forming process is still very relevant even in this millennium. It may be augured that this may be limited within the small and medium scale manufacturers but this group are closer and more accessible to the general populace.

Mechanical forming is one of the conventional methods of metal-forming techniques employed in sheet metal forming process. The mechanical forming process causes permanent bending in sheet metal through the process of plastic deformation. Such permanent deformation may occur by slip, twinning or the combination of both methods during deformation. Mechanically forming a sheet into a shape may be a function of the material property of the sheet and the tooling system comprising of the punch and the die. Metal forming can be classified into different categories under mechanical working process, such as forging, extrusion, rolling, drawing, stretched forming, and other sheet forming processes, casting, powder metallurgy and fiber metal forming, electroforming, rubber, stretched and super-plastic forming [3].

The principle of sheet forming operation involves the application of force to the sheet metal to modify the material geometry; this makes the process of metal forming unique because there is no material removal. The applied force stresses the metal beyond its yield strength, causing the material to plastically deform, but not to fail. By doing so, the sheet can be bent or stretched into a variety of complex shapes. Fig. 1 shows the schematic diagram of stretched forming process. A schematic of a mechanical forming process is shown in Fig. 1.

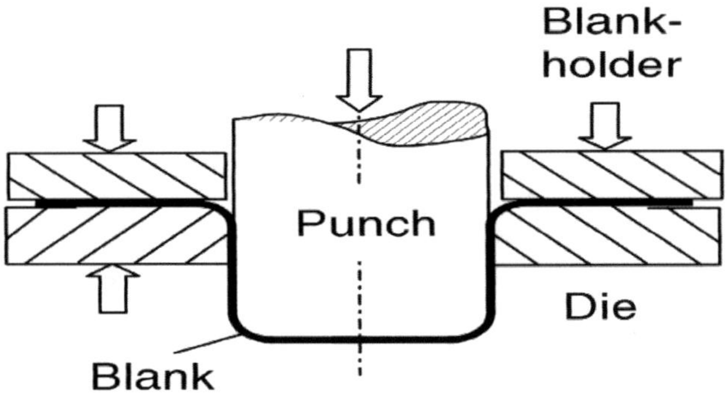

Figure 1: Schematic of Stretched Forming [4]

One of the most widely used press fabrication operation is bending and forming of metallic components. The mechanical forming process employs a punch rammed into a clamped plate positioned across a die. The force is applied to the punch above the plate to be deformed; the type of shape will depend on the punch and the die. The application of force consequently generates internal stresses and displacement thereby causing distortion within the structure of the material subjected to the load. Hence, the geometrical shape of the formed component depends on the load applied to the punch [2].

Over the decades, there had been increasing interest in the preparation and processing of new materials with unique properties and functions. The poor properties of materials such as ceramics and magnesium alloys have presented not only challenges but also great opportunities for metal forming researchers to develop new materials [2]. With this initiative, the real forming techniques such as mechanical forming may be further researched to get a more scientific understanding of the process for better positioning in meeting today's need. The mechanical bending process is by the material properties, and the response during the forming is influenced by its mechanical properties, which include the elongation, the yield point, anisotropy, grain sizes, residual stresses, and spring back.

Virtually all manufacturing processes, such as casting, welding, machining, moulding, and heat treatment, plastic deformation during bending, rolling or forging introduce residual stresses into the manufactured samples [4]. Investigations conducted by researchers on the subject of residual stresses in engineering components found that the most common causes of residual stresses are material plastic deformation, steep thermal gradients and phase transformations, nuclear radiation, as well as the exposure to electromagnetic fields [4]-[6]. On the other hand, a perspective of a researcher was that residual stresses arise as a result of misfits between different regions of the material, component or assembly [7]. Further studies on the subject confirmed that macro residual stresses could arise in engineering samples through the interaction between the misfit of parts within an assembly, and through the generation of chemical, thermal, and plastically induced misfits between different regions within one part [6].

Among other factors known to cause residual stresses can be the development of deformation gradients in various sections of the material. By developing thermal gradients, volumetric changes can

IOP Conf. Series: Materials Science and Engineering **328** (2018) 012012 doi:10.1088/1757-899X/328/1/012012

arise during the solidification or from solid-state transformations, as well as from differences in the coefficient of thermal expansion in samples made from different materials [4].

Residual stresses are often present in sheet metal parts after the process because of the non-uniform deformation of the sheet during forming. They are also referred to as the locked-in stresses and can be defined as those stresses existing within a material in the absence of any external loading or thermal gradients. Virtually all-manufacturing processes, such as casting, welding, machining, moulding, and heat treatment, plastic deformation during bending, rolling or forging induces residual stresses into the manufactured components [8]. This has necessitated the study into the residual stresses developed in sheets during mechanical forming processes, through this; the structural integrity of the formed component can be evaluated.

During past years, many different methods for measuring the residual stresses in different types of samples have been developed. Residual stress measurements may be classified as destructive, semi destructive, or non-destructive according to research and investigation into the subject [9] on residual stress measurement techniques. This is further illustrated with Figure 2, showing the various methods.

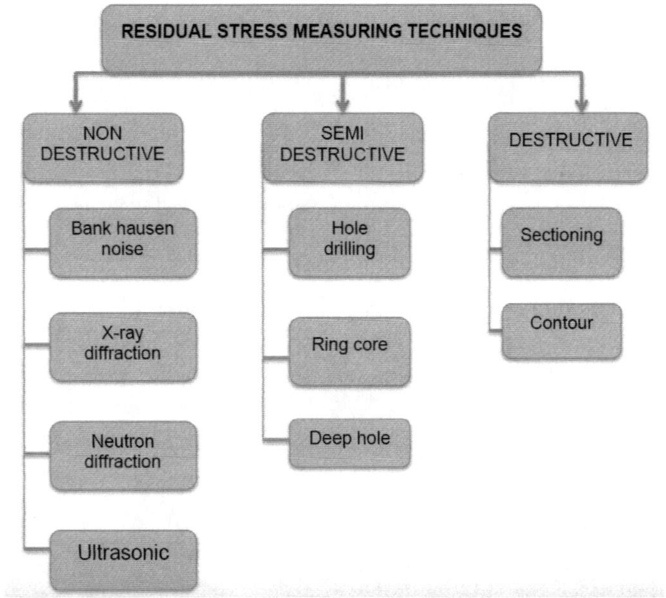

Figure 2: Summary of the Residual stress measurement

The destructive (sectioning technique and contouring) and semi destructive techniques (Hole drilling, Ring core and deep-hole), also referred to as mechanical method, are dependent on inferring the original stress from the displacement incurred by completely or partially relieving the stress when removing some material. These methods rely on the measurement of deformations due to the release of residual stresses upon removal of material from the specimen. Non-destructive techniques on the other hand, usually measure some parameters (strain) that are related to stress and these techniques include the x-ray or neutron diffraction, synchrotron, ultrasonic methods and magnetic methods. These techniques are being increasingly used because of their non-destructive nature.

The engineering properties of materials and structural samples notably fatigue life, distortion, dimensional stability, corrosion resistance, and brittle fracture can all be considerably influenced by residual stresses. Such effects usually bring about additional cost in repairs and restoration of the parts, equipment, and structures. Consequently, residual stress analysis is a compulsory stage in the design of parts and structural elements and in the estimation of their reliability under real service conditions [4], [9]-[11].

IOP Conf. Series: Materials Science and Engineering **328** (2018) 012012 doi:10.1088/1757-899X/328/1/012012

Another investigation of a researcher found that residual stress magnitudes, which do not exceed 80% of the original yield strength of a material contain minimal error [12]. However, if this percentage is exceeded, then the residual stress magnitude should be taken at the 80% yield value. It was also reported that residual stresses cannot be precisely determined and about 10-15% error can be expected when using the hole drilling method [10]. A study under similar experimental conditions further found [11] that an error of about 20% should be allowed when stainless steel plate is used compared to the results of a similar study that reported an error of about 10-15% for a hole drilling method [10].
Similarly, a detailed knowledge of the residual stress distribution remains difficult to gather despite the large number of publications on the investigation of residual stress measurements [11]-[13]. The following methods have been successfully used for residual stress measurements; layer removal, hole drilling, X-ray diffraction, neutron diffraction, synchrotron diffraction, ultrasonic and magnetic method. The most frequently applied among these techniques for metallic samples are the layer removal, the hole drilling technique and the x-ray diffraction method. More closely, X-ray diffraction utilizes the fact that when a metal is under stress, applied or residual, the resulting elastic strains cause the atomic planes in the metallic crystal structure to change their spacing. X-ray diffraction can directly measure this inter-planar atomic spacing; from this quantity, the total stress on the metal can then be obtained [14].

2. Experimental Methodology

The residual stress measurements were conducted at the South African Nuclear Energy Corporation (NECSA) facility and an X-ray diffractometer - Bruker D8 Discovery/Advance was used. The experimental set-up for the residual stress measurement of both parent material and mechanical formed are shown in Fig. 3. (a) and (b) respectively.

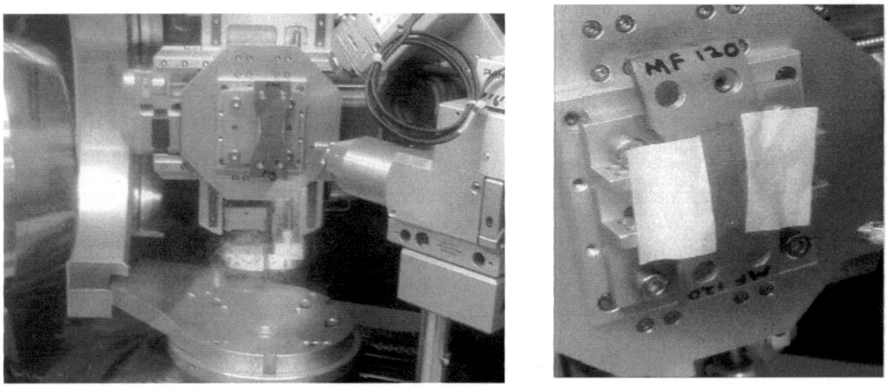

(a) (b)

Figure 3 (a) and (b): Residual Stress measurement setup for Parent and Mechanical Formed in the diffractometer (Photo taken at NECSA during the experiment)

Three sets of parent material and mechanical formed samples each were investigated and average result was presented. The measurements were taken at the three points indicated in Figure. 4. Three points were considered as shown for the measurement.

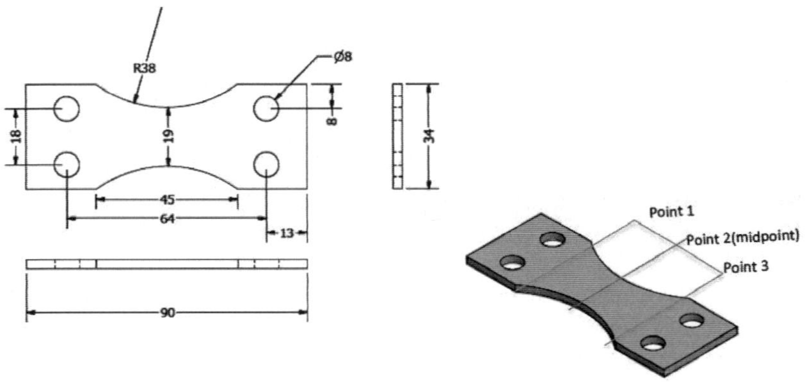

Figure 4: Test-piece isometric and residual stress measurement points

The diffractometer consists of the following samples:

- X-ray source and Graphite monochromator crystal for monochromatic beam on the primary side.
- Goniometer (Eulerian Cradle): Used for the accurate positioning of samples,
- Video microscope and laser for alignment.
- Area detector,
- GADDS software for control, data-acquisition and 2-D diffraction data processing.

The diffractometer is equipped with a standard Bragg-Brentano geometry with Cu tube and a detector. The 2θ scan range was set from 20 to 70° at 10° per step. The measurement for every azimuth angle ϕ has a set of tilt angle ψ between (0-70°) in stepwise stages of 10° each. The beam optics includes the graphite monochromator and a 0.8 mm diameter collimator mounted on the primary side.
The measurements were performed while set at an inclination mode to allow for higher sample tilt [4]. A laser video-alignment system was used for accurate sample positioning and the diffracted data were collected with a high star-area detector. The measurement parameters for the strain determination on the D8 GADDS diffractometer are presented in Table 1. The maximum measurement tilt angle considered in this study was 70° in 2θ. For the purpose of sample shielding, the beam was held at high tilt angles. The peak evaluation method employed is the Pearson VII and the stress model is Biaxial + Shear stresses.

Table 1: Parameters settings on the D8 GADDS diffractometer

Parameters	**Value**
hkl reflection	220
2θ [°]	99.67
Tilt angle ψ [°]	0-70
Frame width, $\Delta\psi$ [°]	7
Azimuth orientation ϕ [°]	0, 45, 90, 180, 225, 270
Sample detector distance [cm]	15

3. Result and Discussion

Three sample sets of parent material and mechanical formed samples were investigated but only one set has been reported because of the compressive nature of both samples. The residual stresses in the two principal axes of transverse σ_{11} and σ_{22} longitudinal for PM and MF, were measured. This is a standard way to present residual stresses of processed materials most especially metals. One of the studies in this field [15] agrees with author's finding that residual stresses and strains in longitudinal direction are tensile and dominant and secondly that with increasing heat input, associated with higher line energy and/or multiple passes, the longitudinal residual strain/stress increases until a threshold is reached. This threshold is due to the material yielding in tension.

The measured frames were analysed using the BRUKER's LEPTOS v6 software. The program slices the diffraction rings into bins of desired width, integrates the slices, and fits the resultant peak. The analysis for strain was conducted using the freestanding 220 reflections. These were the highest reflections with better statistics that could be measured on the D8 GADDS instrument. The spotted patterns originate from the effect of the process to which the material was subjected. The residual stress tensor results measured at the three points (1, 2 and 3) for the parent material and the mechanically formed component are presented in Table 2 respectively.

Table 2: Stress tensor for the mechanically formed component and parent material

Measurement points	**Mechanical formed sample**		**Parent material**	
	σ_{11}	σ_{22}	σ_{11}	σ_{22}
1	-61.7 ± 41.6	--51.1 ± 41.6	-310.7±27.5	-399.5± 27.5
2	-53.9±41.2	11.8 ± 45.9	-299.8± 28.4	-319.6±28.5
3	-43.1 ± 55.9	-54.2 ± 63.6	-297.0±29.2	- 321.8±29.2
HV	116		94	

The stresses considered are in the two principal axes of transverse σ_{11} and longitudinal σ_{22} for the and mechanically formed component.

The parent material was found to be very compressive in nature considering the values of the measured tensors across the three points of measurement (-310.7±27.5, -299.8± 28.4 and -297.0±29.2). Similarly, tensor values of the mechanically formed components were also compressive in nature but observed to be more relaxed from the tensor values (-61.7±41.6, -53.9 ± 41.2 and -43.1±55.9) due to the mechanical forming process. The compressive nature of the stresses is attributed to strain hardening and cold working condition on the material. In addition, the micro Vickers hardness values confirm the exhibited evolved residual stresses of the mechanical formed samples. It was observed that the values of the micro Vickers hardness were enhanced in the formed sample in comparison to the parent material. The improvement in the hardness values is attributed to mechanical work and strain hardening. There are some proportionate relationships between the developed residual stresses and the increasing micro Vickers hardness in the formed samples.

Furthermore, the varying magnitude of the tensor across the three points of measurement can be explained by either of the following or all; the difference in the surface roughness, the effect of stress relaxation on their shape, influence of their grain size distribution, and the effect of machining during fabrication of the test sample.

4. Conclusion

The analysis of the residual stresses was conducted using the x-ray diffraction, and the study revealed the changes in the residual stresses as the samples were formed. The observed variations in the residual tensors from compressive in this study towards tensile in the mechanically formed parts confirm the findings from the literature [15]. Also, the difference in the magnitude in the tensile tensors at the three measurement points is attributed to the flow stress, mechanical and strain hardening. The changes in the micro Vickers hardness values confirm the developed residual stresses. It may be correct to establish that the residual stress can increase proportionately with the hardness value to a certain threshold that should be further determined.

5. Acknowledgement

The authors recognise the support of the Global Excellent Stature (GES) grant award, of the University of Johannesburg.

References

[1] Mechanical stretch forming. [Online] Available: https://www.google.co.za/imgres?imgurl=http://14.139.245.214/mfvlab/images/SheetM%20etal/drawingprocess.jpg&imgrefurl=http://14.139.245.214/mfvlab/SheetMetal.php&h=72%202&w=867&tbnid=crVU2jNkYfl18M:&docid=4Px9NZOZGH4XLM&ei=xuDzVszQMYiSaa%20DwlaAK&tbm=isch&ved=0ahUKEwjMy8bErNnLAhUISRoKHSB4BaQQMwh6KFEwUQ#h=722&imgrc=crVU2jNkYfl18M:&w=867 [Accessed on February 2017]

[2] Kurt, L., 1985, Handbook of Metal Forming. McGraw-Hill Book Company, New York, NY 10020.

[3] Flaman, M. T., 1982, "Investigation of ultra-high speed drilling for residual stress measurements by the Centre-hole method," Experimental Mechanics 22 (1), pp. 26-30.

[4] Sheet metal forming, 2012, [online]. Available: www.customerpartnet.com/wu/sheet-metal-forming, [Accessed February 2016].

[5] Schajer, G.S., 2010, "Advances in Hole-Drilling Residual Stress Measurements," Experimental

[6] Withers, P. J., and Bhadeshia, H. K. D. H., 2001, "Residual stress, Part 2-Nature and origins," Materials Science and Technology. Vol. 17.

[7] Withers, P. J., 2007, "Residual stress and its role in failure," Report of progress in Physics 70, pp. 2211-2264.

[8] Kannatey-Asibu, E., 2009, "Principle of laser materials processing," John Wiley & sons, Inc., New York, N.Y.

[9] Rossini, N. S., Dassisti, Benyounis, K. Y., and Olabi, A. G. "Methods of measuring residual stresses in Samples," *Material and Design*, 35, pp. 572-588, 2012.

[10] Beaney, E. M. and Procter., E. A. "Critical evaluation of the center hole technique for the measurement of residual stresses," *Strain Journal*, pp. 7-15, 1974.

[11] Nikola, W. E. "Post-yield Effect on Centre-Hole Residual Stress Measurement," *Experimental Mechanics*, pp. 126-136, 1984.

[12] Schajer, G. S. "Measurement of Non-uniform Residual stresses using the hole-drilling method, Part I- Stress calculation procedures," *Journal of Engineering materials and Technology*, 1 (110), pp.335-343, 1998.

[13] Schajer, G. S. "Advances in Hole-Drilling Residual Stress Measurements," *Experimental Mechanic,* 50, pp. 159-168, 2010.

[14] Norton, J. H. and "Rosenthal, D. "Stress measurement by x-ray diffraction," *Proc Soc Exp Stress Anal* 1(2), pp. 73–6, 1944.

[15] Stefan, K., Paradowska, A. M., Kirstein O., and Moore, A., 2010, "Investigation of Residual Stress in Laser Forming Steel Plates using Neutron Diffraction," Material Science Forum, Vol. 652, pp. 123-128.

Effect of Bottoming on Material Property during Sheet Forming Process through Finite Element Method

Stephen A Akinlabi[1], Olawale S Fatoba[2], Peter M Mashinini[3] and Esther T Akinlabi[4]

[1&3]Department of Mechanical & Industrial Engineering Technology, University of Johannesburg, South Africa,
[2&4]Department of Mechanical Engineering Science, University of Johannesburg, South Africa

stephenakinlabi@gmail.com , drfatobasamemi@gmail.com , mmashinini@uj.ac.za , etakinlabi@uj.ac.za

Abstract. Metal forming is one of the conventional manufacturing processes of immense relevance till date even though modern manufacturing processes have evolved over the years. It is a known fact that material tends to return or spring back to its original form during forming or bending. The phenomena have been well managed through its application in various manufacturing processes by compensating for the spring back through overbending and bottoming. Overbending is bending the material beyond the desired shape to allow the material to spring back to the expected shape. Bottoming, on the other hand, is a process of undergoing plastic deformation at the point of bending. This study reports on the finite element analysis of the effect of bottoming on the material property during the sheet forming process with the aim of optimising the process. The result of the analysis revealed that the generated plastic strains are in the order between 1.750e00-1 at the peak of the bending and 3.604e00-2, which was at the early stage of the bending.

1. Introduction

Sheet metal forming process is one of the fundamental techniques employed in metalworking. This is one of the common manufacturing processes on the shop floor of a fabrication yard, where sheet metal are often formed and bent into thin and flat pieces. Sheet metal forming processes are classified as a conventional forming process, which applies forces to piece of sheet metal to modify the geometry of the material instead of material removal. The applied load to the material stresses it beyond its yield strength causing the material to plastically deform but not to fail. In this state, the sheet can be bent and stretched into different configuration and shapes [1].
The bending process is applied in the sheet metal forming operation to bend and fabricate curvature shapes such as electronics part. Most of the bending activities are not only employed in the bending process but are found in other many sheet-metal forming manufacturing processes such as the deep drawing process, the U-tube bending and the flanging process. One of the major drawbacks of the bending process is the development of spring back, which is believed to have affected many of the working process parameters such as the material property, bending angle, bending radius, bending stroke [2], [3]. Considerable studies have been conducted with the aim to investigate and reduce the amount of spring-back.

IOP Conf. Series: Materials Science and Engineering **328** (2018) 012013 doi:10.1088/1757-899X/328/1/012013

In the past, sheet metal bending processes are dependent on the designer's experience and involve trials and errors to obtain the desired result [4]. Trial and error can involve adjustments made to the machine tools and process control to compensate for variation in material and unexpected parameters [5]. Many analytical models proposed to study spring-back in bending use simple beam or plate bending theory [6]. Some of the research conducted includes the study into the spring-back of CK67 steel sheet in V-die and U-die bending processes [7]. The studies into the spring-back control of sheet metal air bending process [8] and the investigation of the variation of elastic modulus during plastic deformation and its influence on spring-back [9[. Also the prediction of spring-back in the wipe-bending process of sheet metal using the neural network [10], the determination of spring-back of stainless steel sheet metal in the V-bending die [11] and the study into spring-back prediction, compensation, and optimisation using numerical analysis [12].

Several studies have been conducted on sheet metal forming [8]-[11] by different researchers to date, through this the application of this cut across different industrial applications. It is believed that bending is one of the applications used in metal forming process in many industries and observed that springback as one of the problems associated with bending process [13]. The process of validating the experimental results in qualifying the amount of springback in a U-bending process demand that both experimental and numerical investigation be conducted on the effect of punch speed on the amount of springback in U-bending of auto-body steel sheets [14]. They use SPCC and DP780 steel sheet for both the experimental and finite element analysis and the experimental results show that the amount of springback for SPCC during the bending process decreases but that of DP78 increases as the punch increases.

Furthermore, other research conducted was a study on finite element analysis of punch height effect on the V-bending angle, and the results showed that the effects of punch height on the bending angle could be theoretically clarified based on the material flow analysis and the stress distribution analysis [15]. Considering all the research work conducted in metal sheet bending and forming, the V-bending operation and finite element analysis has been widely investigated experimentally but using U-bending (Figure 1) to study the effect of bottoming on the material property during the sheet forming process is scarcely researched. This paper presents the effect of bottoming on the material property during the sheet forming process using finite element method

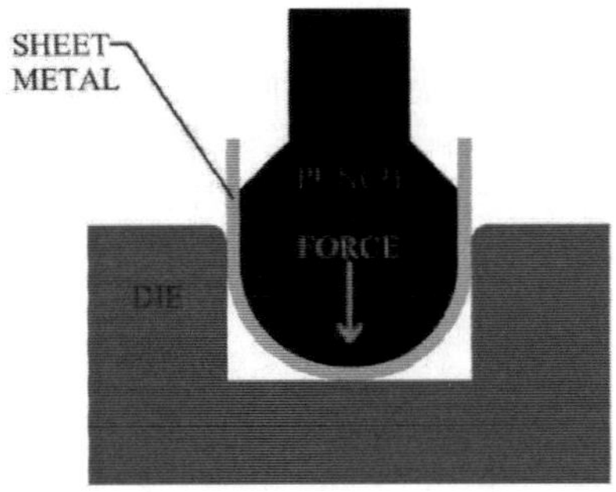

Figure 1: Schematic of U-bending Process [16]

2. Experimental Methodology

The mechanical forming technique is one of the different manufacturing processes employed in the fabrication of sheet metals to form different shapes and part. The process often involved bending and shaping of sheet metals using different technics. Sheet metal bending is the plastic deformation of the work over an axis, creating a change in the part's geometry. In some cases, the bending produces small change in sheet thickness while some instance it does not produce any change in the thickness of the material. It is important to highlight that in the process of making the desired geometry, bending can be used to impart strength and stiffness to sheet metal, change a part's moment of inertia, and to eliminate sharp edges [17].

Sheet metal bending is based primarily on the phenomenon of tension and compression within the materials. Mechanical Engineering principles of metals, especially regarding deformation, are an important factor to understanding sheet metal. The effect that material properties will have in response to the conditions of manufacture will be a factor in sheet metal process design. Usually sheet metal bending is mostly performed at room temperature, however, at times some form of heat are applied to either warm or hot work the material at some temperature other than room condition [17]. The introduction of moderate temperature often improves the material property and help to enhance the bending of some hard materials. Traditionally, sheet bending process mostly employs a punch and die setup for the shaping or bending or forming to be achieved. The different types of punch and die geometry and fixtures have made the bending to be unique to an operation; this may be viewed as a drawback of this process when compared to other modern manufacturing processes. Figure 2 shows a schematic of a representation of a simple sheet metal V-bending configuration.

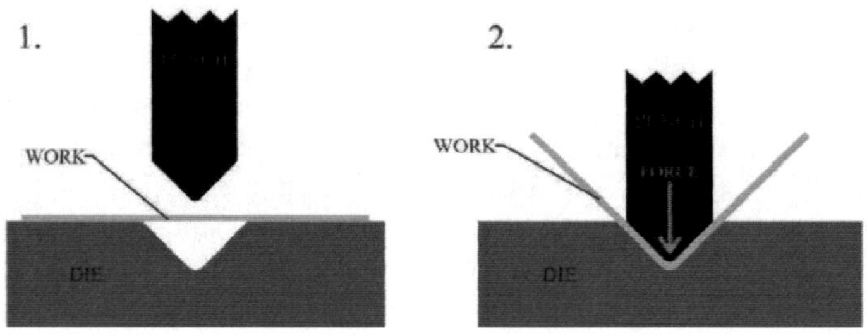

Figure 2: Schematic of V-bending setup [17]

In general, we may consider a whole U-bending process into two steps, loading and unloading. In the loading step, a sheet metal is being bent into the die until the punch moves down completely so that its shape is formed closely to the die shape. During this step, the workpiece undergoes elastoplastic deformation and temperature increases under frictional resistance. Owing to this thermo-mechanical relaxation, the dimension of the final product, particularly the bending angle, becomes different from that of the product before unloading. This dimensional difference is called the elastic recovery phenomenon, and particularly the bending angle alternation is denoted by spring-back. Figure 3 may illustrate the spring-back amount and the negative spring-back is also called spring-go.

Figure 3: Illustration of typical springback phenomenon [13]

Sheet metal bending and forming are long standing manufacturing processes in the fabrication of different shapes despite the drawbacks of spring back. The improvement of the processes over the years has contributed immensely to the relevance of such a traditional manufacturing process to diverse areas of applications. Some of the techniques developed to overcome the effects of spring back include over bending, stretch forming and bottoming. The amount of spring back is quantified, and the sheet metal is over bent to a smaller bend angle than needed. Recovery of the material from spring

IOP Conf. Series: Materials Science and Engineering **328** (2018) 012013 doi:10.1088/1757-899X/328/1/012013

back results in a calculated increase in bend angle. This increase makes the recovered bend angle exactly as what was originally planned. Stretch forming, on the other hand, is a metal bending technique that eliminates most of the spring back in a bend. Subjecting the work to tensile stress while bending will force the elastic region to be plastically deformed [13].

Stretch forming cannot be performed for some complex bends and very sharp angles. The amount of tension must be controlled to avoid cracking of the sheet metal. Stretch forming is a process often used in the aircraft building industry. Bottoming is a bending process where the punch and the workpiece bottom on the die. This makes for a controlled angle with very little spring back. The tonnage required on this type of press is more than in air bending. The inner radius of the workpiece should be a minimum of 1 material thickness [18].

3. Mechanics of Sheet Metal Bending Process

An understanding of the material properties, characteristics and behaviours of metal, is necessary for a good insight into the mechanics of sheet metal bending. The elastic and plastic deformation of the sheet metal is necessary and also that the bending sheet metal produces localised plastic deformation and essentially with no change in the sheet thickness, depending on the type of operations and does not create a metal flow that may affect the region close to the bend. The force required to perform a bend is largely dependent upon the bend and the specific metal bending process because the mechanics of each process can vary considerably. Proper lubrication is essential to control forces and has an effect on the process. In punch and die operations, the size of the die opening is a major factor in the force necessary to perform the bending. Increasing the size of the die opening will decrease the required bending force. As the sheet metal is bent, the force needed will change. Usually, it is important to determine the maximum necessary bending force, to access machine capacity requirements [16].

Some of the important factors influencing the mechanics of bending include material, sheet thickness, width over which bend occurs, the radius of the bend, bend angle, machinery, tooling and specific metal bending process. Bending a sheet will create forces that act in the bend region and through the thickness of the sheet. A schematic of the mechanics of sheet metal bending is shown in Figure 4. The material towards the outside of the bend is in tension, and the material towards the inside is in compression. Tension and compression are opposite forces, therefore when moving from one to the other, a zero region must exist. At this zero region, no forces are exerted on the material. When sheet metal bending, this zero region occurs along a continuous plane within the part's thickness, called the neutral axis. The location of this axis will depend on the different bending and sheet metal factors. However, a general approximation of the location of the axis could be 40 % of sheet thickness, measured from the inside of the bend. Another characteristic of the neutral axis is that because of the lack of forces, the length of the neutral axis remains the same. Fundamentally, to one side of the neutral axis, the material is in tension, to the other side, the material is in compression. The magnitude of the tension or compression increases with increasing distance from the axis [16], [19].

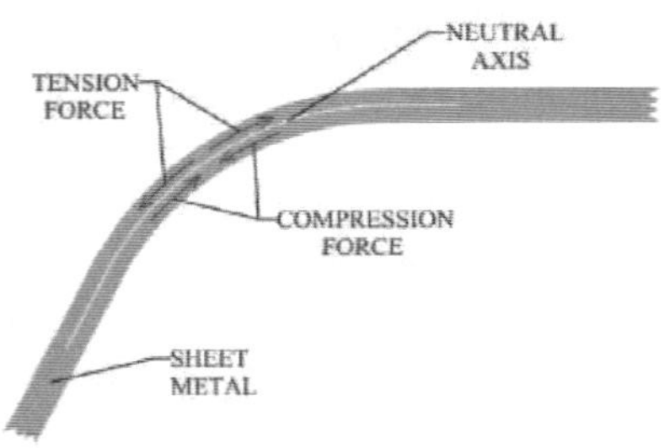

Figure 4: Schematic demonstrating the mechanics of Sheet Metal Bending [16]

4. Finite Element Formulation

The finite element method as applied to the sheet metal bending and U-bending, in particular, is to evaluate the effect of bottoming using a U-bending approach considering the plastic strain and Von-Mises stress developed in the process of and bottoming in particular. Marc software, 2015 version was employed for the FEM analysis. The geometry of the steel was defined in Mentate Marc, a plain strain element was used, and the material properties and boundary conditions were setup. The schematic diagram of the U-bending process and the dimension is shown in Figure 5. The finite element setups of the deformable and rigid bodies are shown in Figure 6 as defined in Marc mentate.

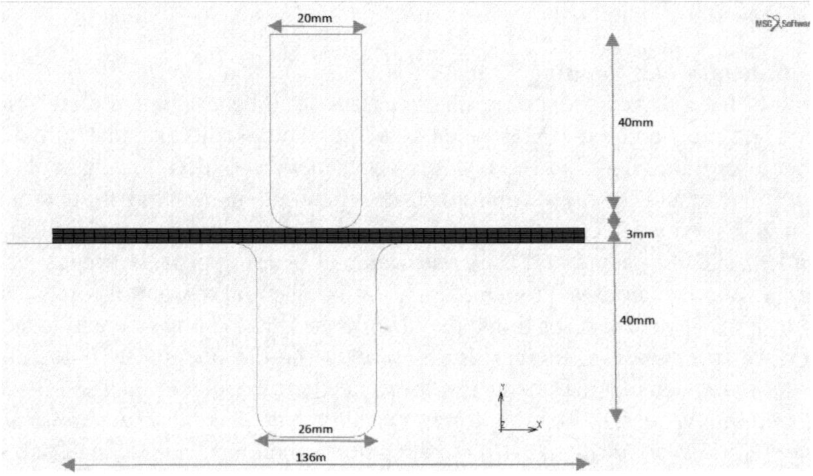

Figure 5: Schematic of U-bending Process and dimensions

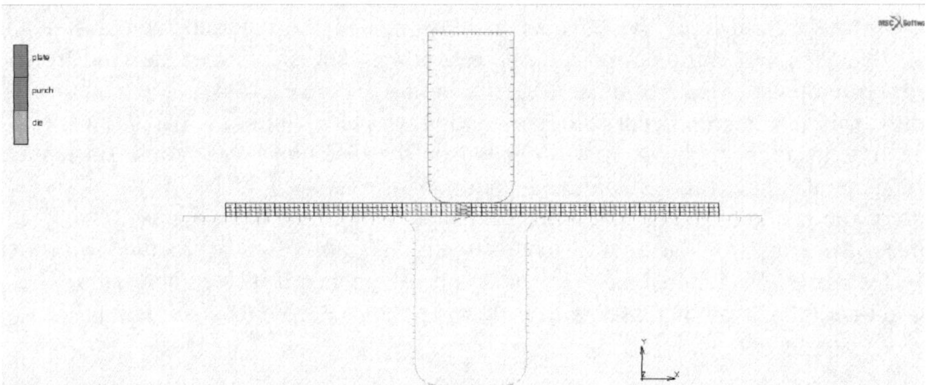

Figure 6: Schematic of the finite element setup of the deformable and rigid body

5. Result and Discussion

It is expected in a bending operation that both tensile stresses and compressive stresses are overcome for the desired shape to be achieved. When the bending is eventually achieved, the locked-in stresses cause the material to spring back towards its original position. Springback phenomena are seriously dependent on the material and the type of bending process employed, but the good story is that spring back can always be compensated for in any bending operation. The static analysis was conducted and completed with plain strain elements in constant time stepping, over thirty steps for loading and ten for unloading. The resulting deformation was described by the movement of the punch into the plate at different stroke lengths (2 mm, 9 mm and 20 mm). The contour plot at the stroke of 2 mm for the plastic strain and the Von Mises stresses is shown in Figure 7. (a) and (b) respectively. Both the equivalent plastic strain and the Von Mises Stress were measured during the bending and bottoming in particular to evaluate the effect on the material property. The punch at a stroke of 2 mm generated a plastic strain of 3.604e-2 and Von Mises stress of 6.040e+4 Pa. At this stage of the process, the plastic strain is compressive in nature while the Von Mises stress is tensile.

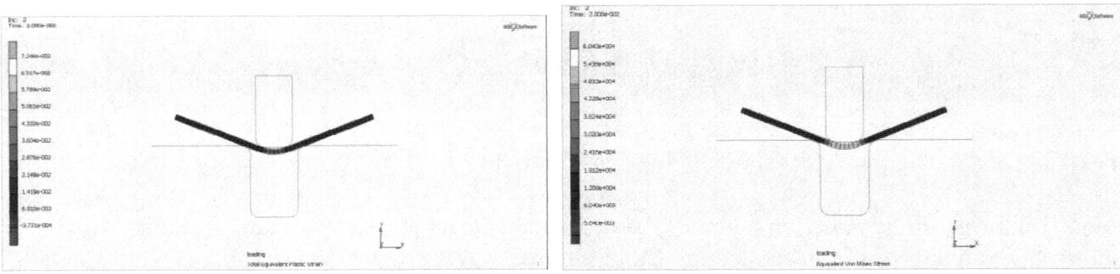

(a) (b)

Figure 7 (a) and (b): Contour Plots showing the Equivalent Plastic Strain generated at a stroke of 2 mm

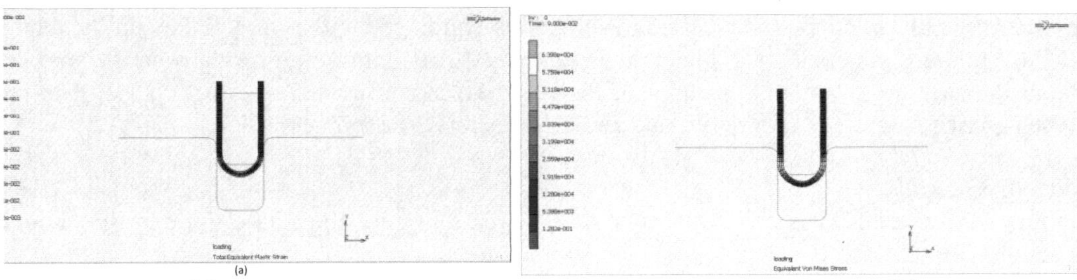

(a) (b)

Figure 8 (a) and (b): Contour Plots showing the Equivalent Plastic Strain generated at a stroke of 9 mm

As the punch stroke increases, there was significant changes in both the plastic strain and Von Mises stress. At a stroke of 9 mm, the plastic strain was still compressive (6.024e-2) but more tensile in nature, the Von Mises stress distribution shown on the contour plot indicated that there is stress distribution along the side wall of the U-bend (5.758+e4Pa) and the bend radius (1.919+4 Pa). Furthermore, It is important to highlight the significant role of the punch in a bending operation because the bend radius depends on the punch, material properties and the thickness of the material. When a sheet material is bent, the sheet stretches in length over the outside edges of the bend in tension and inner bend radius in compression. This consequently induces stresses and strains into the sheet metal as the punch presses, causing a permanent deformation of the material. The Contour plot showing the Equivalent Plastic Strain generated at a stroke of 20 mm is shown in Figure 9. A plastic strain of 1.75e-1 and Von Mises Stress of 5.662e+4 Pa were recorded at the end of the loading with the punch bottoming the material inside the die. The bottoming of the plate in the die was the reason for the increase and the tensile nature of the Von Mises Stress.

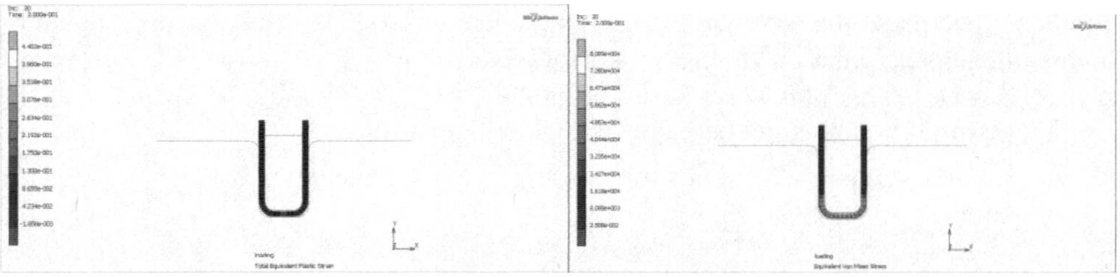

(a) (b)

Figure 9 (a) and (b): Contour Plots showing the Equivalent Plastic Strain generated at a stroke of 20 mm

The significant role of the punch in a bending process indicates that the deformation measured from the bending angle increased the stroke length. More interestingly are the measured strains (Equivalent Plastic Strain) induced into the sheet during the bending, strain being the response of material to the applied load. It is, therefore, important to consider these responses closely.

6. Conclusion

The finite element analysis of the sheet metal forming was conducted and completed using the Marc Mentate, MSC software version 2015. The study established that bottoming during a sheet bending process might increase the strain and the stress. This may have a consequential effect on the material property when a large magnitude of stresses and strain characterise the process.

7. Acknowledgement

The authors recognise the financial support of the University of Johannesburg through the Global Excellent Stature Grant award.

References

[1] Sheet Metal forming http://www.custompartnet.com/wu/sheet-metal-forming Accessed March 2017

[2] Schuler., 1998 Metal forming handbook. Berlin, Heidelberg, New York: Springer-Verlag; p. 366–73.

[3] Lange, K., 1985, Handbook of metal forming. New York: McGraw-Hill Inc., p. 19.1–35.

[4] Thanki, S. J., Raval, H. K., and Dave, A. K., 2001, "Prediction of the punch reversal position under V-plate bending using real material (power-law) behavior," J. Mater. Process. Technol. 114, p. 227–232.

[5] Yang, M., Manabe, K. I., Nishimura, H., 1996, "Development of real-time process control system for precision and flexible V-bending with an on-line database, J. Mater. Process. Technol. 60, p. 249–254.

[6] Forcellese, A., Fratini, L., Gabrielli, F., Micari, F., 1996, "Computer aided engineering of the sheet bending process," J. Mater. Process. Technol. 60,p. 225–232.

[7] Bakhshi-Jooybari, M., Rahmani, B., Daeezadeh, V., Gorji, A., 2009, "The study of springback of CK67 steel sheet in V-die and U-die bending processes. J Mater Des; 30, p. 2410–9.

[8] Wang, J., Verma, S, Alexander, R., Gauc, J., 2008, "Spring-back control of sheet metal air bending process," J Manuf Process10: p.21–7.

[9] Yu, H. Y., 2009, "Variation of elastic modulus during plastic deformation and its influence on springback". J Mater Des; 30, p.846–50.

[10] Kazan, R., Firat, M., Tiryaki, A. E., 2009, 'Prediction of springback in wipe-bending process of sheet metal using neural network," J Mater Des; 30, p.418–23.

[11] Tekaslan, O., Gerger, N., Seker, U., 2008, "Determination of spring-back of stainless steel sheet metal in ''V'' bending dies," J Mater Des; 29, p. 1043–50.

[12] Meinders, T., Burchitz, I. A., Bonte, M. H. A., Lingbeek, R. A., 2008, "Numerical product

design: springback prediction, compensation and optimization," Int J Mach Tool Manuf 48, p. 499–514.

[13] Sutasn, T., and Surasit R., 2008, "Investigation of spring-go phenomenon using finite element method," Materials and Design 29, p.1526-1532.

[14] Min Kuk Choi, and Huh, H., 2014, "Effect of punch speed on amount of springback in U-bending process of auto-body steel sheets," Procedia Engineering 81, p.963-968

[15] Sutasn T., 2010, "Finite element analysis of punch height effect on V-bending angle," Materials and Design 31, P.1593-1598.

[16] Sheet metal bending http://thelibraryofmanufacturing.com/sheetmetal_bending.html Accessed April 2017

[17] Sheet Metal Bending - http://thelibraryofmanufacturing.com/sheetmetal_bending.html Accessed March 2017.

[18] Virtual Lab-Dayalbagh Educational Institute. http://msvs-dei.vlabs.ac.in/SheetMetal.php Accessed April 2017

[19] Cho, J. R., Moon, S., J., Moon, Y. H., and Kang, S. S., 2003, "Finite element investigation on spring-back characteristics in sheet metal U-bending process," Journal of Material Processing Technology 141, pp109-116

IOP Conf. Series: Materials Science and Engineering **328** (2018) 012014 doi:10.1088/1757-899X/328/1/012014

Iridium Oxide pH Sensor Based on Stainless Steel Wire for pH Mapping on Metal Surface

S Shahrestani[1], M C Ismail[1], S Kakooei[1*], M Beheshti[1], M Zabihiazadboni[2], M A Zavareh[3]
[1]Centre for Corrosion Research, Mechanical Department, Universiti Teknologi PETRONAS, Malaysia
[2]F.Z.A Application University & Fouladin Zob Amol Foundry Company, Iran
[3] Department of chemical engineering, faculty of engineering, Tarbiat Modares University, Iran
saeid.kakooei@utp.edu.my

ABSTRACT

A simple technique to fabricate the iridium oxide pH sensor is useful in several applications such as medical, food processing and engineering material where it is able to detect the changes of pH. Generally, the fabrication technique can be classified into three types: electro-deposition iridium oxide film (EIrOF), activated iridium oxide film (AIROF) and sputtering iridium oxide film (SIROF). This study focuses on fabricating electrode, calibration and test. Electro-deposition iridium oxide film is a simple and effective method of fabricating this kind of sensor via cyclic voltammetry process. The iridium oxide thick film was successfully electrodeposited on the surface of stainless steel wire with 500 cycles of sweep potential. A further analysis under FESEM shows detailed image of iridium oxide film which has cauliflower-liked microstructure. EDX analysis shows the highest element present are iridium and oxygen which concluded that the process is successful. The iridium oxide based pH sensor has shown a good performance in comparison to conventional glass pH sensor when it is being calibrated in buffer solutions with 2, 4, 7 and 9 pH values. The iridium oxide pH sensor is specifically designed to measure the pH on the surface of metal plate.
Keywords: pH sensor; iridium oxide; electro-deposition; cyclic voltammetry; thick film, coating

1. INTRODUCTION

pH sensors have been applied in a multi-discipline industries such as chemical, biomedical, electro-biomedical and materials engineering where pH is involved in many control processes [1]. The conventional pH sensors which are basically made up of glass electrode have been widely used in pH measurement applications [2, 3]. In general, the conventional glass-type electrode has several good characteristics like good sensitivity, stability and long lasting. In addition to those advantages, glass-type has property of relatively low electrical resistance and shows no sensitivity to any kinds of redox species, gasses, anions or busser [4]. However, this electrode still has drawbacks in term of fragility and it is not suitable to be used in certain applications in biomedical, clinical and food industries [3]. In biomedical and physiological fields for example, glass electrodes have difficulty of miniaturization, poor response time and defenseless to membrane fouling [4]. In term of preservation, it is hard to maintain the quality and it always failed in harsh environmental condition [5]. Glass type electrodes for pH sensor are high cost, hard to fabricate and sensitive to certain alkaline and fluorhydric acid solution [1]. The fabrication technique for producing iridium oxide pH electrode sensor varies and it depends on the applications of sensor. Several popular fabrication processes which have been used to produce this type of sensor including sputtering [2-6], thermal oxidation [2-4,6] thermal decomposition [5,6], electrochemical growth [2,3,6], sol-gel [3,4] and electro-deposition [2,4,6]. Sputtering iridium oxide film (SIROF) method is excellent in terms of high production output and it is able to fabricate up to sub-micron-size. Sputtering requires expensive materials and instrumentation [2].

SIROF is an efficient way for producing high quality of iridium oxide film and at a high cost [3, 7]. Anodic iridium oxide thin film (AIROF) is an economic approach for producing iridium oxide film. AIROF is just an electrolysis process that basically uses a solution of iridium complexes. Iridium tetrachloride compound is an example of deposition agent [3]. The deposition of efficiency depends on pH value of deposition solution, temperature of solution and current density [3].

On contrary to conventional glass pH electrode, metal/metal-based pH sensor has a quicker pH response. Several solid-state metal oxides which are suitable for pH sensors include PtO_2, IrO_x, RuO_2, OsO_2, Ta_2O_5, RhO_2, TiO_2, PaO_2, Sb/Sb_2O_3, WO_3, PbO_2, Co_3O_4 and SnO_2 [2-5,8]. The most suitable metal oxide of all is IrO_x and shows great stability in any pH condition. Moreover, IrO_x is still able to work in a high surrounded temperature up until 250°C [1-3], pressure up to 270 bars [2, 9]. In addition, IrO_x has a quick response time, minimum chemical interference as well long-term stability [2-4], good response to the hydrogen ion [5] and low in term of impedance, toxicity and reactivity [10].

There are three possible mechanisms that might be involved in pH dependent redox reaction [2, 3] as follows:

$$Ir_2O_3 + 6H^+ + 6e^- \leftrightarrow 2Ir \quad (1)$$

$$IrO^2 + 4H^+ + 4e^- \leftrightarrow Ir + 2H_2O \quad (2)$$

$$2IrO_2 + 2H^+ + 2e^- \leftrightarrow Ir_2O_3 + H_2O \quad (3)$$

The redox potential is defined as follows:

$$E = E° - 2.303\ (RT/nF)\ pH = E° - 0.059pH \quad (4)$$

The value of E° is 926mmV which is the standard electrode potential for a standard hydrogen electrode (SHE). F is for Faraday's constant with a value of $96487 Cmol^{-1}$, n is the number of electrons transfer during redox process. To simplify, RT/F can be replaced with 25.69mV at temperature 25°C. The pH potential sensitivity yields -59mV/pH and this is called as Nernstian response [2, 3].

Iridium oxide film electrode are used as micro-stimulation electrode [11, 12] because Q_{inj} value can be about $4mC/cm^2$ [13-15]. On the other hand, iridium oxide film electrodes can exhibit a more charge storage capacity that it is related to reversible Faradaic reactions which causes good biocompatibility and corrosion resistance. This reaction is between Ir^{3+} and Ir^{4+} redox states [16, 17].

This paper focuses on investigation of pH upon a metal surface that is immersed in a solution. Therefore, one probe must be electrodeposited and one device must be designed and implemented for pH monitoring upon immersed-metal surface in solution.

2. EXPERIMENTAL

A. Solution Preparation

The electro-deposition solution was prepared. 150mg of Iridium (IV) chloride hydrate was added in 100mL of distilled water into a 200mL glass beaker. The mixture of solution was stirred by a magnetic stirrer for 10 minutes to dissolve completely. Next, 360mg of oxalic acid was added to the solution followed by stirring about 10 minutes. Then, 1mL of aqueous 30% hydrogen peroxide was added to the solution and stirred for 15 minutes. Lastly, the pH of solution was first recorded about 5 and it shows that the property of solution is quite acidic. Anhydrous potassium carbonate was gradually added to the solution and stirred for 15 minutes until pH reaches 10.5. At the end of the procedure, the color of solution completely turned to light yellow. The solution was transferred into a bottle and allowed to stabilize for two days in room temperature and its color would gradually change to dark blue.

B. Working Electrode Base Fabrication

The working electrodes for deposition process were fabricated and basically made of stainless-steel wire. Two types of wire diameters were used for making working electrodes (pH sensor probe) with 0.25mm and 0.6mm. These 304-

stainless steel wire were mounted in a mixture of epoxy and hardener. Ratio of 2g hardener and 10g of epoxy were used and followed by stirring the mixture for about 5 minutes. A rotating grinder machine with different size of sandpapers grit were used to grind the working electrodes. The grinding process was done using sandpaper with a grade from 180 to 1500 grit at a speed of 120rpm.

C. Electro-deposition method

This is an important process where the growth of iridium oxide layer took place via cyclic voltammetry method. The cyclic voltammetry method basically consists of several electrodes which are being connected to a potentiostat [18]. The Metrohm PGSTAT 128N potentiostat by Autolab was used in electro-deposition coating of iridium oxide via cyclic voltammetry. Three electrodes were used which primarily consist of Ag/AgCl reference electrode, mesh Pt counter electrode and working electrode. These electrodes were connected to potentiostat with wires which have its own color code, blue (reference electrode), black (counter electrode) and red (working electrode).

The process was started by preparing the lab-made electro-deposition bath which used 100ml beaker to hold the solution. The solution was poured until it reached a point where the tip of all electrodes is fully immersed. To ensure the electro-deposition process could occur, the tip of working electrode which has a polished circular cross section area stainless steel wire should be fully immersed. The growth of iridium oxide layer took place only on the exposed area of stainless steel. Figure 1 shows the electrodes setup for electro-deposition process by cyclic voltammetry method.

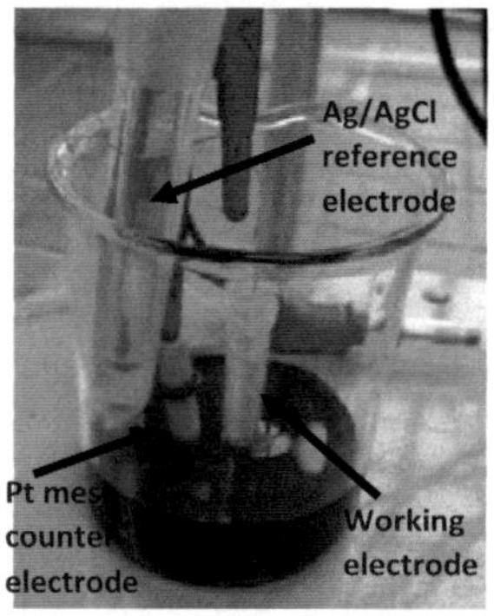

FIGURE 1: Electrodes setup

D. pH Sensor Calibration

The fabricated iridium oxide pH sensors are measured under the open circuit potential to measure the potentiometric response toward several buffer solutions. This process is conducted to study the dynamic range of pH sensor under several buffer solutions. At the end of the process, the calibration curve of pH sensor was obtained. Buffer solutions used are with 2, 4, 7 and 9 pH values. Open circuit potential of iridium oxide pH sensor was measured as a function of pH of buffer solutions respected to Ag/AgCl reference electrode.

E. Prototype Fabrication

The iridium oxide pH micro-sensor was calibrated and tested by mapping the pH value at the surface of a sample. The process is basically the similar to SECM where the probe or sensor is needed to be moved steadily in X and Y axes [19]. Z axis basically controls the distance of the tip of pH sensor to the surface of sample. For this project, X and Y are the main axes that control the movement of the pH sensor in X and Y directions. The position of Z direction was controlled manually using a screw mechanism. This prototype was fully controlled by a microcontroller which mainly works based on C^{++} program. Before writing program codes, several parameters such as the length and the width of sample were determined. The scanning process should be within the range of parameters. Figure 2 shows the X-Y plotter prototype [19].

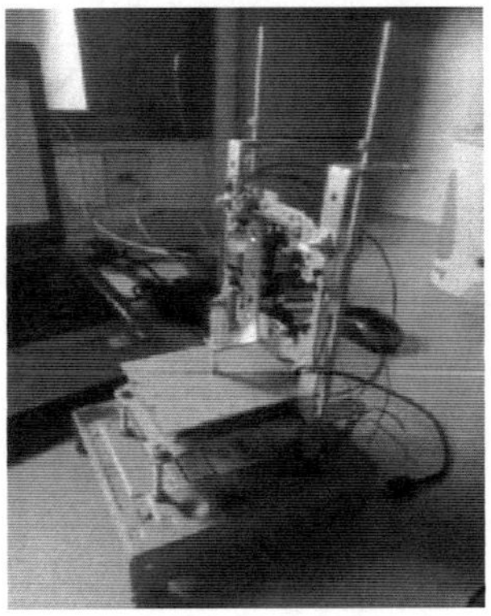

FIGURE 2: X-Y Plotter

The X-Y plotter was controlled by a computer through C++ language program. Therefore, an open-source electronics platform based on hardware and software is required to perform such task. The Arduino electrical board was used as a main processor to translate the input program becoming output which basically the motion in X-Y direction. Of all Arduino products, the Arduino Uno was selected as a microcontroller board for this project. In addition, it has 14 digital input/output pins, 6 analog inputs, USB connection and power jack making it as a suitable platform for this study.

Motor driver is a little current amplifier that primarily takes a low current control signal from microcontroller and turns it into a higher current signal that can drive a motor. Two units of L293D motor driver integrated circuit which has 16 pins were used to control two stepper motors. L293D works based on the concept of H-bridge which is a type of circuit that allows the voltage to flow in two directions. DC motors that can move in discrete steps are called as stepper motors. Stepper motors have multiple coils that are assembled in groups called as "phases". Energizing each phase with a correct sequence, the motor will rotate with a desired rotation. A very precise positioning and speed can be obtained by using a computer controlled stepping such as Arduino Uno. In this study, a screw-type stepper motor was used to generate linear displacement motion in X and Y.

F. pH Mapping

The iridium oxide pH sensor was used along with the X-Y plotter to measure the pH on the surface of metal. The process was conducted in such a way that its probe tip is partially immersed in electrolyte solution. The sample for test must be fully immersed on the 0.1M sodium chloride solution. During the process, tip of pH sensor was not directly contacted to the surface of sample and held with the same or constant distance. Four electrodes systems were used which consist of working electrode (pH sensor), platinum as a counter electrode, Ag/AgCl as a reference electrode and the sample itself as a substrate electrode. Surface generation tip collection mode (SG/TC) was used to measure the voltage and current generated by the surface of sample. Figure 3 shows the electrodes setup for pH scanning experiment.

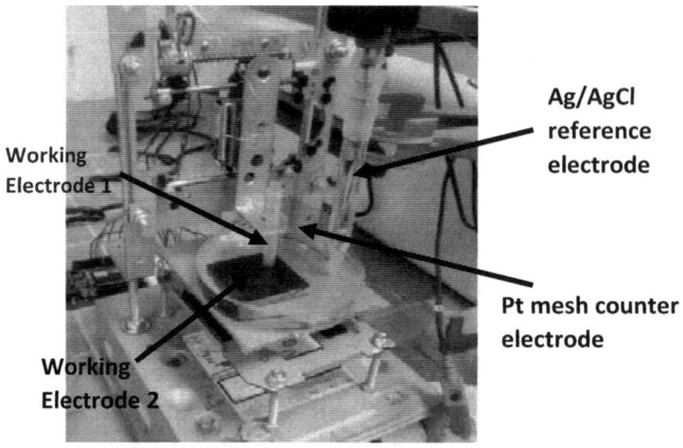

FIGURE 3: pH scanning electrodes setup

3. RESULTS AND DISCUSSION

A. Electro-deposition

The electro-deposition process was conducted by using a cyclic voltammetry method to develop iridium oxide layer on the circular cross section area of stainless steel at the tip of working electrode. Two types of stainless steel wire with diameters of 0.25mm and 0.6mm were electrodeposited separately via cyclic voltammetry method. All parameters for these electro-deposition processes were kept constant at all time. Based on the obtained data, the electro-deposition process was conducted which roughly took 14006 seconds (3 hours and 53 minutes) to complete. In the end of the process, a layer of dark blue, as shown in Figure 4, indicates that the electro-deposition of iridium oxide via cyclic voltammetry was found.

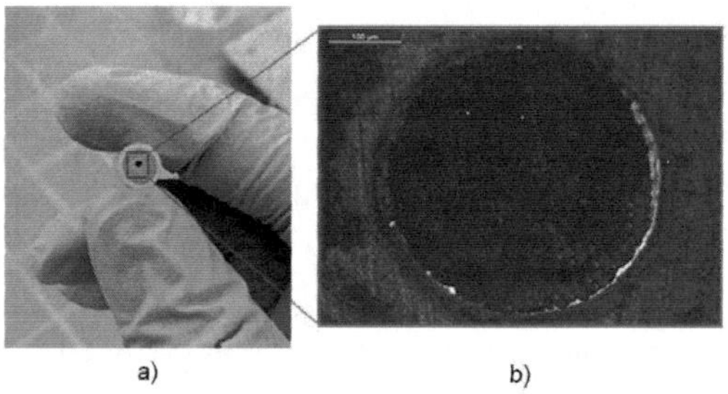

FIGURE 4: (a) Electrodeposited iridium oxide film. (b) Iridium oxide film under optical microscope at 100μm scale.

The cyclic voltammetry process was applied to the working electrodes and the result is shown in the graph. The progress and the pattern of graphs were observed in every 100 cycles. Graphs with potential respective to reference electrode against current were plotted to observe its correlation. In cyclic voltammetry, there are two points that give information about the redox activity during the process and located at the highest and lowest peak of cyclic voltammogram. Based on Figure 5, the cathodic potential peak was recorded at 0.57v with cathodic current peak of 0.141mA. On the other hand, the anodic peak potential was recorded at 0.19V with anodic current peak of -0.502mA.

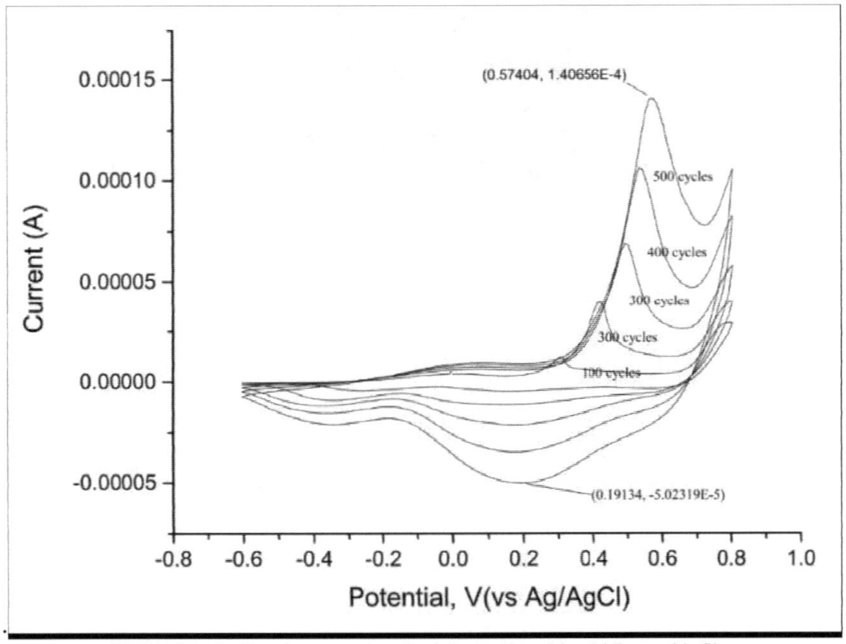

FIGURE 5: Cyclic voltammetry graph of electro-deposition iridium oxide film

B. FESEM Analysis

The microstructure was conducted by using SUPRA 55VP Field Emission Scanning Electron Microscopy (FESEM). FESEM gives a clear topographical and elemental information about the surface with unlimited depth of field. The first microstructure image was taken with 1000X magnification and it shows unevenly distributed growth of iridium oxide film. At certain area, the iridium oxide growth was in a huge lump size while the rest is just average in size, as shown in Figure 6.

FIGURE 6: FESEM image at 1000X magnification

In order to get a more detailed observation, the magnification was increased to 30000X and it has been focused at area that has a rougher iridium oxide surface. Iridium oxide has cauliflower-like appearance and usually occurred in thicker films. Laterally, high roughness of iridium oxide film gives higher surface area and results in high surface exposure to test environment and better response to pH changes [20]. Figure 7 shows the FESEM image of coating at 30000X magnification. The growth of iridium oxide film is perpendicular to the surface of coated-stainless steel wire.

IOP Conf. Series: Materials Science and Engineering **328** (2018) 012014 doi:10.1088/1757-899X/328/1/012014

FIGURE 7: FESEM image of IrO_x at 30000X magnification

C. EDX Analysis

EDX analysis was conducted after observation of microstructure to determine the elemental information about iridium oxide film. From the result of EDX analysis in Figure 8, it shows that the highest element recorded is iridium followed by oxygen. These two elements indicate that the EIROF process is successful. Other types of element such as chromium and iron are also found, they are the main composition of stainless steel.

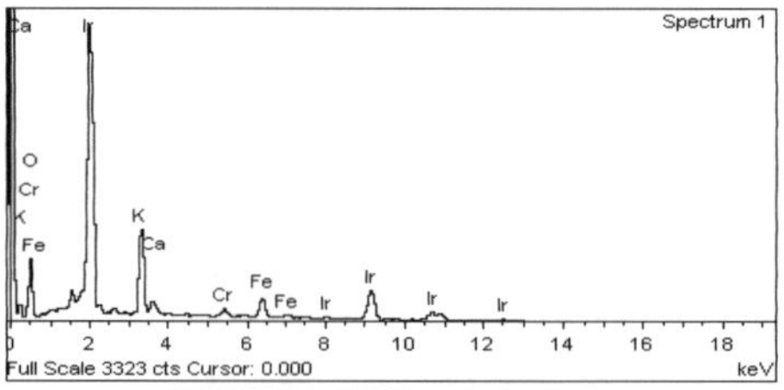

FIGURE 8: EDX analysis result

All elements present on coated-stainless steel wire were analyzed and calculated in term of weight and atomic percentage (Table 1). As shown in Table 1, weight and atomic percentage of oxygen and iridium are significant compared to other materials. This shows that stainless steel wire is coated with IrO_x truly.

TABLE 1: Composition weight and atomic percentage of material for coated- stainless steel wire

Element	Weight	Atomic %
O	23.4	66.79
K	11.14	13
Ca	0.71	0.81
Cr	1.49	1.31
Fe	5.27	4.31
Ir	57.99	13.78
Total	100	

D. Potentiometric Response

The potentiometric response of iridium oxide pH sensor was measured during the open circuit potential experiment by immersing it into buffer solutions with pH of 2,4,7,9. The iridium oxide pH sensor was dipped and left until the reading of average voltage is stable. Figure 9 shows potential, E (Ag/AgCl) versus pH is plotted to obtain the calibration curve. The iridium oxide pH sensor shows a good linear relation, R=0.9986 with super-Nerstian response value of -0.0779V/pH.

IOP Conf. Series: Materials Science and Engineering **328** (2018) 012014 doi:10.1088/1757-899X/328/1/012014

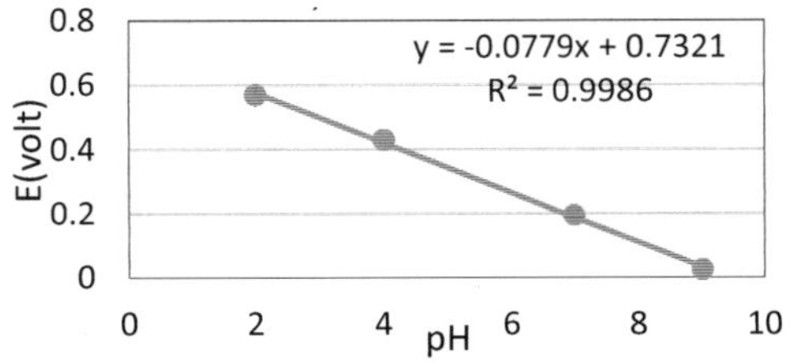

FIGURE 9: Calibration curve of iridium oxide pH sensor E versus pH

E. pH Mapping

The metal plate with 4cm × 4cm dimensions was used for sample as shown in Figure 10. A visual observation after test was conducted and it was found that there are some corrosions on its surface. The sample was further examined to get reading of pH by pH sensor scanning probe. Probe scans the whole surface of substrate step-by-step and collect pH of the surface.

FIGURE 10: 4cm X 4cm metal sample plate

At the end of mapping process, a topographical image of metal surface was obtained showing the distribution of pH as shown in Figures 11-a and 11-b. The Figure shows that the lower region of metal plate has low pH reading which is indicated in red and yellow color. Blue and dark blue of surfaces on the other hand, are due to high pH area of substrate. Therefore, these areas with high pH value are disposed to corrosion. Hence, to prevent of corrosion, these areas must be protected through coating, cathodic protection or other methods. In general, with this method, it can detect pH value upon surfaces that it determines properties and activities of surfaces. As shown in Figure 10, corrosion occurred after test, as pH mapping was predicted. For more detailed analysis, 3D pH contour graph was plotted as shown in Figure 11-b, it is showing that the lowest pH region is on the red part of metal plate.

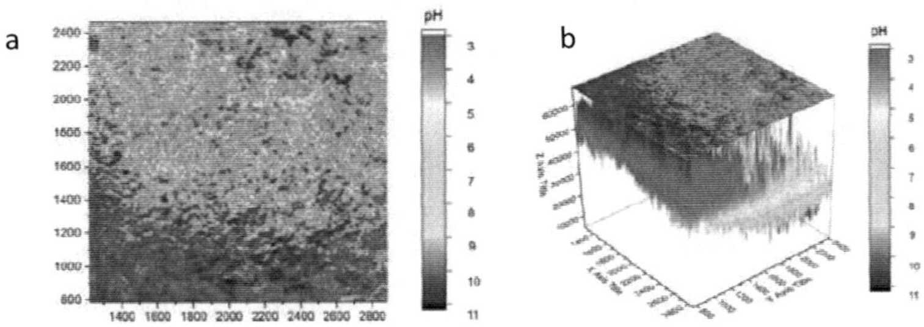

FIGURE 11: a) Topographical and b) 3D topographical images of pH

4. CONCLUSION

The pH microsensor made of stainless steel with coated-iridium oxide film has been successfully developed. Electro-deposition of iridium oxide film (EIROF) is a good and cost-effective technique to produce iridium oxide pH sensor compared to other techniques. Stainless steel is an alternative material to use as substrate for coating. This pH sensor has a good linear response within pH range of 2 to 9 with sensitivity -0.0779V/pH. For the application, the pH sensor was used to record the reading of pH at the metal surface which give a topographical image based on pH scale. It can determine activities of surface in solution, quality of coatings, properties of surface in different solutions, investigation of biomedical and biological knowledge.

REFERENCES

[1] C. C. M. Martinez, R. E. Madrid, and C. J. Felice, 2009, "A pH Sensor Based on a Stainless Steel Electrode Electrodeposited With Iridium Oxide," *IEEE Transactions on Education,* **vol. 52**, pp. 133-136.

[2] T. Y. Kim and S. Yang, 2014, "Fabrication method and characterization of electrodeposited and heat-treated iridium oxide films for pH sensing," *Sensors and Actuators B: Chemical,* **vol. 196**, pp. 31-38.

[3] W.-D. Huang, H. Cao, S. Deb, M. Chiao, and J. C. Chiao, 2011, "A flexible pH sensor based on the iridium oxide sensing film," *Sensors and Actuators A: Physical,* **vol. 169**, pp. 1-11.

[4] X.-r. Huang, Q.-q. Ren, X.-j. Yuan, W. Wen, W. Chen, and D.-p. Zhan, 2014, "Iridium oxide based coaxial pH ultramicroelectrode," *Electrochemistry Communications,* **vol. 40**, pp. 35-37.

[5] J. Park, M. Kim, and S. Kim, 2014, "Surface renewable nano-iridium oxide polymeric composite pH electrodes," *Sensors and Actuators B: Chemical,* **vol. 204**, pp. 197-202.

[6] T. Ryynänen, K. Nurminen, J. Hämäläinen, M. Leskelä, and J. Lekkala, 2010, "pH electrode based on ALD deposited iridium oxide," *Procedia Engineering,* **vol. 5**, pp. 548-551.

[7] M. A. Petit and V. Plichon, 1998, "Anodic electro-deposition of iridium oxide films," *J. of Electroanal. Chemis.,* **vol. 444**, pp. 247-252.

[8] D. D. Zhou, 2008, "CHAPTER 10 - Microelectrodes for in-vivo determination of pH A2 - Zhang, Xueji," in *Electrochemical Sensors, Biosensors and their Biomedical Applications*, H. Ju and J. Wang, Eds., ed San Diego: Academic Press, pp. 261-305.

[9] J. Hendrikse, W. Olthuis, and P. Bergveld, 1998, "A method of reducing oxygen induced drift in iridium oxide pH sensors," *Sensors and Actuators B: Chemical,* **vol. 53**, pp. 97-103.

[10] D.-H. Kim, S.-H. Park, J. Choi, M. H. Yi, and H.-S. Kim, 2015, "Fabrication of iridium oxide nanoparticles supported on activated carbon powder by flashlight irradiation for oxygen evolutions," *Mat. Sci. and Eng.: B,* **vol. 201**, pp. 29-34.

[11] S. Tanghe, K.Najafi, K.D. Wise, 1990, A planar IrO multichannel stimulating electrode for use in neural prostheses, *Sens. Actuators B: Chem.* **1** 464–467.

[12] H. Yang, S.K. Kang, C.A. Choi, H. Kim, D.H. Shin, Y.S. Kim, Y.T. Kim, 2004, An iridium oxide reference electrode for use in microfabricated biosensors and biochips, *Lab. Chip.* **4** pp 42–46.

[13] S.F. Cogan, 2008, Neural stimulation and recording electrodes, *Annu. Rev. Biomed. Eng.* **10**, 275–309.

[14] Y. Lu, Z.X. Cai, Y.L. Cao, H.X. Yang, Y.Y. Duan, 2008, Activated iridium oxide films fabricated by asymmetric pulses for electrical neural microstimulation and recording, *Electrochem. Commun.* **10,** 778–782.

[15] S.F. Cogan, T.D. Plante, J. Ehrlich, 2004, Sputtered iridium oxide films (SIROFs) for low-impedance neural stimulation and recording electrodes, *in: 26th Annual International Conference of IEEE EMBS*, San Francisco, CA, pp. 4153–4156.

[16] R.D. Meyer, S.F. Cogan, T.H. Nguyen, R.D. Rauh, 2001, Electrodeposited iridium oxide for neural stimulation and recording electrodes, *IEEE Trans. Neural. Sys. Reh. Eng.* 92–11.

[17] Yi Lu, TanyuanWang, Zhengxu Cai, Yuliang Cao, Hanxi Yang, Yanwen Y. Duan∗, 2009, " Anodically electrodeposited iridium oxide films microelectrodes for neural microstimulation and recording", *Sensors and Actuators* **B 137** 334–339.

[18] J. R. Kearns and B. J. Little, 1994, *Microbiologically Influenced Corrosion Testing*: ASTM.

[19] M. Beheshti, S. Kakooei, M. C. Ismail, and S. Shahrestani, 2016, "Dedign and implementation of scanning electrochemical microscopy (SECM) for scanning open circuit potential system on damaged coated surfaces" ARPN, *J. of Engin. and appl. Sci.,* **11**(24), 14299-14302.

[20] S. Kakooei, M. C. Ismail, and B. A. Wahjoedi, 2013, "Electrochemical study of iridium oxide coating on stainless steel substrate," *Int. J. Electrochem. Sci,* **vol. 8**, pp. 3290-3301.

Theoretical Development of Biaxial Fabric Prestressed Composites under Tensile or Flexural Loading

Nawras H. Mostafa[a,b] , Z.N. Ismarrubie[a], S.M. Sapuan[a,c], M.T.H. Sultan [d]

[a]Department of Mechanical and Manufacturing Engineering, Faculty of Engineering, Universiti Putra Malaysia, 43400 UPM Serdang, Selangor Darul Ehsan, Malaysia
[b] Department of Mechanical Engineering, Faculty of Engineering, University of Babylon, Babylon Province, Iraq
[c] Institute of Tropical Forestry and Forest Products (INTROP), Universiti Putra Malaysia, 43400 UPM Serdang, Selangor Darul Ehsan, Malaysia
[d] Aerospace Manufacturing Research Centre (AMRC), Level 7, Tower Block, Faculty of Engineering, Universiti Putra Malaysia, 43400 UPM Serdang, Selangor Darul Ehsan, Malaysia

Abstract-In this study, we extend our development of the classical lamination theory of laminated composite with the presence of biaxial fabric prestressed. The aim of this paper is to describe the development of the fibre prestressed composite and its effect on the composite's internal stresses when subjected to tensile or flexural loading. The biaxial fabric prestress of the plain-weave composite could efficiently reduce the overall tensile stress within the composite lamina due to inducing compressive residual stress imparted from releasing the fibre pretension load. Thereby, the fibre prestressed composite could withstand more external tensile or flexural stress than non-prestressed counterparts did.
Keywords: Biaxial prestress; plain-weave fabric; residual stress; tensile and flexural stresses; development of classical lamination theory

1. Introduction

Currently, there is a need for high strength and lightweight materials in widespread applications such as in the aerospace, civil, automotive and sporting goods industries. For that reason, fibre–reinforced composite materials are now of great interest due to their high strength to weight ratio in comparison with most metals. As it is well known, the cost of composites is shared between its materials and its fabrication, for instance, the cost of fibreglass–reinforced polymer is approximately 60% materials and 40% fabrication [1]. Therefore, the focus on improving the fabrication techniques is still reasonable if they can enhance the mechanical behaviour of the composites. Unfortunately, residual stresses are generated within the composites during manufacturing process [2]. These stresses could considerably reduce the mechanical properties of the fibre reinforced composite. Residual stresses can develop in composite materials due to several reasons such as the chemical shrinkage of the polymer matrix, the different thermo–mechanical properties of the constitutions, moisture absorption, and fibre pretension [3]. Residual stresses within the matrix can arise due to the phase change of the resin from liquid to solid state (chemical shrinkage). The mismatch in the coefficient of thermal expansion between the fibre and the matrix will produce residual stresses in the composite when it is cooled from its curing temperature [3,4]. Several methods have been used to minimise the detrimental effects of the induced residual stresses within the fibre reinforced composite such as optimisation of dwell cure cycle [5], curing the composite at low temperatures [6], using electron beam curing (gamma irradiation) [7], using expanding monomers [8], inserting shape memory alloy wires [9], and usingfibre pretension (prestress) method [10].

Fibre pretension could be one of the available options for improving the mechanical properties of polymeric matrix composites (PMCs) without increasing their section dimensions or mass [11]. The pretension was applied to the fibre prior and during the matrix cure. Once the matrix was cured well, the fibre pretension has been released. Up to now, the improvement obtained in the mechanical properties of fibre–reinforced composites due to using the fibre–prestressing method is not well established. However, our early previous studies have effectively confirmed the advantages of fibre prestressed composites for plain-weave composites [12–16]. Fibre pretension (prestressing) during matrix cure has generally a positive effect on the composite mechanical properties as it induces compressive residual stresses in the matrix. According to the authors' opinion, the effect of fibre pretension on the composite's internal stress subjected to a tensile or a flexural external load was not considered yet.
Therefore, the main aim of this research is to investigate the effect of biaxial fibre prestress on the mechanical behaviour of the plain weave composite subjected to tensile or flexural loadings. Theoretical equations of the macro-mechanics lamination theory of the composite are developed to include the fibre prestressing effect.

2. Materials and method

The composite material used in this study consisted of E-glass plain weave fabric/polyester resin system. The mechanical properties of the composite's constituents are listed in Table (1). These properties were measured at a temperature of 25 °C.

Table (1): Mechanical properties of the composite's constituent materials

Property	E-glass plain-weave fabric	Polyester resin
Elastic modulus (E), GPa	70	2.77
Tensile strength, MPa	2200	61
Poisson's ratio (ν)	0.23	0.25
Thermal expansion coefficient (α), 10^{-6} m/(m °C)	120	5.4
Warp density × Fill density of the fabric (ends/m)	285 × 245	-

The analysis of the fibre reinforced composite is based on the classical lamination theory. According to the classical lamination theory, a plane state of stress uses a macro–mechanical analysis using the subscription 1 as the direction of the warp fibres and the subscription 2 as the direction of fill fibres [15], the stress–strain ($\sigma - \varepsilon$) matrix was expressed by:

$$\begin{Bmatrix} \sigma_1 \\ \sigma_2 \\ \tau_{12} \end{Bmatrix} = \begin{bmatrix} Q_{11} & Q_{12} & 0 \\ Q_{12} & Q_{22} & 0 \\ 0 & 0 & Q_{66} \end{bmatrix} \begin{Bmatrix} \varepsilon_1 \\ \varepsilon_2 \\ \gamma_{12} \end{Bmatrix} \tag{1}$$

where the reduced stiffnesses $Q_{11}, Q_{22}, Q_{12}$and Q_{66} were equal to:

$$Q_{11} = \frac{E_1}{1-\nu_{12}\nu_{21}}, Q_{22} = \frac{E_2}{1-\nu_{12}\nu_{21}}, Q_{12} = \frac{\nu_{21}E_1}{1-\nu_{12}\nu_{21}} = \frac{\nu_{12}E_2}{1-\nu_{12}\nu_{21}}, Q_{66} = G_{12} \tag{2}$$

where
G_{12}: Shear modulus of a composite.
E_1 : Elastic modules of a composite along the direction of fibres
E_2 : Elastic modules of a composite transverse to the direction of fibres
ν_{12} : The strain in the transversedirection over the strain in the direction of fibre when the stress is applied along the fibre's direction
ν_{21} : The strain in the direction of fibre over the strain in the transverse direction when the stress is applied along the transverse direction

IOP Conf. Series: Materials Science and Engineering **328** (2018) 012015 doi:10.1088/1757-899X/328/1/012015

The developed residual stress–strain relation of a laminated composite is due to different sources that could be rewritten in the following form:

$$\begin{Bmatrix} \sigma_1 \\ \sigma_2 \\ \tau_{12} \end{Bmatrix}_{res} = \begin{bmatrix} Q_{11} & Q_{12} & 0 \\ Q_{12} & Q_{22} & 0 \\ 0 & 0 & Q_{66} \end{bmatrix} \begin{Bmatrix} \varepsilon_1 \\ \varepsilon_2 \\ \gamma_{12} \end{Bmatrix}_{res} \quad (3)$$

$$\begin{Bmatrix} \varepsilon_1 \\ \varepsilon_2 \\ \gamma_{12} \end{Bmatrix}_{res} = \begin{Bmatrix} \varepsilon_1^{pre} - \varepsilon_1^{ther} \\ \varepsilon_2^{pre} - \varepsilon_2^{ther} \\ 0 \end{Bmatrix} \quad (4)$$

If the composite was cured at a temperature different from the room temperature, thermal strain was induced in the final product. Thermal strain would generate tensile residual stresses within the composite due to the different thermal expansion properties of the composite's constituents. Residual thermal strain was calculated as:

$$\left(\varepsilon_{1,2}^{ther}\right)_{res} = \alpha_{1,2}\,\Delta T \quad (5)$$

where

$\alpha_{1,2}$: Coefficients of thermal deformation in the 1 and 2 directions, respectively
ΔT : Difference between the cure and cool-down temperatures

For the case of a plain–weave fabric pretensioned in the biaxial directions (i.e. the warp and the fill yarns), the induced strain ε_1 and ε_2 due to releasing the fibre pretension load after the matrix has cured well are equal to:

$$\left(\varepsilon_1^{pre}\right)_{res} = \frac{\left(\sigma^{pre}V_f\right)_{1-dir}}{\bar{E}_1} - \frac{\bar{\nu}_{12}\left(\sigma^{pre}V_f\right)_{2-dir}}{\bar{E}_1} \quad (6)$$

$$\left(\varepsilon_2^{pre}\right)_{res} = \frac{\left(\sigma^{pre}V_f\right)_{2-dir}}{\bar{E}_2} - \frac{\bar{\nu}_{21}\left(\sigma^{pre}V_f\right)_{1-dir}}{\bar{E}_2} \quad (7)$$

where

$\bar{E}_1, \bar{E}_2$: Effective elastic modulus in warp and fill directions, respectively
$\bar{\nu}_{12}, \bar{\nu}_{21}$: Effective Poisson's ratio

V_f: Fibre volume fraction

At the micro–mechanical level, the strain in both the fibre and the matrix were equal to the lamina strain due to strain compatibility[15]. Residual stresses in the fibre and the matrix at a ply level could be estimated as follows:

$$\sigma_f^{res} = E_f\left(\varepsilon_{1,2}^{pre} - \varepsilon_{1,2}^{ther}\right)_{res} + \sigma^{pre} \quad (8)$$

$$\sigma_m^{res} = E_m\left(\varepsilon_{1,2}^{pre} - \varepsilon_{1,2}^{ther}\right)_{res} \quad (9)$$

The equivalent elastic properties according to Gay's approach were equal to[15]:

$$\left.\begin{aligned} \bar{E}_1 &\approx \beta E_1 + (1-\beta)E_2 \\ \bar{E}_2 &\approx (1-\beta)E_1 + \beta E_2 \\ \bar{G}_{12} &= G_{12} \\ \bar{\nu}_{12} &\approx \frac{\nu_{12}}{\left(\beta + (1-\beta)\frac{E_1}{E_2}\right)} \\ \bar{\nu}_{21} &\approx \bar{\nu}_{12}\frac{E_2}{E_1} \end{aligned}\right\} \quad (10)$$

where

$$\beta = \frac{n_1}{n_1+n_2} \tag{11}$$

Here n_1 and n_2 denote the number of yarns per meter along the warp and fill directions that are equal to 285 and 245, respectively.

The effective composite thermoelastic properties of a unidirectional composite according to the concentric cylinder approach, assuming fibre behave transversely isotropic properties, were equal to [15]:

$$\left.\begin{aligned} E_1 &= E_f V_f + E_m V_m + \frac{4(\nu_f-\nu_m)^2 V_f V_m}{V_m/k^F + V_f/k^M + 1/G_m} \\ E_2 &= \frac{2(1-\nu_f V_f - \nu_m V_m)\bar{k}E_1}{(E_1+4\bar{k}\nu_{12}^2)} \\ G_{12} &= G_m\left(\frac{G_m V_m + G_f(1+V_f)}{G_f V_m + G_m(1+V_f)}\right) \\ G_{m,f} &= \frac{E_{m,f}}{2(1+\nu_{m,f})} \end{aligned}\right\} \tag{12a}$$

and

$$\left.\begin{aligned} \nu_{12} &= \nu_f V_f + \nu_m V_m + \frac{(\nu_f-\nu_m)(1/k^M - 1/k^F)V_m V_f}{V_m/k^F + V_f/k^M + 1/G_m} \\ \nu_{21} &= \nu_{12}\left(\frac{E_2}{E_1}\right) \\ \alpha_1 &= \frac{\alpha_f E_f V_f + \alpha_m E_m V_m}{E_f V_f + E_m V_m} \\ \alpha_2 &= \alpha_m V_m + \alpha_f V_f + \left(\frac{\nu_f E_m - \nu_m E_f}{\frac{E_m}{V_f} + \frac{E_f}{V_m}}\right)(\alpha_f - \alpha_m) \end{aligned}\right\} \tag{12b}$$

The axial residual stress that induced in the lamina due to the biaxial fabric prestress and thermal residual stresses is obtained from equation (3):

$$\sigma_1{}^{res} = Q_{11}\varepsilon_1{}^{res} + Q_{12}\,\varepsilon_2{}^{res} \tag{13}$$

whereQ_{11} andQ_{12} in the above equation were determined from equation (2) that related the elastic properties of the composite's constituents with the fibre orientation angle; however, the $\varepsilon_1{}^{res}$ and $\varepsilon_2{}^{res}$are the residual strains obtained from equations (3 and 4). Now, if the composite lamina is subjected to external axial tensile stress (σ_{axi}) in the direction of specimen length, the total axial stress (σ_{axi}^{total}) within the composite lamina at that direction would become:

$$\sigma_{axi}^{total} = \sigma_{axi} + \sigma_1{}^{res} \tag{14}$$

The axial tensile stress in the composite resulted from applying external axial load without considering residual stresses is equal to:

$$\sigma_{axi} = \frac{P_{axi}}{A_{total}} \tag{15}$$

where σ_{axi}, P_{axi} and A_{total} represent the axial stress, axial load and the cross sectional area of the composite material, respectively. Consequently, by substituting equation (15) into equation (14), one can find the total internal axial stress in the prestressed composite:

$$\sigma_{axi}^{total} = \frac{P_{axi}}{A_{total}} + \sigma_1{}^{res} \tag{16}$$

In the case of the prestressed lamina subjected to flexural loading, the total internal flexural stress in the lamina is equal to:

$$\sigma_{flex}^{total} = \mp \sigma_{flex} + \sigma_1{}^{res} \tag{17}$$

The $\mp$ sign represents the tensioned and compressed regions located at the lower and upper sides of the neutral axis of the beam under flexural loading. $\sigma_{flex}$is the applied flexural stress in the composite lamina.

3. Results and discussion

The tensile residual stresses in the fabric when it was prestressed with different levels and released after matrix cure are listed in Table (2). It is very clear that increasing the biaxial fibre prestressing level could increase both the tensile residual stress in the fibre and the compressive residual stress within the matrix. On the other hand, the induced residual stresses within the matrix at warp and fill directions of the fabric were not equal due to using unequal yarn number per meter along the principal directions of the used E-glass fabric.

Table (2): Tensile residual stresses in the fibre after matrix cure due to applying different biaxial fabric prestressing levels

Biaxial fabric prestressing (MPa)	Tensile residual stress in the fibre (MPa)	Residual stress in the matrix (warp direction) (MPa)	Residual stress in the matrix (fill direction) (MPa)
25	5.841	-1.14	-1.26
50	11.68	-2.27	-2.53
75	17.52	-3.39	-3.75
100	23.30	-4.52	-5.00

The axial stress in the lamina results from applying axial stress that was developed in equation (14) are shown in Figure (1). The fibre volume fraction of the composite lamina was equal to 11% with cross-sectional area of 3 mm (thickness) x 25 mm (width). The effect of the fibre prestressing was very clear as it could reduce the tensile stresses in the prestressed lamina when subjected to external axial tensile stress. The flexural stress development in the presence of the fibre prestress is shown in Figure (2). The results were obtained from using the developed flexural equation (equation 15). The total tensile flexural stress reduced with increasing the prestressing level; however, compressive flexural stress in the compressed side increased as the fibre prestress was increased.

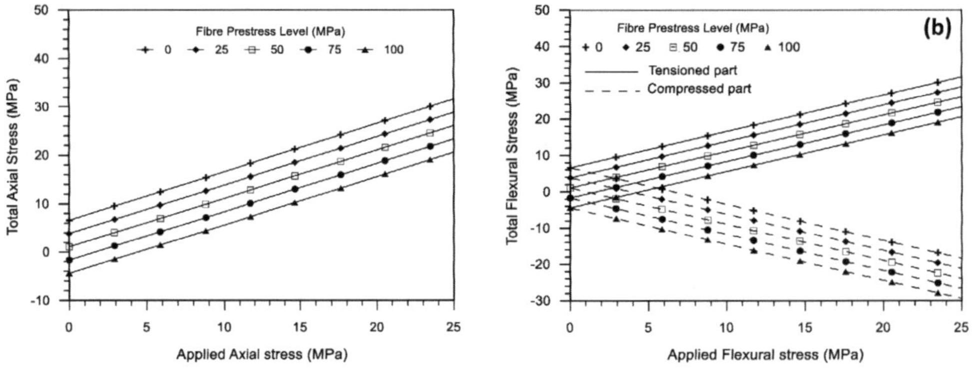

Figure (1): Total internal stress versus applied external stress in the composite lamina (E-glass/Polyester) cured at 50 °C and cooled down to 25 °C and prestressed with different levels. (a) Axial stress and (b) flexural stress.

In general, the tensile stress in the prestressed composite lamina could be reduced by increasing the biaxial fabric prestress [14,16]. This reduction is continued as the prestressing level has been increased, but this behaviour does not always lead to improvement on the mechanical properties of the composite as stated by many experimental studies[3]. The main reason behind this is attributed to the existence of interfacial shearing stress between the fibre and matrix [13], which is not considered in the classical lamination theory.

4. Conclusions

The biaxial fibre prestress of the composite reinforced with plain weave fabric could reduce the tensile residual stresses within the composite that were resulted from different thermo–mechanical properties of the composite's constitutions. The tensile and flexural stresses in the fibre prestressed laminated composite were reduced with increasing the level of the biaxial fibre prestress due to inducing compressive residual stresses within the lamina. This could improve the internal stress state of the composite subjected to external tensile or flexural loading.

References

[1] Ashby MF, Jones DRH. Engineering Materials 1: An Introduction to Properties, Applications, and Design. Fourth edi. The Boulevard, Langford Lane, Kidlington, Oxford: Elsevier; 2012.
[2] Safarabadi M, Shokrieh MM. Residual stresses in composite materials. 2014.
[3] Krishnamurthy S. Prestressed advanced fibre reinforced composites: fabrication and mechanical performance. PhD thesis, Engineering System Department, Defence College of Management and Technology, Cranfield University, UK, 2006.
[4] Shokrieh M. Residual stresses in composite materials. Cambridge: Woodhead Publishing Limited; 2014.
[5] White SR, Hahn HT. Cure Cycle Optimization for the Reduction of Processing-Induced Residual Stresses in Composite Materials. J Compos Mater 1993;27:1352–78.
[6] Gopal AK, Adali S, Verijenko VE. Optimal temperature profiles for minimum residual stress in the cure process of polymer composites. Compos Struct 2000;48:99–106.
[7] Korenev S. Electron beam curing of composites. Vacuum 2001;62:233–6.
[8] Fu J, Liu W, Liu X, Tuladhar SL, Wan Q, Wang H. Properties of a new dental photocurable resin based on the expanding monomer and three-component photoinitiator system. J Wuhan Univ Technol Sci Ed 2014;29:384–90.
[9] Naghashian S, Fox BL, Barnett MR. Actuation curvature limits for a composite beam with embedded shape memory alloy wires. Smart Mater Struct 2014;23:1–10.
[10] Motahhari S, Cameron J. Measurement of Micro-Residual Stresses in Fiber-Prestressed Composites. J Compos Mater 1997;16:1129–37.
[11] Graczykowski C, Orlowska A, Holnicki-Szulc J. Prestressed Composite Structures - Modeling, Manufacturing, Design. Compos Struct 2016;151:172–82.
[12] Mostafa NH, Ismarrubie ZN, Sapuan SM, Sultan MTH. Effect of fabric biaxial prestress on the fatigue of woven E-glass/polyester composites. Mater Des 2016;92:579–89.
[13] Mostafa NH, Ismarrubie ZN, Sapuan S, Sultan M. Fibre prestressed polymer-matrix composites: a review. J Compos Mater 2017;51:39–66.
[14] Mostafa NH, Ismarrubie ZN, Sapuan S, Sultan M. The influence of equi-biaxially fabric prestressing on the flexural performance of woven E-glass/polyester-reinforced composites. J Compos Mater 2016;50:3385–93.
[15] Mostafa NH, Ismarrubie ZN, Sapuan SM, Sultan MTH. Fibre prestressed composites: Theoretical and numerical modelling of unidirectional and plain-weave fibre reinforcement forms. Compos Struct 2017;159:410–23.
[16] Mostafa NH, Ismarrubie ZN, Sapuan S, Sultan M. Effect of equi-biaxially fabric prestressing on the tensile performance of woven E-glass/polyester reinforced composites. J Reinf Plast Compos 2016;35:1093–103.

The Effects of Sn Addition on the Microstructure and Surface Properties of Laser Deposited Al-Si-Sn Coatings on ASTM A29 Steel

Olawale S. Fatoba[1*]; Stephen A. Akinlabi[2]; Esther T. Akinlabi[1]

[1]Department of Mechanical Engineering Science, Faculty of Engineering and the Built Environment, University of Johannesburg, South Africa.

[2] Department of Mechanical and Industrial Engineering Technology, Faculty of Engineering and the Built Environment, University of Johannesburg, South Africa.

Corresponding author*: drfatobasameni@gmail.com

Abstract-Aluminium and its alloys have been successful metal materials used for many applications like commodity roles, automotive and vital structural components in aircrafts. A substantial portion of Al-Fe-Si alloy is also used for manufacturing the packaging foils and sheets for common heat exchanger applications. The present research was aimed at studying the morphology and surface analyses of laser deposited Al-Sn-Si coatings on ASTM A29 steel. These Fe-intermetallic compounds influence the material properties during rapid cooling by laser alloying technique and play a crucial role for the material quality. Thus, it is of considerable technological interest to control the morphology and distribution of these phases in order to eliminate the negative effects on microstructure. A 3 kW continuous wave ytterbium laser system (YLS) attached to a KUKA robot which controls the movement of the alloying process was utilized for the fabrication of the coatings at optimum laser parameters. The fabricated coatings were investigated for its hardness and wear resistance performance. The field emission scanning electron microscope equipped with energy dispersive spectroscopy (SEM/EDS) was used to study the morphology of the fabricated coatings and X-ray diffractometer (XRD) for the identification of the phases present in the coatings. The coatings were free of cracks and pores with homogeneous and refined microstructures. The enhanced hardness and wear resistance performance were attributed to metastable intermetallic compounds formed.
Keywords: Solidification; Al-Sn-Si coating; wear resistance; micro-hardness; crystallization

1. Introduction

The increasing interest in attempting to improve wear, corrosion resistance and hardness of surfaces that are exposed to harsh surroundings and are in contact with abrasives has accelerated the development of several techniques for creating suitable and appropriate protective coatings [1]. Area of failure or degradation for engineering components often begins on the surface. This is because the intensity of external stress and environmental attacks are often highest at the surface during mechanical and chemical/electrochemical interaction with the surrounding environment [2].

Aluminium based alloy provides good combination of strength and corrosion resistance. They are light in weight, economically viable, amenable for production by various processing techniques and possess high strength and good corrosion resistance. They are widely used in industrial applications due to better tribological properties. Silicon is good in metallic alloys used for casting. This is because it increases the fluidity of the melt, reduces the melting temperature, decreases the contraction associated with solidification and is very cheap as a raw material. Silicon also has a low density (2.34 g cm^{-3}), which may be an advantage in reducing the overall weight of the cast component. Silicon has a very low solubility in aluminium; it therefore precipitates as virtually pure silicon, which is hard and hence improves the abrasion resistance. Tribological properties of Al-Si alloys are affected by the shape and distribution of silicon protection. Because of the excellent tribological properties of Al-Si alloys, they are used in engineering application, especially in plan bearing, internal combustion engine, and hardness increases when silicon is available as coarse polyhedral particles. Al-Si cast alloys have been widely employed to produce automotive components working at ambient and fairly high temperature (up to 200 °C) due to excellent characteristics such as low-cost manufacturing, excellent castability, high specific strength and recyclability [3, 4]. Al–Sn based alloys are widely used as sliding bearing materials in automobile and shipbuilding industry due to their good compactibility, wear resistance, thermal conductivity, and sliding properties [5]. The alloys also excel in high temperature stability. In these alloy systems, tin is a necessary soft phase in the aluminum matrix. Due to its low modulus, low strength and the excellent anti-welding characteristics with iron, tin phase in Al–Sn bearing materials can provide suitable friction properties and shear surface during sliding [6]. However, literature on Al-Sn-Si alloy coatings through laser surface alloying (LSA) technique is very limited.

LSA modifies the surface morphology and near surface structure of components and its alloys with perfect adhesion to the interface of the bulk steel and the distinctive advantages of the LSA technique for surface modification include the refinement of the grain size because of rapid quench rates and the generation of meta-stable structures with novel properties that are not feasible by competing methods [7, 8]. With optimum laser processing parameters, a reliable coating that is free of cracks and pores can be produced on the matrix. The present study investigates the microhardness property and wear resistance performance of Al-Sn-Si coating on ASTM A29 steel.

2. EXPERIMENTAL DETAILS

2.1. Materials Specifications and Sample Preparation Method

The substrate material used in the present investigation was ASTM A29 steel with the chemical composition (wt%) 0.42 Mn, 0.030 S, 0.010 P, 0.180 C and Balance Fe. The SEM micrograph and XRD spectrum of the ASTM A29 steel are shown in Figure 1. The substrate was cut, and machined into dimensions 75 x 75 x 4 mm^3. Prior to laser treatment, the substrates (ASTM A29 steel) were sandblasted, washed, rinsed in water, cleaned with acetone and dried in hot air before exposure to laser beam to minimize reflection of radiation during laser processing and enhance the absorption of the laser beam radiation. Al (99.9 purity), Sn (99.9% purity) and Si (99.9% purity) reinforcement metallic powders were used as alloying powders mixed in 50:20:30 (A_1) and 60:15:25 (A_2) ratios, respectively, in a shaker mixer

(Turbular T2F; Glenn Mills, Inc.) for 18 hours at a speed of 49 rpm to obtain a homogeneous mixture. The particle shape of the powder used was spherical with 45-90 μm particle sizes. Prior to the characterization of the laser surface alloyed materials, samples were prepared by cutting to rectangles of 15 x 15 mm^2, and cold mounted in clear thermosetting Bakelite resin for optical micrographs and black conductive thermosetting resin for SEM and energy dispersive spectroscopy (EDS) analysis. Specimens for SEM (JSM-7600F; JOEL, Ltd) were prepared by cutting samples in such a way to reveal the transverse section of the coatings. Wear tests were performed on the deposited sample at room temperature using the reciprocating tribometer (CERT UMT-2; Bruker Nano Inc., Campell, CA) under dry reciprocating conditions with continual recording of the dynamic coefficient of friction values. The normal load applied on the samples was 25 N at a frequency of 5 Hz and 2 mm stroke length using tungsten carbide (WC) counter material.
Laser surface alloying was performed using a 3-kW continuous wave (CW) Ytterbium Laser System (YLS) controlled by a KUKA robot which controls the movement of the nozzle head and emitting a Gaussian beam at 1064 nm. The nozzle was fixed at 4 mm from the steel substrate. The admixed powders were fed coaxially by employing a commercial powder feeder instrument equipped with a flow balance to control the powder feed rate. The metallic powder was fed through the off-axes nozzle fitted onto the Ytterbium fibre laser and it was injected simultaneously into a melt pool formed during scanning of the ASTM A29 steel by the laser beam. An argon gas flowing at a rate of 2.5 L/min was used as a shielding gas to prevent oxidation of the sample during laser surface alloying. Overlapping tracks were obtained by overlapping of melt tracks at 50%. To determine the best processing parameters, optimization tests were performed with the laser power of 600 to 900 W and scanning speed varied from 0.6 to 1.0 m/min. The final selection criteria during optimization tests were based on surface having homogeneous layer free of porosity and cracks determined from SEM analysis. The optimum laser parameters used was 750 W power, a beam diameter of 4 mm, gas flow rate of 2.5 L/min, powder flow rate of 2.5 g/min and scanning speeds of 0.6 m/min and 0.8 m/min respectively. X-ray diffraction analyses were carried out utilizing the X'Pert Pro model diffractometer to identify the phases present on the resulted alloyed layer. A Cu Kα radiation fitted on the X'Pert Pro diffractometer set at 40 kV and 20 mA was used to scan in a range between 10° and 80° two theta (2θ) with a step size of 0.02° degrees.

3. Results and Discussion
XRD spectrum of ASTM A29 steel clearly shows a high peak of α-Fe (ferrite phase) at 44.5°C and low peak of F_3C (cementite) at 67.8°C. The high α-Fe phase than the peak of F_3C confirms the substrate to be A29 steel. The XRD spectra of admixed powder Al-Sn and alloyed 65Al-15Sn revealed major diffraction peaks with known inter-planar distances. The phases at these peaks are $Al_{13}Fe_4$, Sn, $Al_{0.4}Fe_{0.6}$, and $FeSn_2$ and $Al_{13}Fe_{14}$ as shown in Figure 4. The various phases formed in the laser alloyed samples depend on the laser power, composition of the alloy, and the laser scanning speed. Hard intermetallic phases such as $Al_{0.4}Fe_{0.6}$, $FeSn_2$, and $Al_{13}Fe_4$ were formed in the laser alloyed Al-Sn samples. This may be attributed to the reaction between elemental Al, Sn, and the Fe in the A29 steel substrate.

IOP Conf. Series: Materials Science and Engineering **328** (2018) 012016 doi:10.1088/1757-899X/328/1/012016

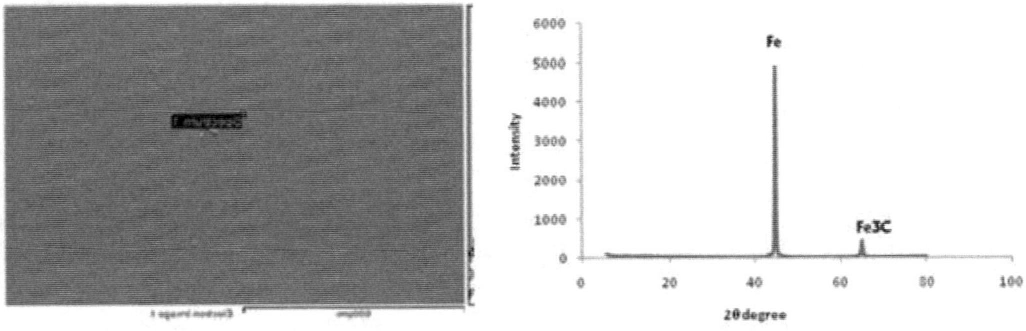

Figure 1: SEM micrograph and XRD spectrum of ASTM A29 steel

3.1 Hardness property of the samples after laser surface alloying.

Microhardness measurements were conducted across the cross section of the deposited hybrid coating tracks (Figure 2), using digital microhardness tester. All the tests were conducted in accordance to the ASTM standards.

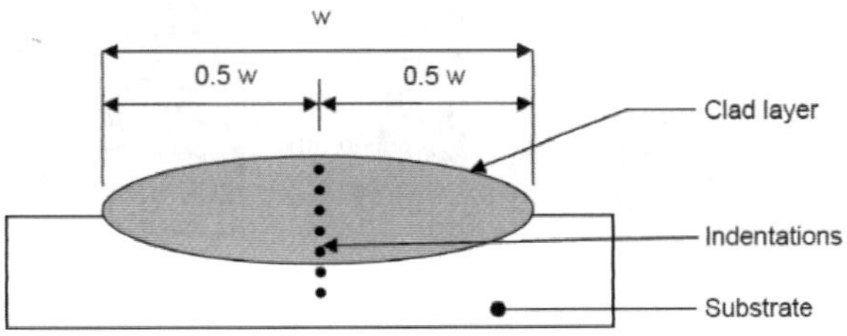

Figure 2: Placement of indentations

The results of the hardness values are shown in Figure 3. The hardness values are highly dependent in the laser scanning speed. It is worthy nothing that as the laser speed increases from 0.6 to 0.8 m/min the hardness values also rise. The high hardness values of the sample at laser speed of 0.8 m/min could be attributed to the smaller grain sizes formed after the laser alloyed process as shown in Figure 5. Increase in hardness value is caused by the obstacle posed to the movement of dislocation by fine grain sizes [9]. Grain size is generally smaller at a faster cooling rate, but the grain size will not reduce further by increasing the cooling rate over a certain rate [10]. Large grain size is the main reason that causes the decrease of hardness and fracture toughness [11].

The lowest and the highest measured microhardness value on the deposited layer across all the 4 samples were found to be 530 and 942 HV respectively. The results showed that the laser alloying process enhances the hardness value of the substrate as shown in Figure 3. The hardness values of 118, 530, 635, 830, and 942 HV were obtained for A29 steel, Al-20Sn-30Si-0.6, Al-20Sn-30Si-0.8, Al-15Sn-25Si-0.6, Al-15Sn-25Si-0.8 respectively. A raise of 85.78 and 87.47% in hardness values above that of the substrate at Al-15Sn-25Si-0.6, and Al-15Sn-25Si-0.8 respectively as shown in Figure 3. This increased hardness values are attributed to the hard phases of iron-aluminium ($Al_{13}Fe4$), iron-tin ($FeSn_2$) formed after the laser alloying process as evident by the XRD spectrum and SEM image shown in Figures 4, 5 and 6. The EDS of the coated sample Al-20Sn-30Si at 0.8 m/min is examined (Figure 6), Fe, Al, Sn, Si, O existed in the coating. From chemical composition of A29 steel, we deduce that Fe is from the matrix itself, Al, Sn and Si are from main reinforcement powders.

It can also be observed that as the laser speed increases from 0.6 to 0.8 m/min there are smaller grain sizes of the various phases formed. This could be attributed to the fact that increases in laser speed led to faster cooling of the melt pool which resulted in the fine grain sizes as shown in Figure 5. Average grain size increases on decreasing laser scanning speed [11]. The grain size becomes larger when the scanning speed is further decreased. Moreover, increase in number of scan changes to type of heat treatment and produces strain hardening in material causing the grain sizes to be reduced as laser scans increases [12]. In addendum, increase of the scanning speed results in finer microstructure due to the larger cooling rate during solidification [13].

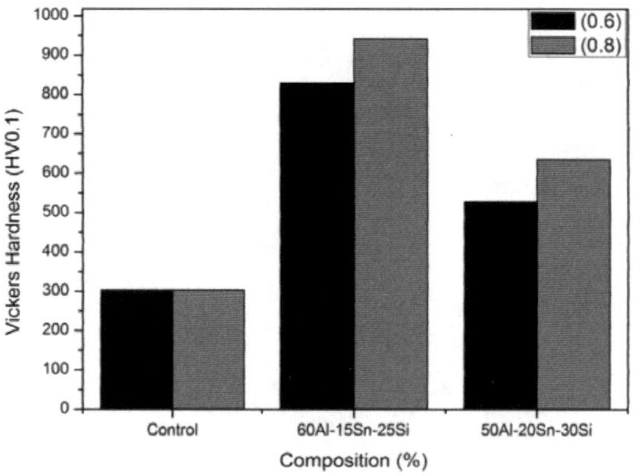

Figure 3: Variation of Hardness values with Al-Sn-Si Coating Compositions

3.2 Microstructural evolution of Al-Sn-Si alloy

The results from the experiment was obtained and it was found there was a presence of α Si phase where the Al-Si alloy depend on the cooling rate and composition of Fe and Si. The Si contents in Si-rich layer increases with increasing Sn content as indicated in 50Al-20Sn-30Si coating. Increasing Sn content from 15 to 20 % caused an increase in nucleation temperature, drop in growth temperature and significant increase in the crystallization temperature of the primary Si [14].

$$L \rightarrow \alpha + \beta(eutectic)$$

(1)

L = Liquid phase

α = Predominantly aluminium

β = Predominantly silicon

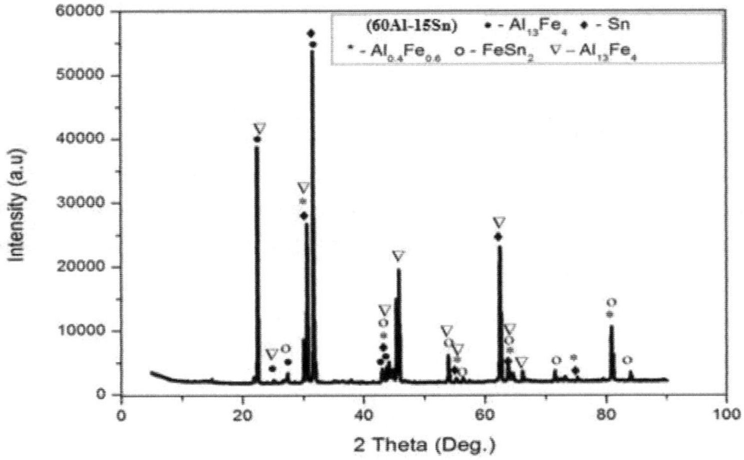

Figure 4: XRD Spectrum of Al-15Sn-25Si Coating at 0.8 m/min scanning speed

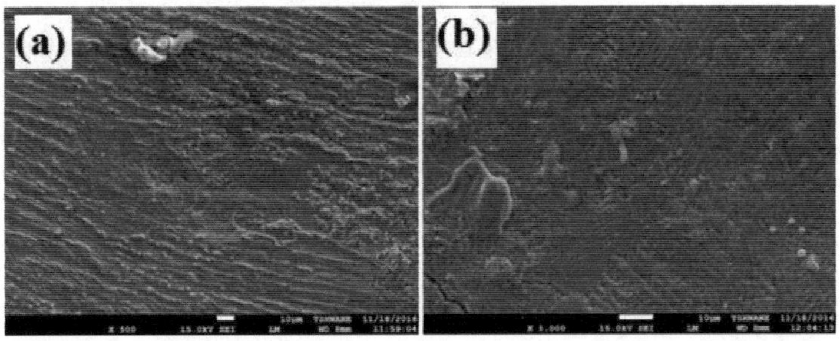

Figure 5: SEM images of Al-15Sn-25Si Coatings at 0.8 m/min (a) x 500 (b) x 1000

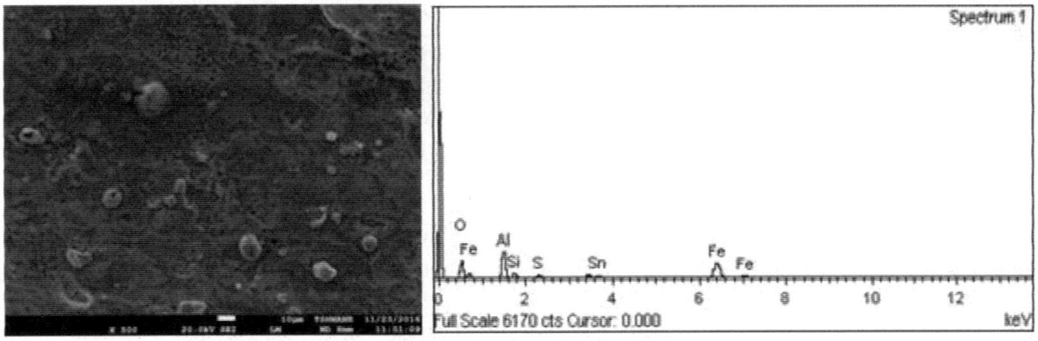

Figure 6: SEM/EDS image of Al-20Sn-30Si Coatings at 0.8 m/min scanning speed

Figure 7 shows the equilibrium phases diagram for the Al-Si system showing metastable extensions of the liquidus and the solidus lines [15]. Al-Si phase diagram is a simple binary eutectic with limited solubility of aluminum in silicon and limited solubility of silicon in aluminum. The solubility of silicon in

IOP Conf. Series: Materials Science and Engineering **328** (2018) 012016 doi:10.1088/1757-899X/328/1/012016

aluminum reaches a maximum 1.5 at. % at the eutectic temperature and the solubility of silicon increases with temperature to 0.016% Si at 119^0 C [15].

The authors reported that at higher cooling rates, the system behaves as though the eutectic point is shifted to the higher silicon contents and the eutectic temperature is depressed. The eutectic on this phase diagram contains much more than α and Si and the mixture (α+Si) is expected to be mainly α according to the lever rule at the eutectic temperature. Primary Si forms first, depleting the liquid of Si until eutectic composition is reached and the remaining solidification will follow the eutectic reaction. Most Al-Si alloys always have a near-eutectic composition since it reduces the melting point. Addition of a small quantity of a ternary element like Sn causes modification of the microstructure. The carbon added from the substrate during laser alloying formed carbide to produce Al-Si-C metal matrix composite (MMC) with better thermal expansion performances. Primary Si can be separated from the Al-Si-Sn melt pool and can accumulate to form a Si-rich layer. The content of primary Si in Si-rich layer increases with increasing of Sn content as reported Murray and McAlister [15]. Sn addition tends to enlarge the crystallization temperature range and enhance the separation efficiency.

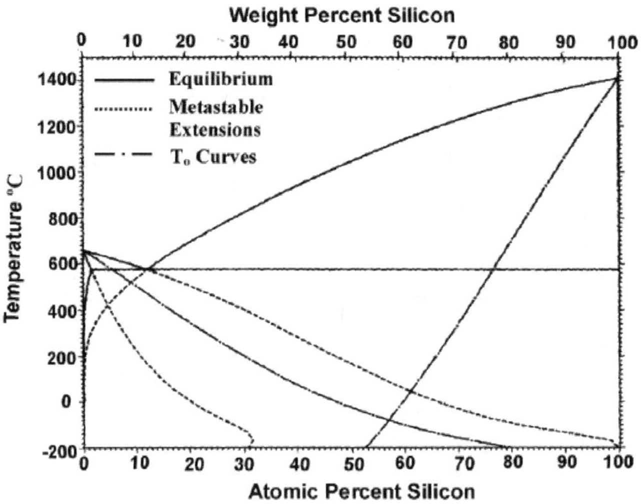

Figure 7: The equilibrium phase diagram for Al-Si system [15].

The results have shown that the use of silicon as an alloying element because of its good properties such as good castability (high fluidity, low shrinkage), the low density of silicon (2.34g/cm) is an advantage as it is able to reduce the overall weight of the component. It also reduces the expansion of Al-Si alloys. Al-Si can be divided into categories: Hypoeutectic (< 12 wt% Si), Eutectic (12-13 wt% Si) and Hypereutectic (14 Wt% Si). Si is the main and most important alloying element of Al-Si alloy. The more Si an alloy contains, the lower is its thermal expansion coefficient. Si is a very hard phase; thus, it contributes significantly to an alloy's wear resistance. Si combines with other elements to improve an alloy's strength and to make alloys heat treatable. Al-Si alloys solidify by a primary precipitation of dendrites. In hypoeutectic Al-Si alloys primary aluminium solidifies dendritically and grows in direction. Dendrites are often drawn having four secondary arms growing around the primary stem at each junction which is true for cubic structures [15]. The undercooling depends on the cooling rate, the concentration of the alloying element in the melt and the type of the alloying element. It is well established that the undercooling increases with increasing cooling rate and increasing concentration of the alloying element [16].

The cooling rate refinement of eutectic Si-particles has been described based on the surface energy of the Al-Si solid interface [17]. This theory is one of the widely accepted theories for quench modification. The rate of advance of the solidification interface depends on a balance between the rate of heat flow from the liquid to the solid through the interface and the latent heat of fusion released during solidification. The thermal conductivities of Al and Si in their pure form are 205 and 83 W/(mK) respectively, and their latent heats of fusion are 396 and 1411 J/g respectively. Since the difference between the magnitude of the thermal conductivity of pure Al and pure Si and the difference between the magnitude of the latent heat of fusion of pure Al and pure Si are large, Al will solidify much faster than Si. Thus, Al gains a lead during solidification of the eutectic.

3.3 Wear property of the samples after laser surface alloying.

The results indicated that the COF of the control sample decreases with fluctuations in the first 150 s while the ternary coating also exhibited fluctuation in the steady state throughout the sliding time. The fluctuation might be coming from debris formed. The control sample exhibited highest friction coefficient of 0.65. The coefficient of friction drastically reduced with Al-Sn-Si ternary coating. Lack of surface coating is responsible for the poor wear resistance displayed by the control sample with high coefficient of friction. Overall, all the Al-Sn-Si coatings at scanning speed of 0.6 and 0.8 m/min respectively showed remarkable improvements in wear resistance and anti-friction behaviour. Wear resistance is enhanced by increasing hardness according to Archard's equation [18]. Al-15Sn-25Si-0.8 alloy with highest microhardness (920 HV) has the lowest coefficient of friction as shown in Figures 8 and 9. There are two opposite factors influencing the wear rate. First one is the hardness. The higher the hardness, the lower the material removal during the wear test. The second factor affecting the wear rate is the ductility. Material with higher ductility (and usually lower hardness) has higher uniform wear but chipping of bigger particles is limited. Improved wear resistance performance of Al-20Sn alloy was due to high microhardness which increased the load-carrying capacity of the alloy [19]. Wear volume was calculated by measuring the width of the wear scar (transverse to the sliding direction) and the wear depth obtained from the UMT-2 software according to the equation formulated [20]. The wear volume loss graph is shown in Figure 9 indicating that the wear volume losses decreased with increasing laser scanning speed at the same operating conditions. The physical and mechanical properties of surface films are what determine the tribological properties of self-lubrication composites. The chief factor responsible for the excellent tribological properties of these materials is the ability to form surface films during sliding. A new microstructure was induced by the reaction between Fe and Sn which played a fundamental role in the wear process [21]. It is capable of resisting wear which led to a 50% wear reduction. This improvement was attributed to intermetallic $FeSn_2$. This intermetallic $FeSn_2$ is a very hard structure as shown in Figure 4. $FeSn_2$ intermetallic may have contributed to these properties. Enhancement in hardness and compressive strength of Al-matrix composite was due to the presence of iron in the composite. The strengthening mechanism was related to the grain refinement of the matrix, and the formation of $Al_{13}Fe_4$ intermetallic [22].

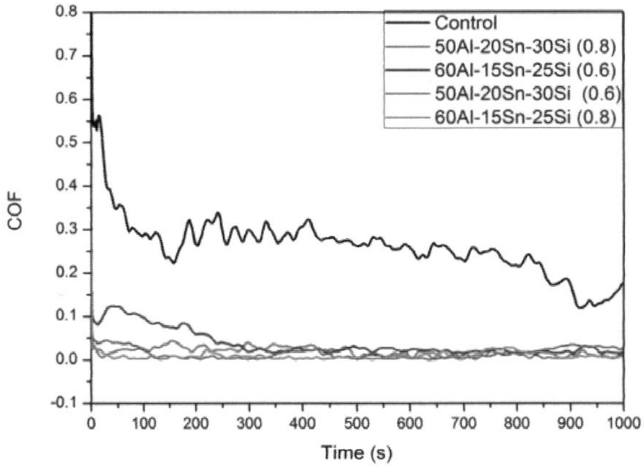

Figure 8: Variation of the Friction Coefficient with Time for the Control and Al-Sn-Si Ternary Coatings.

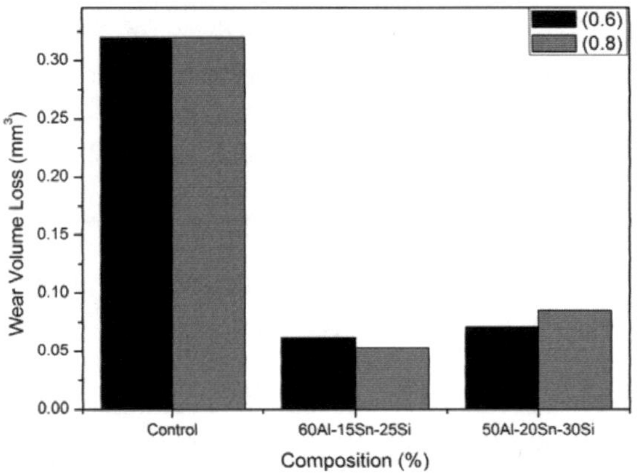

Figure 9: Wear Volume Losses of the Control and Al-Sn-Si Ternary Coatings at 25N Load under Dry Sliding Condition

Moreover, uniform distribution of fine $Al_{13}Fe_4$ intermetallic formed and grain size refinement in aluminium matrix were responsible for enhanced hardness and modulus increase [23-26]. The effects of intermetallic compound on the worn surface of the alloy contribute to the reduction in friction coefficient and wear volume loss. At high temperature reached during wear process, a protective lubricant oxide layer is generated, which prevents the surface from severe damage [27]. The generated oxide provides good layer adhesion which ensures cohesive forces for the contact surfaces [28]. This indicates an outstanding decrease in adhered layer and surface dislocation. Silicon has a very low solubility in aluminium; it therefore precipitates as virtually pure silicon, which is hard and hence improves the abrasion resistance.

4. Conclusion

- Al-Sn-Si coatings were synthesized by continuous wave (CW) Ytterbium Laser System via laser alloying technique to improve the surface property of ASTM A29 steel. The composition proportion of mixed powders has a great influence on the phase structure of the laser deposited coatings.

- The friction coefficient of coated samples indicated remarkable improvement in wear resistance performance compared to the control sample, with 84.41% and 85.29% reduction in coefficient of friction of Al-15Sn-25Si-0.6, and Al-15Sn-25Si-0.8 respectively. All the ternary coatings samples showed decrease in plastic deformation indicating an outstanding decrease in adhered layer and surface dislocation.
- Enhanced hardness property compared to the unprotected substrate comprised a significant increase in hardness from 118 to 942 HV. A raise of 85.78 and 87.47% in hardness values above that of the substrate at Al-15Sn-25Si-0.6, and Al-15Sn-25Si-0.8 respectively.

References

[1] Dobrzanski, L.A., Bonek, M., Hajduczek, E. & Klimpel, A. 2005. Alloying the X40CrMoV5-1 Steel Surface Layer with Tungsten Carbide by the use of a High-Power Diode Laser. *Applied Surface Science*, **247(1)**, 328-332.

[2] Razavi, R. S. & Gordani, G. R. 2011. Recent Trends in Processing and Degradation of Aluminium Alloys [Online].

[3] Handbook A, New York. ASM International, 1985.

[4] Davies J.R Aluminium and aluminium alloys, ASM international, 1993.

[5] Lepper K, James M, Chashechkina K, Rigneya, D.A. (1997). Sliding behaviour of selected aluminium alloys. *Wear*, **203-204**, 46-56.

[6] Desaki T, Goto Y, Kamiya S. (200). Development of the Aluminium Alloy Bearing with Higher Wear Resistance, *Soc. Autom. Engr. of Japan Rev.* **21**, 321-325.

[7] Kwok, C.T., Cheng, F.T. and Man, H.C. (2006). Cavitation erosion and corrosion behaviour of laser –aluminized mild steel. *Surface and Coatings Technology*. **200**, 3544-3552.

[8] Dobrzanski, L.A., Piec, M., Bonek, M., Jonda, E. and Klimpel, A. (2007). Mechanical and tribological properties of the laser alloyed surface coatings. *Journal of Achievements in Materials and Manufacturing Engineering*. **20(1-2)**, 235-238.

[9] Rahman M.J, Sen S.R., Moniruzzaman M. and Shorowordi K.M. (2009). Morphology and properties of electrodeposited Zn-Ni alloy coatings on mild steel, *J. Mech. Eng. Trans. Mech. Eng. Div. Inst. Eng.* **40**, 9-12.

[10] Liu X., Warren A.P., Nuhfer N.T., Rollett A.D., Coffey K.R. and Barmak K. (2014). Comparison of Crystal Orientation Mapping-Based and Image Measurement of Grain Size Distribution in a Thin Aluminium Film. *Acta Materialia,* **79**, 138-145.

[11] Shuai C. Feng, P. Zhang L. Gao C. Hu H. Peng S. and Min A. (2013). Correlation between Properties and Microstructure of Laser Sintered Porous β-TriCalcium Phosphate Bone Scaffolds, *Sci. Technol. Adv. Mater.* **14**, 1-10.

[12] Akinlabi E.T. and Akinlabi S.A. (2012). Effect of Heat Input on the Properties of Dissimilar Friction Stir Welds of Aluminium and Copper. *Amer. J. Mater. Sci.* **2**, 147-152.

[13] Gong X. Lydon J. Cooper K. Chou K. (2014b). Beam Speed Effects on Ti-6Al-4V Microstructures in Electron Beam Additive Manufacturing, *J. Mater. Res.* **29**, 1951-1959.

[14] Zou Q.C. Jie J.C. Dong Y. Li T.J. Effect of Sn addition on the electromagnetic separation of primary Si from Al-30Si alloy during solidification. 8th International Conference on Electromagnetic Processing of Materials, Oct 2015, Cannes, France. EPM2015.

[15] Murray J.L. and McAlister A.J. (1984). Alloy Phase Diagrams, *Bull.* **5**, 74-84.

[16] Fredriksson H. and Åkerlind U. (2012). Solidification and Crystallization Processing in Metals and Alloys, John Wiley & Sons Ltd., Chichester, 2012.

[17] Makhlouf, M. and Guthy H. (2001). *Journal of Light Metals*, **1**, 199-218.

[18] Archad A. (1953). Contact and Rubbing of Flat Surfaces. *J. Appl. Phys*, **24**, 981-988.

[19] Liu X., Zeng M.Q., Ma Y. Zhu M. (2008). Wear behaviour of Al-Sn alloys with different distribution of Sn dispersoids manipulated by mechanical alloying and sintering. *Wear*, **265,** 1857-1863.

[20] Qu J. and Truhan J.J. (2006). An efficient method for accurately determining wear volumes of sliders with non-flat wear scars and compound curvatures. *Wear*, **261**, 848–855.

[21] Kossowsky R. and Singhal S.C. (2012). Surface Engineering, *Surface Mod. Mater. Technol. Eng.* **764**, 1-8.

[22] Fathy A., El-Kady O. and Mohammed M.M.M. (2015). Effect of iron addition on microstructure, mechanical and magnetic properties of Al-matrix composite produced by powder metallurgy route. *Trans. Non-Ferrous Met. Soc. China*, **25**, 46-53.

[23] Lee I.S. Kao P.W. and Ho N.J. (2008). Microstructure and mechanical properties of Al-Fe in-situ nanocomposite produced by friction stir processing. *Intermetallics*, **16**, 1104-1108.

[24] Makhatha, M.E., Fatoba, O.S. and Akinlabi, E.T. (2017). Effects of rapid solidification on the microstructure and surface analyses of laser-deposited Al-Sn coatings on AISI 1015 steel. *Int J Adv Manuf Technol.* 1-15. https://doi.org/10.1007/s00170-017-0876-y

[25] Fatoba O.S., Akinlabi E.T. and Makhatha M.E. (2017). Effect of silicon on the microstructure, mechanical and surface properties of laser metal deposition Al-Si based coatings, Fiber Laser, Dr. Subbarayan Sivasankaran (Ed.), InTech, http://dx.doi.org/10.5772/intechopen.71698

[26] Fatoba O.S. Akinlabi E.T. and Makhatha M.E. (2017). Effect of Process Parameters on the Microstructure and Tribological Property of Zn-Sn-Ti Coatings on AISI 1015 Steel: Laser Alloying Technique. *International Journal of Surface Science and Engineering.* **11 (6),** 489-511.

[27] Zhu M. and Gao Y. (2000). Improvement of the wear behaviour of Al-Pb alloys by mechanical alloying. *Wear*, **242**, 47-53.

[28] Su Y.L. and Kao W.H. (2003). Tribological behaviour and wear mechanism of MoS_2-Cr coatings sliding against various counter-bodies. *Tribol. Int.* **36**, 11-26.

Post-Impact and Open Hole Tensile Of Kenaf Hybrid Composites

S. Yunus[1,a], Z. Salleh[2], N.R.N.M. Masdek[2], Y.M. Taib[2], I.I.S. Azhar[3], K. M. Hyie[4]
[1] Mechatronics Technology, Industrial Electronics Department, German Malaysian Institute (GMI), 43000 Kajang, Selangor.
[2] Faculty of Mechanical Engineering, Universiti Teknologi MARA (UiTM), 40000 Shah Alam, Selangor.
[3] Fakulti Kejuruteraan Mekanikal UiTM Cawangan Johor, Kampus Pasir Gudang, 81750 Masai, Johor.
[4] Faculty of Mechanical Engineering, Universiti Teknologi MARA (Pulau Pinang), 13500 Permatang Pauh, Pulau Pinang

Email: [a] sya_mechys@yahoo.co.uk

Abstract. Nowadays, kenaf hybrid glass composites has been used for a vast field of study throughout the globe. There are several compositions and orientation of kenaf hybrid glass composites that has been studied. With regards to the study that has been done, this study will be focussing on a 90FG/0/90/90/0/90FG orientation of kenaf hybrid glass composites. Polyester resin is used as a matrix to these hybrid composites. Impacted and open hole specimens were then analyzed through tensile test. All specimens were fabricated by using the cold press hand lay-up technique. The results revealed that the hybrid composites were hardly affected by the impact up to 6J. After 6J the impacted specimens experienced a significant damage for both strength and modulus. The same goes to open hole specimens where the same trend of tensile properties were observed as impacted specimens.

Keywords: kenaf hybrid glass composites, polyester resin, cold press hand- lay-up, impact, open hole, tensile test

1. Introduction

Kenaf (*Hibiscus cannabinus*, L. family Malvacea) has been found to be an important source of fiber for composites and other industrial applications [1,2]. It has been domestically grown in Africa and spread throughout the globe [3]. Later, it was spread to nurture in India and China. Presently, kenaf is commercially grown in the United States of America (USA). There are a lot of advantages in using kenaf in the research field such as economical, recyclable and environmentally harmless. The ecological advantages of the kenaf are their grow condition and also their growth rate speed [4]. Furthermore, the optimum weather condition for kenaf to grow optimally is during high weather temperature which is up to 30°C where the growth rate speed up to 80% faster compared to the low temperature condition [5]. Kenaf has a right future in the composite material field as it can be used in many applications such as in the automotive sector. However, it also has disadvantages that make it less attractive to be used such as having low mechanical properties (eg. strength, compressive, bending etc.) compared to other synthethic fibres (fiberglass, Kevlar, carbon etc.). To hybrid it with fiberglass is one of the popular methods to improve its mechanical properties [6].

Composites material are very sensitive to impact where it can create damage and break the materials. Impact can happen anytime and sometimes undetectable damage cannot be seen such as flying debris, falling tools etc. Thus, it needs to give more attention before it can be fully utilized. There is also another factor that need to be taken care of when using the composite material, as for example when drill hole into it. The damage can happen as the hole can initiate the crack thus material failure will happen. This

IOP Conf. Series: Materials Science and Engineering **328** (2018) 012017 doi:10.1088/1757-899X/328/1/012017

is where the studying and analyzing of the impact and hole to the composite materials are important, hence the prediction of the reliability and robustness to that area can be done.
Thus, this study will investigate the hybridizations of kenaf with fibreglass in order to improve the mechanical properties of nature fibre composites. The fabrication technique used to prepare this composite is a hand lay-up process as well as cold press method. In this study, 90FG/0/90/90/0/90FG orientation is chosen and polyester resin is used to hold this hybrid composites. To study its mechanical properties, there are two types of samples prepared before it is going through tensile testing. The first sample, the specimens will undergo many numbers of low velocity impact which were from 2J, 4J, 6J, 8J and 16J. Another samples is open hole specimens which were drilled with the diameter of 2mm, 4mm, 6mm, 8mm and 16mm. The damages resulted from this test were then discussed in the following section.

2. Methodology

Sample preparations

Hybrid composites are made up of long non-treated kenaf fiber (used as received) and woven fibreglass. Both fibres were supplied by a local company. As for the matrix resin, a standard unsaturated polyester resin and polyester hardener which acted as a catalyst are used. Both resins were bought from Mostrong Industries Sdn. Bhd. The sample orientation is shown in Figure 1 and the testing standard used is referred to the ASTM D5766.

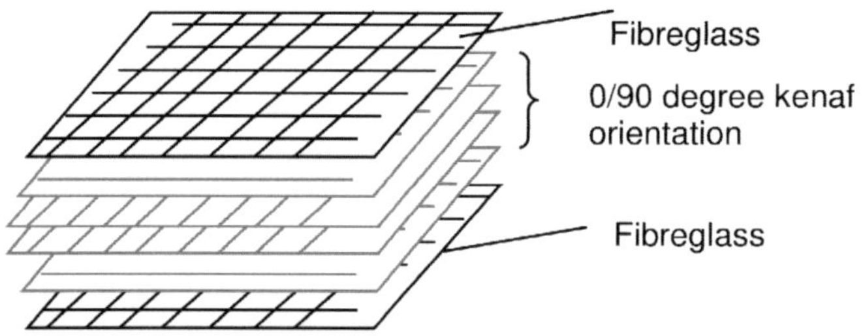

Figure 1: 90FG/0/90/90/0/90FG hybrid composites orientation

Open hole and post impact specimens

The test was done according to the ASTM D5766 standard. For the open hole specimens, before the specimens undergo the tensile test, it was drilled at the center with a hole size ranging from 2,4,6,8 and 16 mm as shown in Figure 2 (a). For the post impact specimens, the specimens were impacted at 2,4,6,8 and 16J as shown in Figure 2(b). All specimens were then mounted on the grip of the Instron 3382 universal tester machine with a gauge length of 100 mm. A sandpaper was wrapped around both ends of the specimen to reduce slipping to happen during the testing.

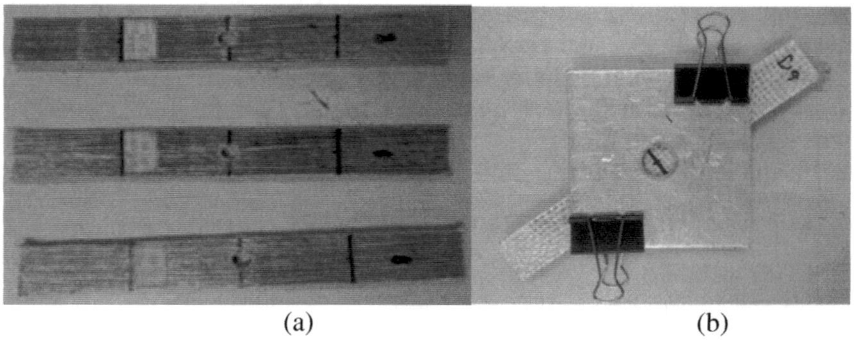

(a) (b)

IOP Conf. Series: Materials Science and Engineering **328** (2018) 012017 doi:10.1088/1757-899X/328/1/012017

Figure 2: (a) Open hole specimens, (b) post impact specimen

3. Result and Discussion

The mechanical properties of the specimens such as the tensile modulus and tensile strength were determined by referring to the ASTM D5766 standard. There were two types of samples prepared, which were post-impact and open hole tensile specimens. For the post-impact tensile test, a total of 15 to 25 specimens were conducted by using an impactor with a hemispherical nose of diameter 12.7 mm as the drop weight and later the specimens were tested using an Intron 3382 universal tester machine. Meanwhile, for the open hole tensile test, a hole was drilled on the centre of the specimens and later the specimens were tested using an Intron 3382 universal tester machine. The test was carried out at a room temperature and a set of at least three readings for each parameter were taken from both tests.

Figure 3 and Figure 4 shows the result obtained from the open hole tensile test. Based on Figure 3, the tensile modulus of the specimens was decreased gradually but the gap between them increased tremendously after 6 mm hole diameter and decreased after 8 mm hole diameter. The 2 mm hole diameter exhibited the highest values among all of the others with an average value of 3690.667 MPa while the lowest value is at 1706 MPa for the 16 mm hole diameter. The value for tensile modulus was dropped to about 50% of its original value when the hole diameter was at 8 mm. As for Figure 4, the tensile strength also showed the same trend where the value drop was the highest after 6 mm and decreased tremendously after 8 mm hole diameter. The highest value of the tensile strength was at 60.803 MPa for 2 mm hole diameter and the lowest at 18.677 MPa for the 16 mm hole diameter. The tensile strength value dropped to about 40% of its original value when the hole diameter was at 8 mm. This behaviour showed that both properties were affected by the hole diameter in the same manners. It can be concluded that the stress can be distributed to a larger part of the cross-section for the small hole diameter, meanwhile for the larger diameter, there is lesser remaining cross-section part which the stress can be distributed to support the stress [7]. Thus, this behaviour can be simplified that when the load increases, due to stress concentration the damage initiates near the hole and then propagate through the cross-section area near the hole site as the stress or strain until reaches a certain value. [8]

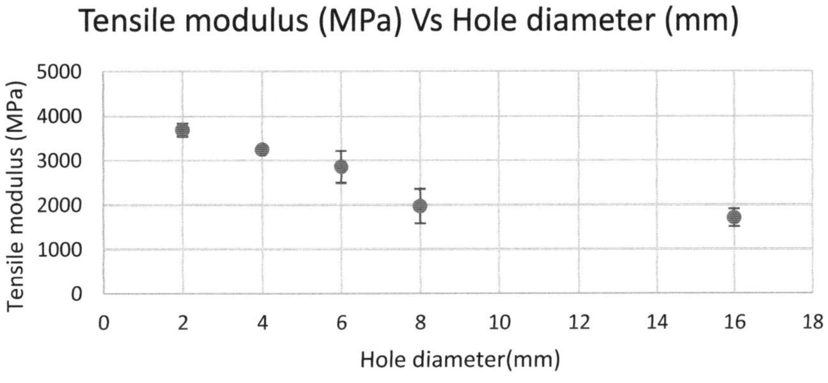

Figure 3: The average value of tensile modulus

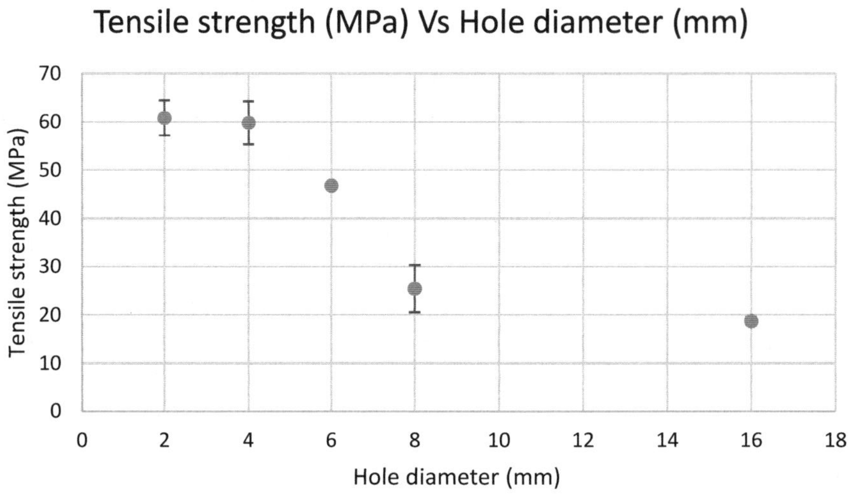

Figure 4: The average value of tensile

Figure 5 and Figure 6 shows the results for a post-impact tensile test. It showed that the tensile modulus and tensile strength drop dramatically after 6 J of impact energy. The behavior was the same as the open hole tensile test result. Thus, it could be said that the specimens were physically impaired after the impacted of 6 J of impact energy. In Figure 5, the maximum tensile modulus for 2 J of impact energy was at 3055 MPa and the lowest was at 1157 MPa for 16 J of impact energy. When the impact energy reach to 8 J the tensile modulus, it was reduced to about 54% of the original value and as for the tensile strength, it was reduced by about 25%. When the specimens were subjected to 8J of impact energy, matrix cracking, fiber-matrix debonding and delamination could have transpired on the specimens which lead to the drastic drop in the physical properties [9]. Thus, the failures that happened on the specimen contributed to the specimen fracture.

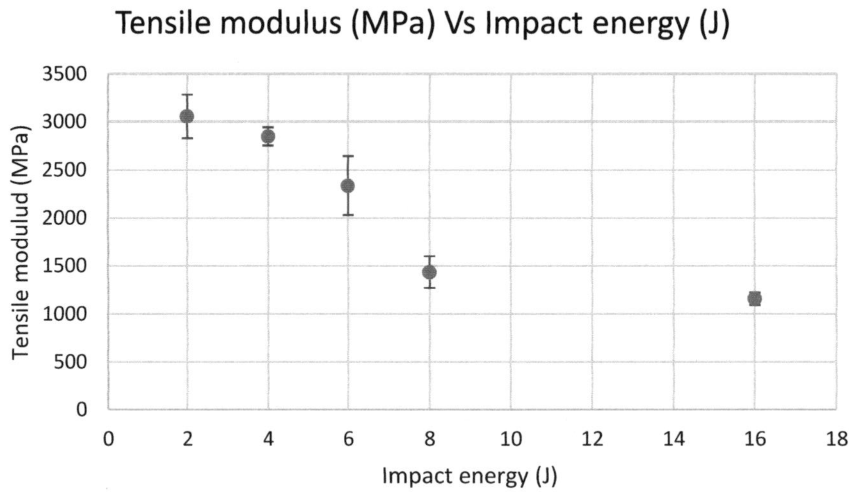

Figure 5: The average value of tensile modulus (MPa)

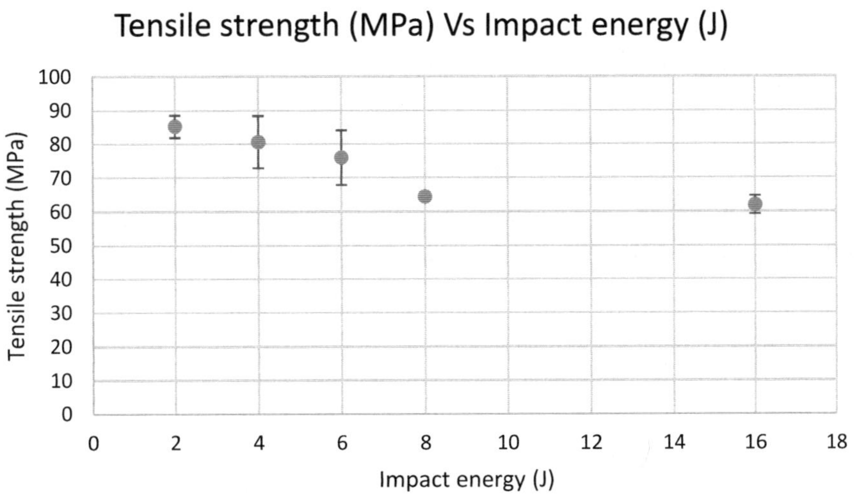

Figure 6: The average value of tensile strength

From Figure 7, it was showed the microstructure where the fracture happen at the hole diameter. As we can see, the specimen experience debonding, fibre pull off and there is also a glass fiber pull off. As for the matrix structure, it has undergone a matrix fracture which results in the failure of the matrix structure to withstand further stress. The specimen that is subjected to a higher strain-rate loading shows large areas of interlaminar splitting (delamination) and numerous fiber breakages and pull-out, which could be caused by changes in the local interfacial bonding [10].

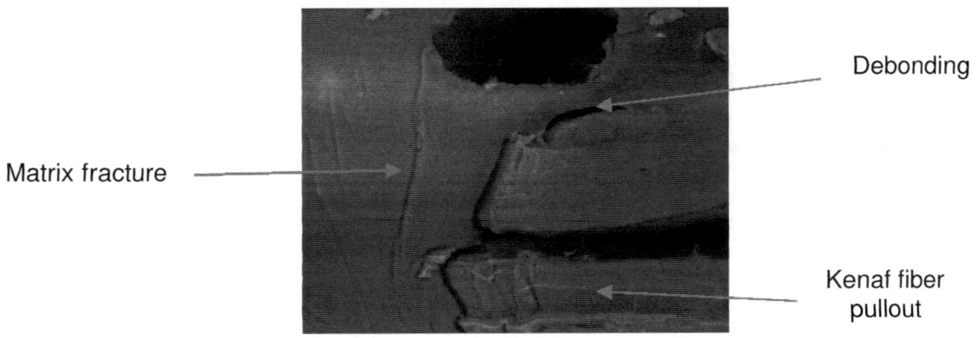

Figure 7: The microstructure of open hole tensile specimen

4. Conclusion

From the test results that have been analysed and discussed, the kenaf hybrid composite that was fabricated using the hand-lay-up technique and cold press method displayed great mechanical properties. With regard to the open hole tensile and post impact test, we can conclude that the specimen was affected by the hole diameter and impact energy and upon a hole diameter of 8mm and 8 J of impact energy, the specimens were drastically impaired which resulted in the drop of tensile modulus and tensile strength. Furthermore, the results of the mechanical properties could be enhanced by controlling one of the specimen parameters such as a uniform thickness of the kenaf hybrid fibreglass composite.
echnical detail that it is necessary to include, but that interrupts the flow of the article, may be consigned to an appendix. Any appendices should be included at the end of the main text of the paper, after the

acknowledgments section (if any) but before the reference list. If there are two or more appendices they should be called appendix A, appendix B, etc. Numbered equations should be in the form (A.1), (A.2), etc, figures should appear as figure A1, figure B1, etc and tables as table A1, table B1, etc.

5. References

[1] Vignesh. S, S. V Alagarsamy, H. Saravanan. A review on natural fibre reinforced polymer composites. International Journal of Engineering Sciences and Research Technology. ISSN :22775-9655, (10) Oct 2016.

[2] R. Karnani, M. Krishnan, R. Narayan (1997). Biofiber Reinforced Polypropylene Composite. Polym Eng Sci, 37 (1997), pp 476-483.

[3] E. Alexopoulou, D. Li, Y. Papatheohari, H. Siqi, D. Scordia, G. Testa (2014). How Kenaf (Hibiscus Cannabinus L.) Can Achieve High Yields In Europe and China. Industrial Crops and Products. Volume 68, June 2015, pp 131-140

[4] H. M. Akil, M. F. Omar, A. A. M. Mazuki, S. Safiee, Z. A. M. Ishak, A. Abu Bakar (2010). Kenaf Fiber Reinforced Composite: A Review. Materials & Design. Volume 32, Issues 8-9, September 2011, pp 4107-4121

[5] Ochi S. (2008). Mechanical Properties of Kenaf Fibers and Kenaf/PLA Composites. Mech Mater, 446–52.

[6] K.L. Pickering, M.G. Aruan Efendy and T.M. Le. A review of recent developments in natural fibre composites and their mechanical performance. Part A 83 (2016), pp 98-112.

[7] P. K. Govindan, R. B. Nageswaran, and V. K. Srivastawa. Notched Tensile Strength of Randomly Oriented E-glass Composites laminates. Materials Science and Engineering (2000), pp 59-66.

[8] Z. Salleh, M. N. Berhan, Koay Mei Hyie, Y. M. Taib, A. Latip, N. R. Nik Roselina (2013). Open Hole Tensile Properties of Kenaf Composite and Kenaf/Fibreglass Hybrid Composite Laminates. The Malaysian International Tribology Conference 2013, MITC2013. Procedia Engineering 68 (2013) 399-404.

[9] Z. Salleh, Koay Mei Hyie, M. N. Berhan, Y. M. D. Taib, M. K. Hassan, D. H. Isaac (2013). Effect of Low Impact Energy on Kenaf Composite and Kenaf/Fiberglass Hybrid Composite Laminates. Applied Mechanics and Materials Vol 393 (2013), pp 228-233.

[10] Sung-Choo Woo, Tae-Won Kim (2013). High-Strain-Rate Impact in Kevlar-Woven Composite and Fracture Analysis Using Acoustic Emission. Composite Part B: Engineering. Volume 60, April 2014, pp 125-136

Acknowledgments

The authors would like to express their gratitude to the Faculty of Mechanical Engineering (UiTM) for the facilities and technical support. The authors also grateful acknowledge the financial support from Research Acculturation Grant Scheme (Ref: 600-RMI/DANA 5/3/LESTARI (111/2015) and 600-RMI/RAGS 5/3 (18/2015)) for the successful implementation of this project. Lastly, special thanks to the final year project student, Shahrul Izwan Bin Adul Ghani for his assistance in completing this study.

Tensile and Flexural Test on Kenaf Hybrid Composites

Z. Salleh[1,a], S. Yunus[2], N.R.N.M. Masdek[1], Y.M. Taib[1], I.I.S. Azhar[3], K. M. Hyie[4]
[1] Faculty of Mechanical Engineering, Universiti Teknologi MARA (UiTM), 40000 Shah Alam, Selangor.
[2] Mechatronics Technology, Industrial Electronics Department, German Malaysian Institute (GMI), 43000 Kajang, Selangor.
[3] Fakulti Kejuruteraan Mekanikal UiTM Cawangan Johor, Kampus Pasir Gudang, 81750 Masai, Johor.
[4] Faculty of Mechanical Engineering, Universiti Teknologi MARA (Pulau Pinang), 13500 Permatang Pauh, Pulau Pinang

Email: [a] a_kzue@yahoo.com

Abstract. The widely use of synthetic materials like carbon and fiberglass in various industries such as automotive and *aircraft has lead to human health and environment problems. Therefore, the use of natural fibres such as kenaf has received higher attention as reinforcement. Kenaf or the scientific name is Hibiscus Cannabinus. L is one of the group of Malvecea plant which in the early days, the application of kenaf served only rope and canvas. However, it has more advantages than synthetic materials such as; widely availaible, renewable, lightweight, non-abbrasiveness during processing, high specific strength, free from health hazard and biodegradeable. This study was carried out to investigate the effects of different arrangement of kenaf and fiberglass composites on Young's Modulus. The material composite was hardened with polyester resin and their properties was characterized. The tensile and the flexural properties is determined using an Instron universal tensile testing machine and carried out by following ASTM D3039 for tensile and ASTM D790 for a flexural test. The experimental program was designed to correlate the flexural and tensile Young's Modulus of kenaf and fiberglass composite under the same load condition but different arrangement of kenaf and fiberglass on the mold . The resistance to change in shape was described by the behavior and characteristic of the composite materials. The stiffness or the elastic modulus of the composite material was determined at the end of the experiment. The results obtained show that the [±90FG/0/90/90/0/±90FG] kenaf/fiberglass composite arrangement has the highest elastic value.*

Keywords: Kenaf fiber, Woven Fiberglass, Tensile, Flexural, Young's Modulus, Mechanical Properties.

1. Introduction

In the last decade, an extensive research work was carried out on natural fiber reinforced composite materials. Natural fibers can be used to reinforce polymers to obtain light and strong materials. Nowadays, natural fibers from plants are beginning to find their way into commercial applications such as in the automotive industries and household applications [1]. Due to the relatively high cost of synthetic fibers such as, glass, plastic, carbon and Kevlar used in fiber reinforced composites, and the health hazard issue of asbestos fibers, it has become necessary to explore natural fibers as a substitute. Natural fibers are produced from renewable resources; and are biodegradable and relatively inexpensive compared to the traditionally used synthetic fibers. Fibers of this type, for instance, hemp and flax, are successfully used as packaging material, interior panels in vehicles, and building components, among others. Also, natural fibers like banana, sisal, hemp, flax, jute, coconut husk, local fibers and oil palm [4-11] have attracted scientists and technologists for applications in consumer goods, low-cost housing and other civil structures.

IOP Conf. Series: Materials Science and Engineering **328** (2018) 012018 doi:10.1088/1757-899X/328/1/012018

The fiber/matrix interface has an important role in the micromechanical behavior of composites. One difficulty that has prevented the use of natural fibers is the lack of good adhesion to polymeric matrix. In particular, the large moisture sorption of natural fibers adversely affect adhesion to hydrophobic matrix material leading to a premature ageing by degradation and loss of strength. To prevent this phenomenon, fiber surface properties have been modified in order to promote adhesion [3]. To reinforce thermoplastic composites, blue agave fibers were modified via esterification reaction[12]. The mechanical properties characterization of these fibers showed a change in the elastic modulus and an improvement in the impact resistance [2]. There are, however, several separations of fiber processing techniques such as mechanical or chemical pulping, whereby the lignin is degraded and dissolved, leaving most of the cellulose and hemicelluloses in the form of fibers. This generally has an important effect on both mechanical and chemical properties of the fibers.

Recently, natural fibers were proved to be an effective reinforcement that serve as simple fillers in thermoplastic and thermoset matrix composites for automotive sectors [11]. A number of investigations have been conducted on several types of natural fibers to study the effects of these fibers on the mechanical properties of composite materials. This work was carried out to investigate the flexural and tensile properties of kenaf/ fiberglass composite. Polyester resin acts as adhesive which sticks woven fiberglass and natural long kenaf fiber. In this study, four different types of arrangement of kenaf fiber with woven fiberglass composite were made, and its mechanical properties were investigated. In addition , the young modulus of flexural and tensile test was determined at the end of the experiment.

Formatting the title, authors and affiliations

Please follow these instructions as carefully as possible so all articles within a conference have the same style to the title page. This paragraph follows a section title so it should not be indented.

2. Methodology

Figure 1 shows a process flow chart which was followed in this study. Meanwhile Figure 2 shows a composite arrangement and designation for each configuration.

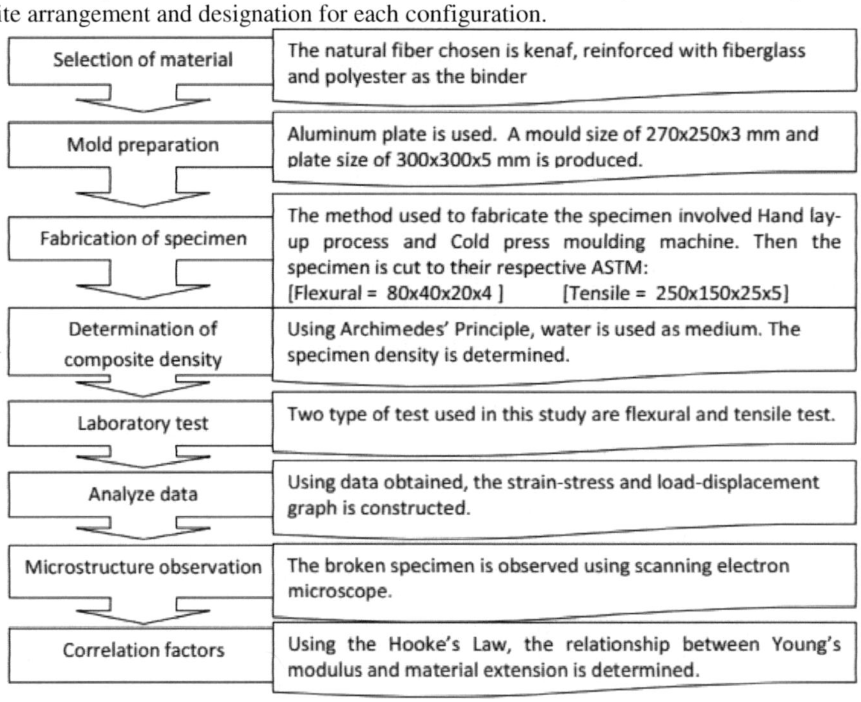

Figure 1: Process flowchart

IOP Conf. Series: Materials Science and Engineering **328** (2018) 012018 doi:10.1088/1757-899X/328/1/012018

Composite materials configuration/arrangement	Description
1 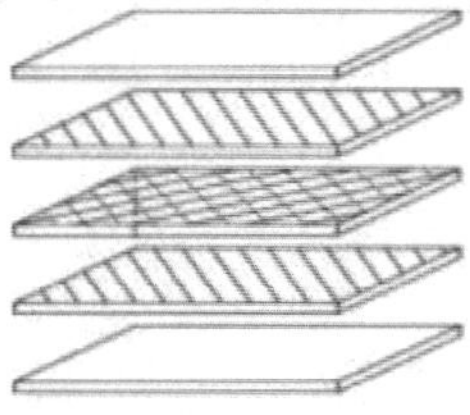 $[0/90/\pm90_{2FG}/90/0]$	Layer of composite are as follow : 1) Horizontal (0^0) kenaf fiber 2) Vertical (90^0) kenaf fiber 3) Double layer of woven fiberglass 4) Horizontal (0^0) kenaf fiber 5) Vertical (90^0) kenaf fiber
2 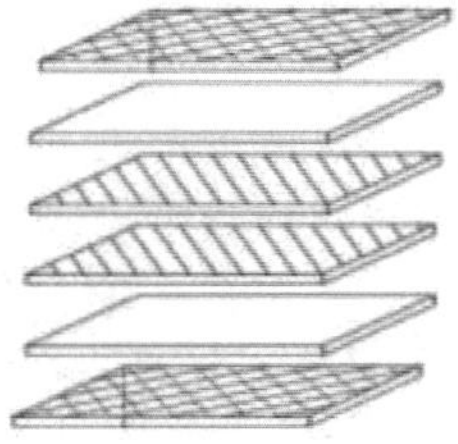 $[\pm90_{FG}/0/90/90/0/\pm90_{FG}]$	Layer of composite are as follow : 1) Single layer of woven fiberglass 2) Horizontal (0^0) kenaf fiber 3) Vertical (90^0) kenaf fiber 4) Vertical (90^0) kenaf fiber 5) Horizontal (0^0) kenaf fiber 6) Single layer of woven fiberglass
3 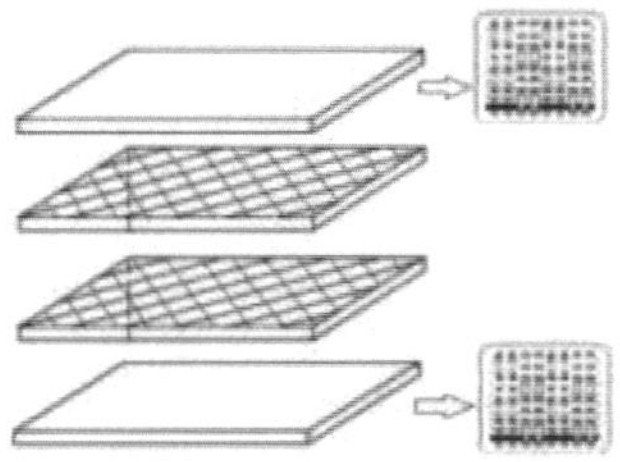 $[\pm90_{w}/\pm90_{2FG}/\pm90_{w}]$	Layer of composite are as follow : 1) A layer of woven kenaf fiber 2) Double layer of woven fiberglass 3) A layer of woven kenaf fiber
4 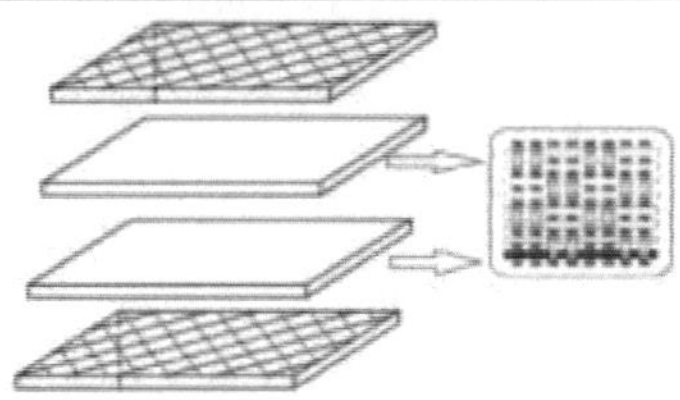 $[\pm90_{FG}/\pm90_{2w}/\pm90_{FG}]$	Layer of composite are as follow : 1) Single layer of woven fiberglass 2) Double layer of woven kenaf fiber 3) Single layer of woven fiberglass

Figure 2: Composite arrangement and designationLine thickness

3. Result and Discussion

This section describes the experimental results after the various composite samples were broken using a universal tester and a three-point bending machine. All results were plotted in terms of stress versus strain of the sample produced by the electronic universal tester machine. In each case, there is a reasonably good correlation in the results.

Mechanical properties such as modulus, flexural and tensile strength were determined by the ASTM Test; methods D790 and D3039 respectively. For the flexural test, the composite beam specimen of

IOP Conf. Series: Materials Science and Engineering **328** (2018) 012018 doi:10.1088/1757-899X/328/1/012018

rectangular cross section was loaded in a three-point bending mode. The tensile test was carried out on laminate composite specimens by applying a tensile load on the specimen which was mounted on an Instron universal tester machine. A set of five readings were taken from both experiments of composites and the average values were obtained. Tests were carried out at a room temperature and as per the standards mentioned.

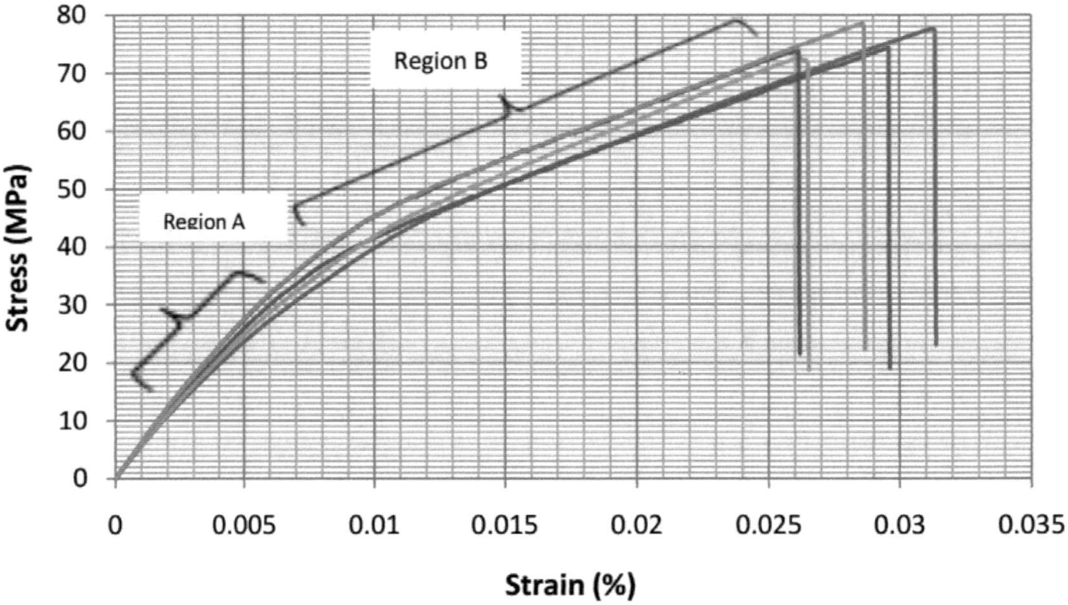

Figure 3: Tensile stress-strain graph of [±90FG/±90 2w /±90FG] woven kenaf/fiberglass composite

Figure 3 shows the tensile stress-strain graph of [±90FG/±90 2W /±90FG] woven kenaf/fiberglass composite for each specimen. The curves indicate that the pull force acting on the specimens were finally breaks after it reachs the maximum tensile stress that the specimen can withstand. The individual breaks are formed at the weaker flaws along the fiber as the load increase. At the point of fiber failure, the [±90 FG/±90 2w /±90FG] woven kenaf/fiberglass interface debonds with the fiber sliding against the matrix. At the region A on tensile stress-strain curve corresponds to an elastic matrix. While the region B corresponds to the plastically yielding matrix. In addition, the graph also shows that the elastic region leaves at approximately 15~20% of the maximum stress. In this case, the elastic region is from 0 to 0.005 mm/mm of a tensile strain. Fibers are assumed to behave elastically up to failure [8].

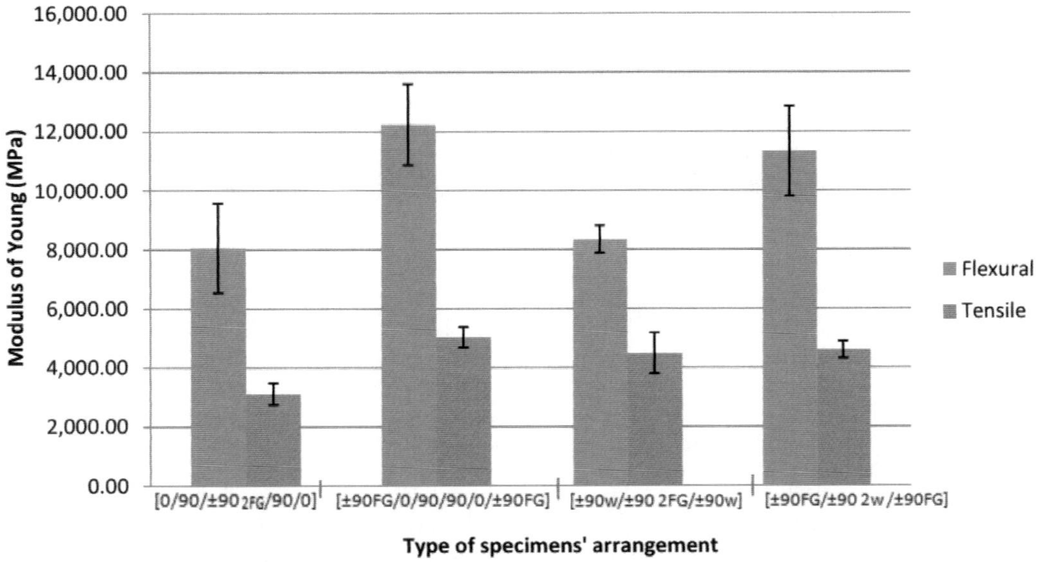

Figure 4: Average Young's Modulus of flexural and tensile properties

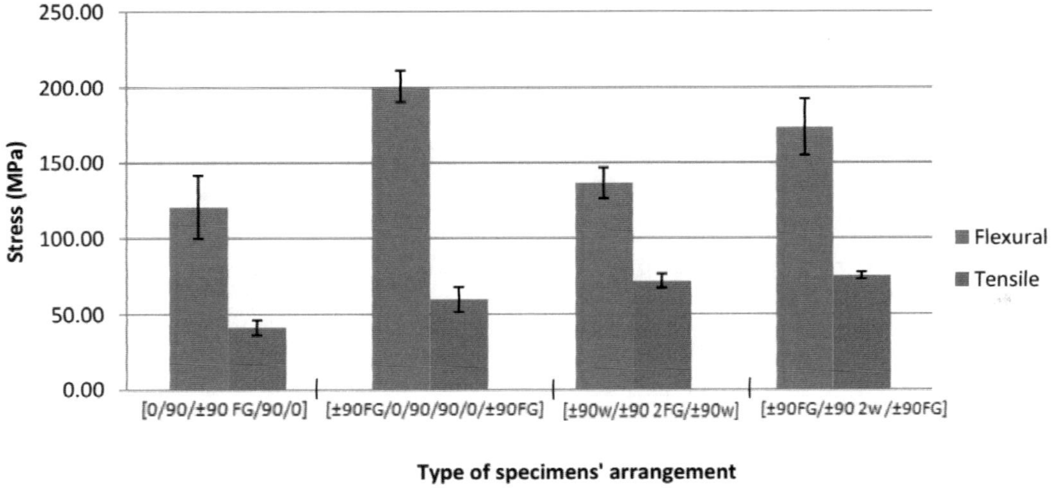

Figure 5: Average ultimate flexural and tensile strength

From both figures, the bar chart pattern shows similar consistency with acceptable variability errors of less than 10%. According to Figure 4, [±90FG/0/90/90/0/±90FG] dominates the result of Young's Modulus on both experiment which is 12.233 GPa in flexural and 5029.67MPa in tensile test compared than other arrangements of laminate composite. This indicate the maximum elastic modulus is available in that composite arrangement. Meanwhile, [0/90/±90 FG/90/0] shows the worst result which is 8050.11MPa in flexural and 3114.29MPa in tensile. As for flexural strength, it can be seen clearly in Figure 5, [±90FG/0/90/90/0/±90FG] with 200.54 MPa has the highest flexural strength compared to other arrangements of laminate composite. [±90FG/±90 2w /±90FG] is 27MPa slightly lower than [±90FG/0/90/90/0/±90FG] and followed by [±90w/±90 2FG/±90w] which is 37MPa higher. However, [0/90/±90 FG/90/0] shows the lowest flexural strength which is 120.93MPa. As for tensile strength,

[±90FG/±90 2w /±90FG] has 75.38MPa which is higher compared to other arrangements of laminate composite. [±90w/±90 2FG/±90w] is slightly lower which is 71.82MPa and followed up by [±90FG/0/90/90/0/±90FG] which is 59.67MPa. However, [0/90/±90 FG/90/0] is once again placed at the bottom having a value of 41.23MPa only. The varying results between flexural and tensile happen due to the inconsistency in thickness for each specimen due to a manual hand lay-up process. Even a slight increase in thickness of the fiber composite without sacrificing much on the weight will affect the increasing/decreasing of the flexural and tensile properties [9].Equations and mathematics

4. Conclusion

This research focus on the mechanical properties and the elastic modulus of the kenaf/fiberglass composite. The kenaf/fiberglass composite is fabricated with combination of a hand lay-up and cold press method which shows good mechanical properties. From the two mechanical tests that were conducted, flexural and tensile testings, the results and data obtained from all these experiments clearly show that [±90FG/0/90/90/0/±90FG] composite arrangement has better mechanical properties and Young's Modulus value compared to the other arrangements. This implies that the kenaf/fiberglass composite is best suit using that arrangement. Moreover these mechanical properties of kenaf/fiberglass composite could be improved by optimizing the structural parameter like thickness of the composite specimens.

echnical detail that it is necessary to include, but that interrupts the flow of the article, may be consigned to an appendix. Any appendices should be included at the end of the main text of the paper, after the acknowledgments section (if any) but before the reference list. If there are two or more appendices they should be called appendix A, appendix B, etc. Numbered equations should be in the form (A.1), (A.2), etc, figures should appear as figure A1, figure B1, etc and tables as table A1, table B1, etc.

Acknowledgments

The authors would like to express their gratitude to the Faculty of Mechanical Engineering (UiTM) for the facilities and technical support. The authors are also grateful the financial support from Research Acculturation Grant Scheme (Ref: 600-RMI/DANA 5/3/LESTARI (111/2015) and 600-RMI/RAGS 5/3 (18/2015)) for the successful implementation of this project. Finally, special thanks to the final year project student, Wan Muhammad Shafiq Bin Wan Yusof for his assistance in completing this study.

References

[1] William D. Callister, David G. Rethwisch, Materials Science and Engineering : An Introduction, 8th Edition

[2] Fabrics, Fibers, and Nonwoven – Engineered for Performance. Retrieved from http://www.dupont.com/product-and- services/fabrics-fiber

[3] Shinji Ochi (2010). Tensile properties of kenaf fiber bundle. Materials Science (Vol.2010). 1-6

[4] H.M. Akil, M.F. Omar, A.A.M. Mazuki, S. Safiee, Z.A.M. Ishak, A. Abu Bakar (n.d.) Kenaf fiber Reinforced composites

[5] M. Faizal. The Effect of Kenaf Fiber as a Cushioning Material in Packaging , 10-14. Retrieved from http://eprints.uthm.edu.my/4757/1/mohd_faizal_esa.pdf

[6] Hidekazu Ichikawa, Takashi Mikuniya, Hideaki Motohashi, Mutsuo Musha (1998). Loudspeaker diaphragm for use in electroacoustic transducer containing Kenaf fibers. Patents no. EP 0613321 B1

[7] Mohd Yuhazri, Phongsakorn, Haeryip Sihombing, Jeefferie A.R., Puvanasvaran Perumal, Kamarul, A.M., Kannan Rassiah. (2011). Mechanical Properties of Kenaf/polyester Composites (Vol.11). No. 01, 106-110

[8] M. A. Ghani, Z. Salleh, Koay Mei Hyie, M.N. Berhan, Y.M.D. Taib. M.A.I. Bakri (2012).

Mechanical Properties of Kenaf/Fiberglass Polyester Hybrid Composite. International Symposium
on Robotics and Intelligent Sensors 2012 (IRIS 2012). Procedia Engineering 41 (2012). 1654-1659.

[9] Chemical Industry Education Centre (2013). Composite: Phase in composite. Department Chemistry University of York.

[10] Srinivas Nunna, P. Ravi Chandra, S. Shrivastava and AK Jalan (2012). A review on Mechanical Behaviour of Natural Based Hybrid Composites, Reinf Plastics and Composites (Vol 31) 759-756.

[11] Armstrong, K. B., & Barrett, R. T. (1998). Care and repair of advanced composites. Society of Automotive Engineers.

[12] A. Shalwan, B.F. Yousif (2012) Mechanical and Tribological Behavior of Polymeric Composites Based on Natural Fibers. Materials and Design (Vol.45). 1893-1899

IOP Conf. Series: Materials Science and Engineering **328** (2018) 012019 doi:10.1088/1757-899X/328/1/012019

Reduction of Defects on Microstructure Aluminium Nitride Using High Temperature Annealing Heat Treatment

Z Tanasta[1], P Muhamad[1], N Kuwano[2], H M Y Norfazrina[3], M H Unuh[1]

[1]Intelligent Dynamics & System Research Lab, Malaysia-Japan International Institute of Technology, Universiti Teknologi Malaysia, 54100 Kuala Lumpur Malaysia [2]Nano-characterization, Structural Control and Processing Malaysia-Japan International Institute of Technology, Universiti Teknologi Malaysia, 54100 Kuala Lumpur Malaysia [3]Department of Mechanical Engineering, Kulliyah of Engineering, International Islamic University Malaysia, 50728 Kuala Lumpur, Malaysia

Email:zaimitanasta@gmail.com

Abstract. Aluminium Nitride (AlN) is a ceramic 111-nitride material that is used widely as components in functional devices. Besides good thermal conductivity, it also has a high band gap in emitting light which is 6 eV. AlN thin film is grown on the sapphire substrate (0001). However, lattice mismatch between both materials has caused defects to exist along the microstructure of AlN thin films. The defects have affected the properties of Aluminium Nitride. Annealing heat treatment has been proved by the previous researcher to be the best method to improve the microstructure of Aluminium Nitride thin films. Hence, this method is applied at four different temperatures for two hour. The changes of Aluminium Nitride microstructures before and after annealing is observed using Transmission Electron Microscope. It is observed that inversion domains start to occur at temperature of 1500 °C. Convergent Beam Electron Diffraction pattern simulation has confirmed the defects as inversion domain. Therefore, this paper is about to extract the matters occurred during the process of producing high quality Aluminium Nitride thin films and the ways to overcome this problem.

1. Introduction

AlN was discovered in 1862 by F. Briegler and A. Geuther. However, the potential of this material for functional devices component only been applied after 1980 [1]. Since then, the demand for this 111-nitride material gained high attention among the industries due to costing, easy to produce and easy to machine compared to other ceramic materials. Most recently, AlN has been used widely in electronic components such as Light Emitting Diode (LED), Laser Diode (LD) and photo detector for ultraviolet region [2]. Until today, more than 20 companies worldwide are developing AlN technology.

AlN is a 111-nitride ceramic material produced from the covalent bonding between Aluminium (Al) and Nitrogen (N). The arrangement of the atomic positions in AlN microstructure placed non-tetrahedral or non-centro symmetric positions. The positions called Wurtzite crystal structure. Due to that, the polarity exhibits the (0001) axis give AlN the piezoelectric properties [3]. AlN also good in emitting lights with shorter wavelength of 210 nm and high band gap up to 6 eV.

Other promising properties by this material are high thermal conductivity, good dielectric, low thermal expansion coefficient and non-reactive to normal semiconductor process chemical and gaseous [4]. Thermal conductivity of AlN is $140W/mK$ to 180 W/mK. However, previous researcher reported that the thermal conductivity of polycrystalline AlN at room temperature is in the range of $100\ W/mK$ to $260\ W/mK$ and for single crystal can reach until $300\ W/mK$ [4]. The different of thermal conductivity due to the density of oxygen impurities presence along the microstructure of AlN. In other words, thermal conductivity level of material is inversely proportional to the presence of defects in their microstructure [5].

The growth process of AlN thin films on sapphire substrates simultaneous to the existence of defects such as inversion domains and dislocations. This conditions becoming the major problems among the manufacturers. J. Jasinski et al identified the formed of Al polarity inversion domains on AlN layer growth on sapphire substrate [6]. The columnar inversion domain formed at boundaries that deviated $2\ \pm 0.5°$ from c-axis in V-like shaped. Figure 1 and Figure 1.1 shows the dark field TEM images of AlN layer recorded with a g-vector perpendicular and parallel to c-axis.

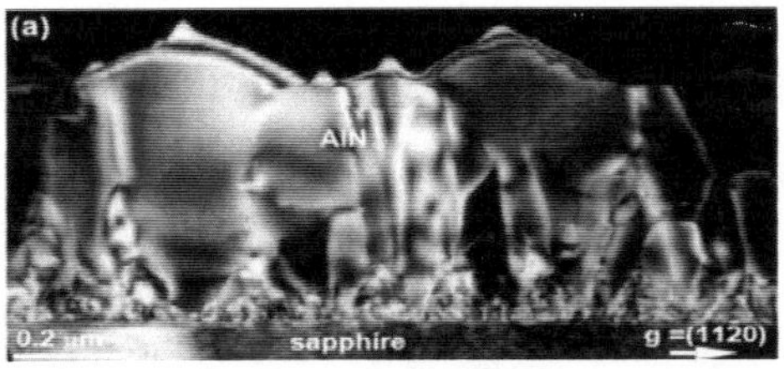

Figure 1. Dark-field TEM images of the AlN layer recorded with a g-vector perpendicular to c-axis [6]

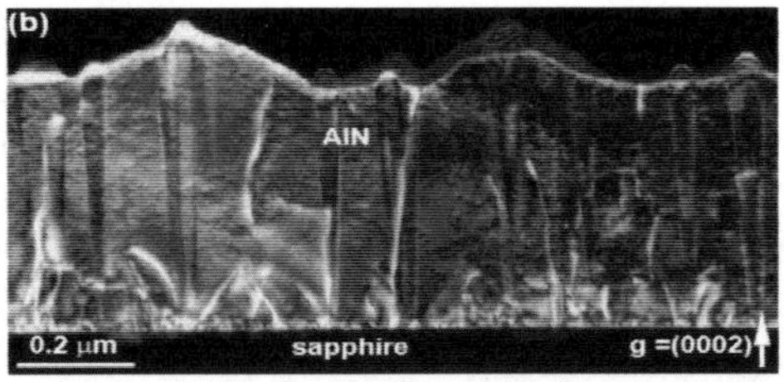

Figure 1.1 Dark-field TEM images of the AlN layer recorded with a g-vector parallel to c-axis [6]

From the images the boundaries of inversion domain along the AlN boundary shows the strong strain contrast. This is indicated that there is a displacement associated with these boundaries along the c-axis. Before this, the treatment using annealing already done to overcome this problem at the temperature less than 1500 °C. However, the defects exist in the AlN microstructure after annealing process still considers as high percentage. Therefore, this paper will be focused on annealing temperature of 1500 °C, 1550 °C, 1600 °C and 1650 °C. The temperature set is still below the melting point of AlN (T_{AlN} = 2200 °C) and this material is very stable at the high temperature [6].

2. Experimental Procedure

2.1 Growing process of AlN on the sapphire substrate

Before going further with annealing heat treatment process, AlN thin films must be growth on (0001) plane of sapphire substrate. On the (0001) plane of sapphire, the c-axis of AlN is aligned normal to the substrate surface, while on sapphire (0112), the c-axis of AlN is inclined at about 28 °C to the film normal. This situation called to be the different in lattice mismatch between both AlN thin films and sapphire that lead to the existence of defects on the microstructure of AlN thin films. The existence of defects on the AlN microstructure affected the effectiveness of this films when applied in the functional devices. Growing of AlN thin films on sapphire substrates done in the Metal Organic Vapor Phase Epitaxy Reactor (MOVPE) machine. Basically, this machine process based on the chemical deposition principle. Trimethylaluminium (TMAl) were supplied in the machine and after 10 seconds Ammonia (NH_3) introduced. During the reaction take places, Hydrogen gas (H_2) were flowed as a gas carrier under a low-pressure of 13.3 kPa. Along the process, the concentration of these substances controlled to get the desired alloy decomposition over the whole range with velocity $100\ ms^{-1}$. Lastly, AlN thin film with thickness 4.5 μm produced. However, this thickness is beyond the limitation to be observed under the Transmission Electron Microscope (TEM).

2.2 Focused Ion Beam (FIB)

To comply the TEM requirement the material must be thicken less than $200\ nm$. Focused Ion Beam (FIB) HITACHI ML-4000L was introduced for cross sectioning the AlN thin film. Firstly, AlN thin film on sapphire substrate is attached on the thicker glass by using a wax. Sample then cut using cutter into $2\ mm\ \times 4\ mm$ and $2\ mm\ \times 3\ mm$ respectively. Next, the samples coated with carbon for 20 times in 10 s until the coating thickness reach $20\ mm$ to $30\ mm$. Noted that the carbon used was already heated for $30\ s$ at $5\ V$. Before furthering the process, FIB were setup then the samples loaded using rod. After that, the process of sloping, etching and deposition on the samples taken place. Lastly, the samples picked up by using W needle. During this process, precaution must be concern to avoid needles clashes with the stages. Noted that the needle must be at their eucentric positions when picked up the samples.

2.3 Annealing heat treatment process

Annealing heat treatment is a process to heat the microstructure of AlN thin film. Previous research has proved that annealing heat treatment has ability in improving the microstructure of AlN thin film besides increasing the mechanical and electrical properties [9]. The process involved is to heated AlN thin films until reaches it critical temperature but lower than it melting point, $T_{AlN} = 2200$ °C. AlN thin films on sapphire substrates annealed t temperatures of 1500 °C, 1550 °C, 1600 °C and 1650 °C for two hours.

2.4 Transmission Electron Microscope (TEM)

The final stage for this research is observing and analyzing the conditions on the microstructure of AlN thin films after applying the annealing heat treatment. As mentioned above, the thickness of AlN thin films must be less than $200\ nm$ in order to get the better imaging quality. So, an electron beam can easily passes through the cross sections of AlN thin film. The thickness of specimens set to be $50\ nm$ and $100\ nm$. An electron beam shoot with the high velocity at the accelerating voltage of $300\ kV$.

3. Results and Discussion

3.1 $T_{annealing} = 1500$ °C

Figure 2 is bright field image TEM results of AlN thin film microstructure after facing the annealing heat treatment at temperature of 1500 °C. The thickness of specimen is $50\ nm$. From the observation, cone shaped like of inversion domain was formed along the $[\bar{1}2\bar{1}0]$ zone axis. The line of defect also can be seen clearly in the image. At this stage the density of inversion domain boundaries is high due to the lattice mismatch between AlN thin film and sapphire substrate (0001) during the growth process.

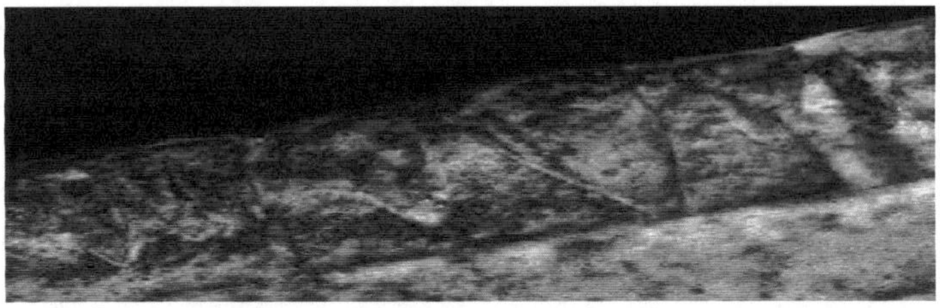

Figure 2. Bright field image TEM of AlN thin film microstructure after annealed at temperature of 1500 °C.

Other than inversion domain, dislocation defect type also found along the boundary of AlN microstructure. Convergent Beam Electron Diffraction (CBED) pattern has confirmed the defects through the inverted polarity region between region A and region B as shown in Figure 2.1. The CBED pattern analyzed the AlN microstructure along $[2\bar{1}\bar{1}0]$ zone axis. Both region also shows different polarity. Region A shows more on Al polarity while region B is N polarity. The quality of AlN thin film at this condition is very poor.

This is due to the low thermal conductivity that affected by the inversion domain and not suitable for functional devices.

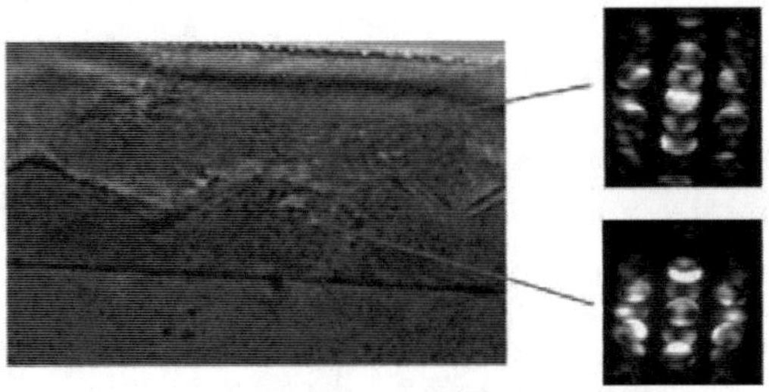

Figure 2.1. CBED pattern of AlN microstructure along $[2\bar{1}\bar{1}0]$ zone axis.

3.2 $T_{annealing} = 1550$ °C

Figure 3 is bright field image TEM of AlN thin films microstructure after annealed at 1550 °C. The specimen thickness is 100 nm. The formation of inversion domain on the microstructure of AlN thin film shows some changed compared to the annealed at 1500 °C. The cone shaped of inversion domain is reduced into the wavy shaped. Other than that, the surface boundary also seen to be sharper. The formation of wavy shaped boundary due to the incline in X and Y axis. The sharp boundary is because it is parallel with the incident beam. At this stage the density of inversion domain boundary still consider as high percentage for efficient function. Besides that, the polarity mechanisms from CBED pattern analysis shows only small changes between both region A and region B.

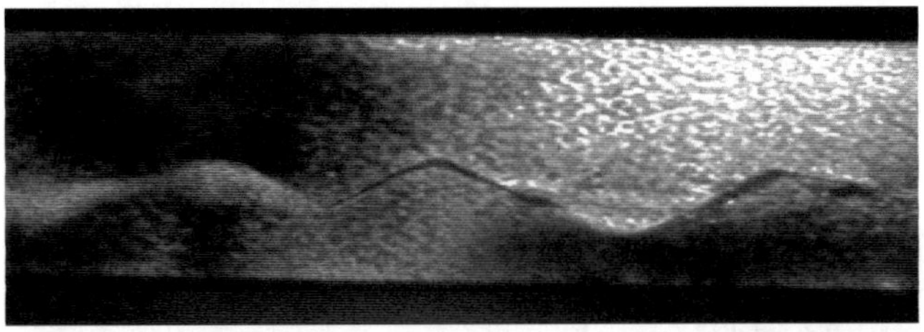

Figure 3. Bright field image TEM of AlN thin film microstructure after annealed at temperature of 1550 °C.

3.3 $T_{annealing} = 1600$ °C

Figure 4 is the bright field image TEM of AlN thin film microstructure after annealed at temperature of 1600 °C. The thickness set for this specimen is 100 nm. At this level of temperature, the microstructure of AlN become flatter and smoother compared to before. The length of wavy shaped is increased compare to before and the sharp boundary also less visible. That condition shows the decreasing for density of inversion domain boundaries along the AlN microstructure.

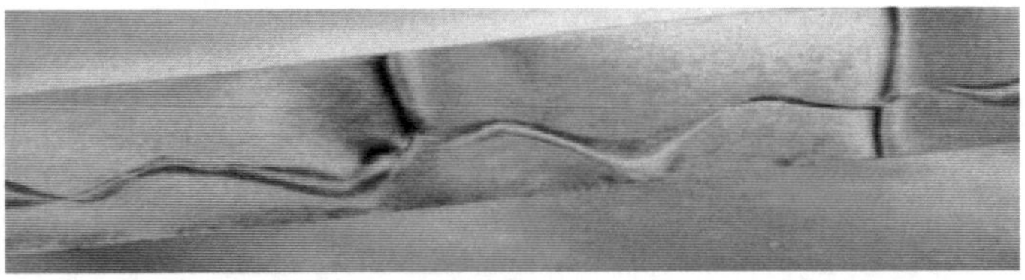

Figure 4. Bright field image TEM of AlN thin film microstructure after annealed at temperature of 1600 °C.

3.4 $T_{annealing} = 1650$ °C

Lastly, AlN thin film annealed at the temperature of 1650 °C. From the observation of TEM image as shown in Figure 5, the surface of AlN thin film microstructure become fine and the grain boundary seem to be smoother. At this stage the density of inversion domain boundary is stable for functional devices.

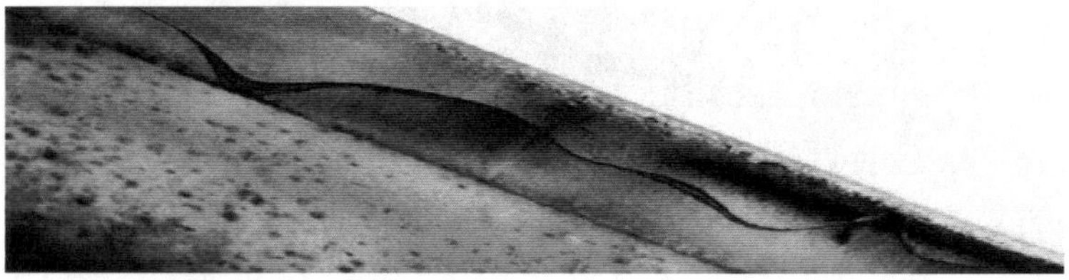

Figure 5. Bright field image TEM of AlN thin film microstructure after annealed at temperature of 1650 °C.

4. Conclusion

The existence of defects inside the AlN microstructure was the major effect in producing the high quality of thin films. The poor thermal conductivity of AlN thin film is due to the high density of inversion domain boundaries found along it microstructure. The defects started to occurred during growth process of AlN thin films on the sapphire substrate (0001) due to the lattice mismatch. Annealing heat treatment at different temperature (1500 °C, 1550 °C, 1600 °C *and* 1650 °C) can reduced the defects inside AlN microstructure. The defect mechanism formation has shown through the TEM imaging techniques. At temperature of 1500 °C, inversion domain can be seen clearly and become less visible until 1650 °C. The inverted polarity from CBED pattern between Al and N along the AlN microstructure has confirmed the defect type is inversion domains. Therefore, annealing heat treatment is effective in improving the AlN microstructure. Hence, this method contributed in developing the high quality of AlN thin films then producing the high efficiency functional devices for customer satisfaction.

References

[1]. Ying Da Yu, Aase Marie Hundere, Ragnvald Hoier, Rafal E. Dunin-Borkowski, Mari-Ann Einarsrud. 2002. Microstructural Characterization and Microstructural Effects on the Thermal Conductivity of AlN(Y2O3). *Journal of the European Society*. **22(2)**:247-252.

[2]. K. M. Taylor and Camille Lenie. 1960. Some Properties of Aluminium Nitride. *Journal of the Electrochemical Society*. **107**:641-647.

[3]. Masataka Imura, Ujjal Gautam, Kiyomi Nakajima, Yasuo Koide, Hiroshi Amano and Kenji Tsuda. 2013. Analysis of Broken Symmetry in Convergent-Beam Electron Diffraction along $\langle 11\bar{2}0\rangle$ and $\langle 1\bar{1}00\rangle$ Zone-Axes of AlN for Polarity Determination. *Japanese Journal of Applied Physics*. **52(8S)**:08JE15.

[4]. Y. Kurokawa, K. Utsumi and H. Takamizawa. 1988. Development and Microstructural Characteristics of High-Thermal-Conducting AlN Ceramics. *Journal of American Ceramic Society*. **71**:588-594.

IOP Conf. Series: Materials Science and Engineering **328** (2018) 012019 doi:10.1088/1757-899X/328/1/012019

[5]. K. Watari, K. Ishizaki and T. Fujikawa. 1992. Thermal Conduction Mechanism of AlN Ceramics. *Journal of Material Science*. **27**:2627-2630.

[6]. J. Jasinski, Z. Liliental-Weber, Q. S. Paduano and D. W. Weyburne. 2013. Inversion Domains in AlN grown on (0001) Sapphire. *Applied Physics Letters*. **83**:2811.

[7]. Z. W. Shan, Raja. K. Mishra, S. A. Syed Asif, Oden L. Warren and Andrew M. Minor. 2007. Mechanical Annealing and Source Limited Deformation in Submicrometer Diameter Nickel Crystal. *Nature Materials*. **7**:115-119.

[8]. Masayoshi Adachi, Kenji Tsuda, Masashi Sugiyama, Junji Lida, Akikazu Tanaka and Hiroyuki Fukuyama. 2013. High-Quality AlN Layer Homoepitaxially Grown on Nitrided a-Plane Sapphire Using a Ga-Al Flux. *Applied Physics Express*. **6**:091001.

[9]. Fude Liu, Ramon Collazo, Seiji Mital, Zlatko Sitar and Gerd Dusher. 2007. The Mechanism for Polarity Inversion of GaN via a Thin AlN Layer. *Applied Physics Letter*. **92**:203115.

[10]. Chihiro Iwamoto, Xu-Qiang Shen, Hajime Okumura, Hirofumi Matsuhata and Yuuchi Ikuhara. 2002. Structure Analysis of GaN Thin Film with Inversion Domains by High Voltage Atomic Resolution Microscopy. *Material Transaction*. **43**:1542-1546.

[11]. Fude Liu, Ramon Collazo, Seiji Mita, Zlatko Sitar, Stephen J. Pennycook and Gerd Duscher. 2008. Direct Observation of Inversion Domain Boundaries of GaN on c-Sapphire at Sub-Angstrom Resolution. *Advanced Materials*. **20**:2162-2165.

IOP Conf. Series: Materials Science and Engineering **328** (2018) 012020 doi:10.1088/1757-899X/328/1/012020

Parametric study of guided ultrasonic wave propagation in carbon-fiber composite plates

N A Ibrahim[1], M A Kamarudin[1], M H F M Jurimi[1] and B I S Murat[1]
[1]Faculty of Mechanical Engineering, Universiti Teknologi MARA (UiTM), 40450 Shah Alam, Selangor, Malaysia

E-mail: faten.asyikin@gmail.com

Abstract. The aim of this work is to study the guided ultrasonic wave (GUW) behaviour in composite plates using 3D Finite Element Analysis (FEA). Two types of composite models are chosen: plates with and without damage. The damage is modelled as a circular-shaped delamination inside the plate, representing one kind of low-velocity impact damage. Parameters such as excitation frequency, monitoring directivity, plate thickness, delamination size and shape were used to investigate the influence of these parameters on the GUW propagation and scattering behaviour. The models were constructed and coded in Matlab platform, while the simulations were performed in ABAQUS Explicit. From the results, the received signals have shown a strong dependency on the parameters. Significant scattering from the models with delamination were also observed, which indicates the possibility of using GUW for rapid non-destructive monitoring of composite panels and structures.

1. Introduction

The use of composite materials has grown rapidly from the occasional application for a non-structural part to the construction of complete airframes. Previous research has shown that composite materials became favourable in various form because of its toughness, stiffness and high strength to weight ratio [1]. In general, composite laminates consist of layers of polymer matrix reinforced with high strength fibers. The combination of these two materials produces a material with characteristics different from the individual constituents. A composite laminate is typically made up of several plies with different fiber orientation. This laminate is weak when subjected to impact loading. The impact loading on a composite laminate can lead to barely visible impact damage that could potentially result in a catastrophic failure if exposed to repetitive loading [2]. Cracks in the polymer matrix, debonding between fibers and separation of laminae (called delamination) are the common damage modes observed in composite structures. Among those damage modes, attention is focused on the delamination, as it reduces the load-carrying capacity of a structure which might lead to a catastrophic failure.

In order to maintain the quality and reliability of a composite structure, non-destructive testing (NDT) is usually used. Visual inspection, ultrasonic testing, acoustic emission, X-ray radiography, liquid penetrant and eddy-current are amongst the NDE methods employed in aerospace inspection [3]. In the present trend for aircraft maintenance, the maintenance of aircraft must be accomplished within a scheduled time. However, for the large aircraft structures, most methods are very time consuming, costly and they normally interrupt the service of the aircraft. This indicates a need for rapid inspection

IOP Conf. Series: Materials Science and Engineering **328** (2018) 012020 doi:10.1088/1757-899X/328/1/012020

and cost-effective methods for monitoring large composite structures. One possible method, the guided ultrasonic waves (GUW), is chosen to be further explored in this study.

Using low excitation frequency, GUW can propagate over long distance with limited energy loss [4]. From a single location, the guided wave can cover large areas, which can help to reduce the inspection time. The reflection of the propagating wave at defects enables rapid detection of defects in large structures [5]. This method has been used successfully for the detection of defects in large metal plates and long pipes, i.e., corrosion and crack detection. However, the behaviour of the guided waves is somewhat more complicated in composite structures due to the physical properties of composites that are generally inhomogeneous and anisotropic in nature [6]. For waves propagating in composite laminates, the wave interaction depends on many factors such as the excitation frequency, geometry of the structure, material properties, direction of propagation, and inter laminar conditions [7], [8]. Knowledge of the group velocity and its direction are needed when considering the propagation path of a wave packet for calculating the arrival time, defect location and area coverage of the inspection.

In general, when a guided wave is incident on a structural discontinuity, it is scattered in all directions. Scattering characteristics such as arrival time, amplitude, frequency content, attenuation, reflected and transmitted waves are normally used to explain the wave propagation in the defective area [9]–[12]. Strong interaction between the incident wave and defect is important for sizing and localizing the defect [13]. However, the interactions are strongly dependent on the mode of the incident wave and the size of the damage [10], [13], [14]. The most important key in using guided waves successfully is to understand the relationship between the characteristic of a defect and the scattered waves.

Building from the literature, it can be understood that the capability of the guided waves for the NDE of composite structures is still under investigation. Therefore, the objective of this research is to investigate the use of GUW for detecting typical impact damage, such as delamination that can occur on composite plates. This study aims to achieve a better understanding of the guided wave propagation in composite plates and its interaction with impact damage.

2. 3D Finite Element Modelling

A large cross-ply composite plate was constructed in MATLAB as detailed in [15] which was then simulated in ABAQUS/Explicit software. The plate consists of 8 layers through the thickness with the same 0°/90° layup direction. The material properties of each layer were modelled with properties of unidirectional composite plate as shown in table 1. The size of the model was set to 500 mm x 500 mm x thickness (t), where t is varied with 2 mm, 4 mm and 6 mm. Damaged plates were modelled with circular shape delamination at the center of the plates. An out-of-plane excitation was set to generate an A_0 Lamb wave mode propagating along the plate. The excitation frequency was chosen to be 100 kHz, where there was a good trade-off between lower wave attenuation and good transducer sensitivity. The excitation point was located at location 150 mm in x-direction, 250 mm in y-direction and half-thickness direction. The out-of-plane displacement was monitored using a line scan across the plates and a circular scan with 30 mm radius around the center of the plates. Parametric inputs and damage definition were controlled by using MATLAB which then saved as the input file for simulations in ABAQUS/Explicit. Figure 1 shows the illustration of the model with excitation point (EP) and circular crack.

IOP Conf. Series: Materials Science and Engineering **328** (2018) 012020 doi:10.1088/1757-899X/328/1/012020

Table 1. Material properties of the unidirectional plate; 2 MHz characterization frequency [16].

Stiffness constants of Carbon epoxy [0/90] (GPa)	
C11	12.15 + i0.34
C12	8.39 + i0.88
C13	7.73 + i0.65
C22	70.87 + i9.19
C23	5.60 + i0.60
C33	64.24 + i10.1
C44	4.70 + i0.28
C55	3.06 + i0.19
C66	2.97+ i0.20

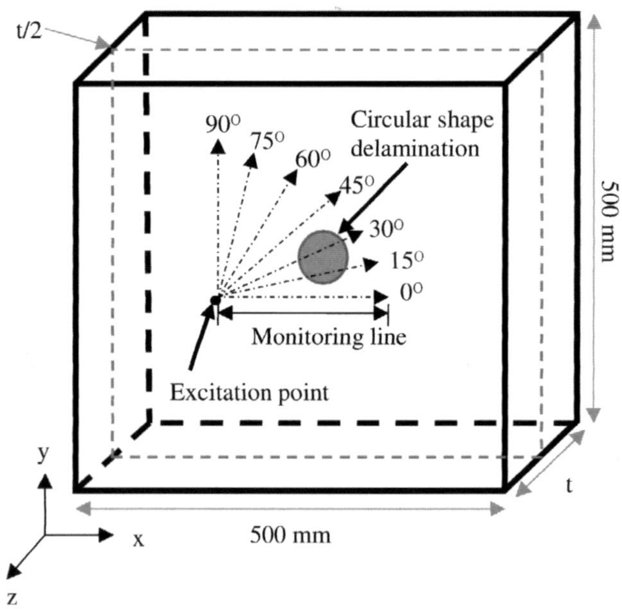

Figure 1. Illustration of the carbon-fiber plate model (500 mm x 500 mm x t) with 200 mm monitoring points in different directions (0° to 90° with 15° interval); excitation point (150 mm, 200 mm, t/2); center of the circular-shaped damage located at the plate center.

3. Parametric study on undamaged carbon-fiber plates

3.1 Influence of excitation frequency

This part presents the simulation results for the A_0 mode propagation in undamaged laminate composite plates. Figure 2 shows the comparison between the received signals in order to investigate the influence of different excitation frequencies (50 kHz, 100 kHz and 150 kHz). Repeatable amplitude profiles can be observed, especially from the line scans in the 0° direction, although the amplitude values showed some great discrepancies. A similar observation can also be seen from the circular scan, where the 150 kHz excitation frequencies produced the lowest signal amplitudes while showing a very minimal steering effect. Meanwhile, a higher dispersion can be seen from the 50 kHz result (figure 2b). Since different frequency components in a wave packet travel at different speeds, the shape of the wave packet is expected to have small distortion while traveling due to the differences in the arrival times of each frequency component. Using the 100 kHz frequency of the excited wave packet, the signal amplitudes are very strong and have less steering effect, which means the wave packet can retain its shape as it travels. This feature is desirable for the NDE of composites, and supports the decision to choose 100 kHz as the nominal excitation frequency for later simulations.

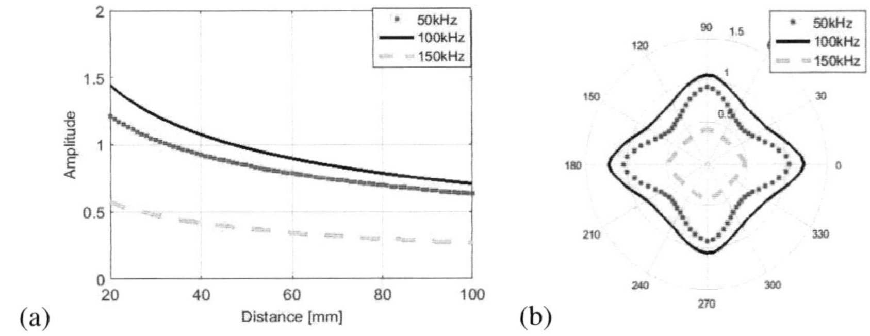

Figure 2. Maximum amplitude of signal envelope in 2 mm cross-ply plate (a) Line scan measured every 1mm step along 0° direction, (b) Polar plot of amplitude at 30 mm radius around the excitation point; different excitation frequencies.

3.2 *Influence of plate thickness*

Figure 3 shows the amplitude behaviour in composite plates with different thicknesses (2 mm, 4 mm and 6 mm). The GUW behaviours show that the amplitude decreased when the plate thickness was increased. The reason for this behaviour could relate to the increase of energy absorption. Study conducted by [16] showed thicker material leads to increase of energy absorption. This result may be explained by the fact that when energy absorption is high, the damping will increase and the wave intensity becomes low.

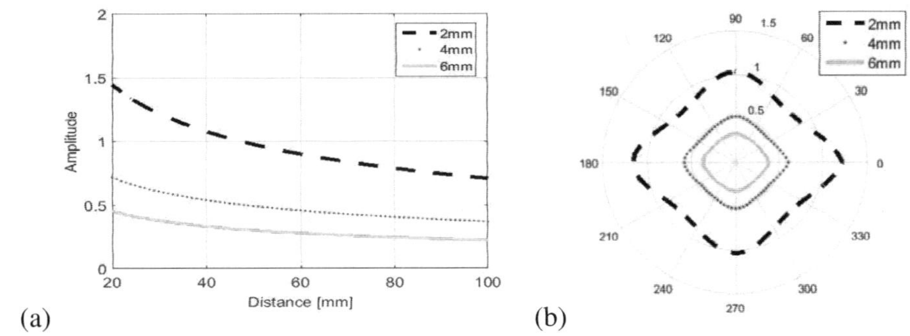

Figure 3. Maximum amplitude of signal envelope in composite plates for (a) Line scan measured every 1 mm step along line of 0° of monitoring point direction; b) Polar plot of amplitude at 30 mm radius around excitation point for different thicknesses; 100 kHz frequency.

3.3 *Influence of angular directivity*

Results for wave propagation at different angular directivities (0°,15°, 30°, 45°, 60°, 75°, and 90°) in composite plate is as shown in figure 4. Amplitude profiles of 0° and 90° directions show higher amplitude than others. Both directions are the same direction with layup orientation in the composite plate. This situation could be related to the steering effect [6]. Steering effect is a phenomenon caused by the anisotropy of the material resulting in different directions of phase and energy velocity vectors. Nonetheless, 90° direction has slightly less value than 0° direction. This discrepancy may be causes by outer layer composite plate that has material orientation at 0° direction which is related to lower bending stiffness [16]. Angular directions of 30°, 45° and 60° show the lowest amplitude profile. This behaviour signifies that these directions slowing down the wave propagation. The amplitude for 15° and 75° directions are in between higher and lower amplitude because they are the closest direction to 0° and 90°.

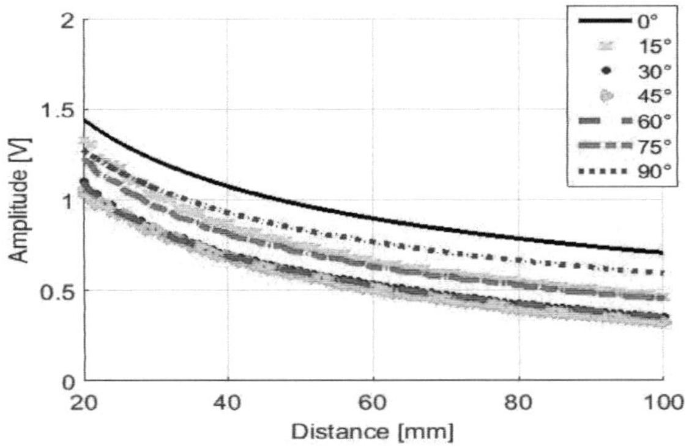

Figure 4. Maximum amplitude of signal envelope in cross-ply composite plates models; line scan measured every 1 mm step along line for different angle of monitoring point directivity; 2 mm thick; 100 kHz frequency.

4. Parametric study on damaged carbon fiber plates

4.1 Effects of angular dependency on wave scattering

Figure 5 shows the simulated signals on 2 mm thick composite plates with 10 mm radius circular-shaped damage, monitored in different directions (from 0° to 90° with 15° interval). The location of excitation point and damage are shown in figure 1. From figure 5, it can be seen that there are disturbances in the amplitude profile monitored in directions of 15° and 30°. A very minimal disturbance can also be observed from the monitoring line in 0° direction. Meanwhile the rest monitoring lines remain unchanged. About a similar reflection has been seen from a previous study [6], but on a composite plate with a rectangular-shaped damage. A possible explanation for this behaviour is the damage detected in between 0°, 15° and 30° directions of monitoring line. Throughout these findings, only monitoring line across the damage could give the accurate location of the damage.

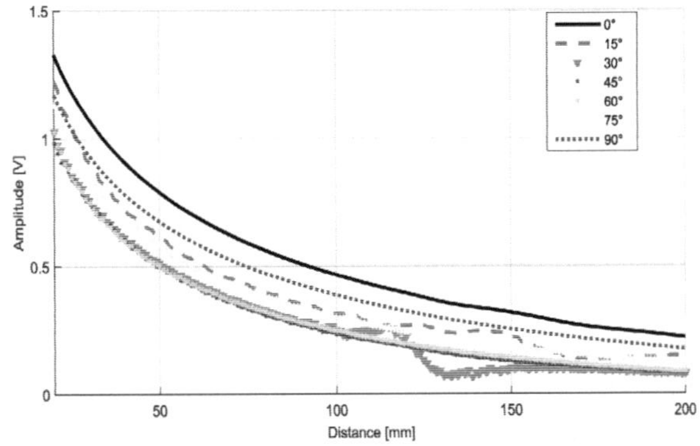

Figure 5. Amplitude profiles in layered composite plate with a circular-shaped delamination; 200 mm line scan measured every 1 mm step along in different direction; circular damaged (radius: 10 mm) at 1 mm depth from the surface; 100 kHz excitation frequency.

4.2 Effects of damage size and depth on wave scattering

Figure 6(a) shows the amplitude profiles of GUW on composite plates with different damage sizes (10 mm, 20 mm, 30 mm, 40 mm, 50 mm and 60 mm). The amplitude profiles shows inconsistency for different delamination sizes. Amplitudes over the intact part of the model (before the delamination) were exactly the same as the baseline data. A small fluctuation in the amplitude profile for all case studies, about 7% compared to the baseline data, can be seen within 10 mm to 20 mm distance before the delamination entrance. This is due to the reflected waves were interfering with the incident waves. Significant changes in amplitudes levels can be observed in the defective area, for all case studies. This could be due to the newly generated modes that travel with different speeds, as well as interferences between each group of new modes that are being reflected within the delamination sideways [15].

Behind the delamination, the amplitudes for all case studies started to show different patterns and much bigger discrepancies compared to the baseline data. This finding may support the previous observations on the analysis of the large defect wave fields, where a part of several groups of waves with dissimilar wave speeds are reflected at the delamination exit and transmitted out from the delamination. This event created different amplitude profiles compared to the baseline data even though the waves now travelled in the area behind the delamination. Figure 6(b) shows the scattered field for the different damage sizes effect. Some small variations were observed in the forward (0° direction) scattered amplitude. The results for each different damage sizes show different amplitude propagation. This unexpected behaviour reveals that different damage sizes propagate at different directions for circular scans.

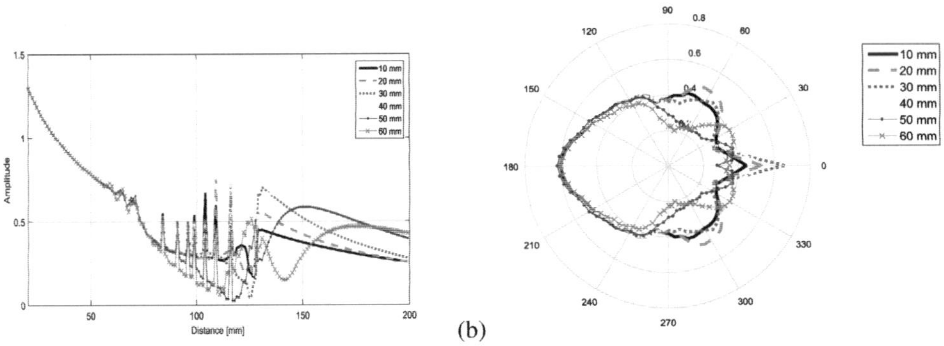

(a) (b)

Figure 6. Comparison between different size of damage (diameter: 10 mm, 20 mm, 30 mm, 40 mm, 50 mm and 60 mm) at 1 mm depth: (a) Amplitude of signal envelope over defective area of composite plate, (b) Scattered amplitude around various size of damage; 100 kHz excitation frequency; 2 mm thick.

Figure 7 shows the amplitude profiles for effect of different damage depth (0.25 mm, 0.50 mm, 0.75 mm and 1 mm) for each damage diameter size (10 mm, 20 mm, 30 mm, 40 mm, 50 mm and 60 mm). The amplitude pattern becomes higher as the depth going towards 1 mm (center of plate) where it was the closest distance with the excitation point. The closer the damage to the excitation point, the stronger the GUW reflection towards the damage.

The scattered waves around delamination were investigated and the results are as shown in figure 8. The left side of the graph shows before GUW entered the damage and the other side was after damage. The 1 and 0.75 mm depth amplitudes were about the same where they were consistently high. This could be due to the nearest damage distance with the excitation point. Meanwhile, 0.5 and 0.25 mm amplitudes did not show a clear pattern. Surprisingly, 0.50 mm damage depth shows the lowest amplitude, not at 0.25 mm depth. This behaviour could be related to wave velocity in different layup. Layup in 0.50 mm depth has fiber arranged in 90° direction. Thus, the velocity reduced as the monitoring line was in 0° direction. It is supported by the data in table 2 which shows the effect of additional 90°layup reduces the velocity and lower the amplitude.

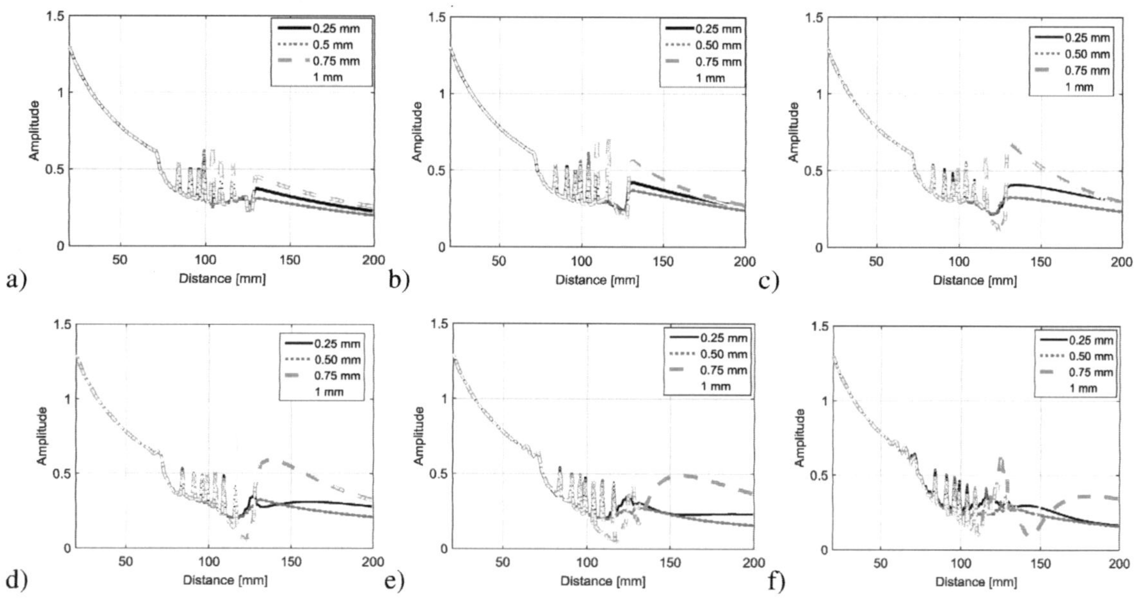

Figure 7. Amplitude of signal envelope across various damage circular diameter: a) 10 mm, b) 20 mm, c) 30 mm, d) 40 mm, e) 50 mm and f) 60 mm; at different depths (0.25 mm, 0.5 mm, 0.75 mm and 1 mm); 2 mm thick; 100 kHz excitation frequency.

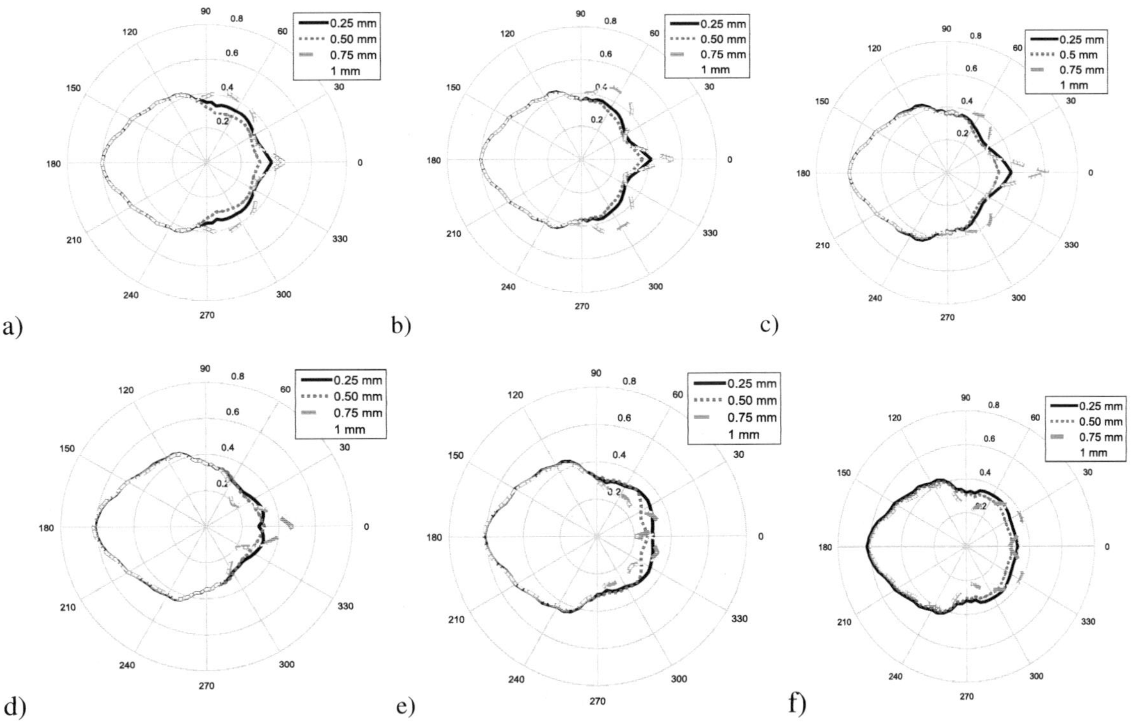

Figure 8. Scattered amplitudes around various damage diameter: a) 10 mm, b) 20 mm, c) 30 mm, d) 40 mm, e) 50 mm and f) 60 mm; at different depths; measured every 5° at 30 mm radius around center of the damage; 2 mm thick; 100 kHz frequency.

Table 2. Wave dispersion attenuation and angular dependency data comparison between FEM and DISPERSE for different layup orientations [16].

Layup orientation (plate's thickness)	Group velocity, C_g (m/s)		Phase velocity, C_p (m/s)		Attenuation, (dB/mm)
	FEM	DISPERSE	FEM	DISPERSE	
0/90 (0.5mm)	1062	1064	1042	600	0.3567
0/90/0/90 (1mm)	1246	1255	1235	866	0.1604
0/90/0/90/90/0 (1.5mm)	1366	1385	1333	1074	0.1180
0/90/0/90/90/0/90/0 (2mm)	1372	1388	1351	1134	0.1098

5. Conclusion

This paper studied the behaviour of the wave propagations towards damaged and undamaged carbon fiber plates. Based on the undamaged composite plate study, 100 kHz frequency was found as the most effective frequency where it has strong signal amplitude and less steering effect. Meanwhile, a thicker material increased the damping, which resulting in a slower wave intensity. For damaged plate study, the analysis found that the additional of 90° layup in composite plates decrease the velocity and the amplitude of the GUW.

IOP Conf. Series: Materials Science and Engineering **328** (2018) 012020 doi:10.1088/1757-899X/328/1/012020

References

[1] González C, Vilatela J J, Molina-Aldareguía J M, Lopes C S, and LLorca J, "Structural composites for multifunctional applications: current challenges and future trends," *Prog. Mater. Sci.* **89** 194–251 2017
[2] Murat B I S and Rahman A A A, "Study of impact damage behavior in woven carbon fiber plates," *Procedia Eng.* **170** 47–54 2017
[3] Diligent O, "Interaction between fundamental lamb modes and defects in plates," *Mech. Eng.*, 1–148 2003
[4] Rose J L, "Ultrasonic guided waves in structural health monitoring," *Key Eng. Mater.*, **270–273** 14–21 2004
[5] Alleyne D N, Pavlakovic B, Lowe M J S, and Cawley P, "Rapid, long range inspection of chemical plant pipework using guided waves," *Key Eng. Mater.* **270–273** 434–441 2004
[6] Putkis O, Dalton R P, and Croxford A J, "The anisotropic propagation of ultrasonic guided waves in composite materials and implications for practical applications," *Ultrasonics* **65** 390–399 2016
[7] Murat B I S, Khalili P, and Fromme P, "Impact damage detection in composite panels using guided ultrasonic waves," *AIP Conf. Proc.* **9064** 906419 2014
[8] Castaings M and Hosten B, "Guided waves propagating in sandwich structures made of anisotropic, viscoelastic, composite materials," *J. Acoust. Soc. Am.* **113** 2622–2634 2003
[9] Qing-Tian D and Zhi-Chun Y, "Propagation of guided waves in bonded composite structures with tapered adhesive layer," *Appl. Math. Model.* **35** 5369–5381 2011
[10] Ng C T and Veidt M, "Scattering analysis of fundamental anti-symmetric lamb wave at delaminations in composite laminates," *Aust. J. Mech. Eng.* **8** 197–206 2011
[11] Sohn H *et al.*, "Delamination detection in composites through guided wave field image processing," *Compos. Sci. Technol.* **71** 1250–1256 2011
[12] Castaings M and Lowe M, "Finite element model for waves guided along solid systems of arbitrary section coupled to infinite solid media," *J. Acoust. Soc. Am.* **123** 696–708 2008
[13] Chakrapani S K and Dayal V, "The interaction of Rayleigh waves with delaminations in composite laminates," *J. Acoust. Soc. Am.* **135** 2646–2653 2014
[14] Liu Z, Yu H, He C, and Wu B, "Delamination damage detection of laminated composite beams using air-coupled ultrasonic transducers," *Sci. China Physics, Mech. Astron.* **56** 1269–1279 2013
[15] Murat B I S, Khalili P, and Fromme P, "Scattering of guided waves at delaminations in composite plates," *J. Acoust. Soc. Am.* **139** 3044–3052 2016
[16] Murat B I S and Fromme P, "Propagation and scattering of guided waves in composite plates with defects," *Int. J. Automot. Mech. Eng.* **13** 3728–3741 2016

IOP Conf. Series: Materials Science and Engineering **328** (2018) 012021 doi:10.1088/1757-899X/328/1/012021

Investigating the effects of rock porosity and permeability on the performance of nitrogen injection into a southern Iranian oil reservoirs through neural network

M S Gheshmi[1], S M Fatahiyan[2], N T Khanesary[2], C W Sia[1], M S Momeni[1]

[1] University technology PETRONAS, Department of Petroleum Engineering
[2] Petroleum University Technology of Ahwaz, Department of Gas engineering

Abstract. In this work, a comprehensive model for Nitrogen injection into an oil reservoir (southern Iranian oil fields) was developed and used to investigate the effects of rock porosity and permeability on the oil production rate and the reservoir pressure decline. The model was simulated and developed by using ECLIPSE300 software, which involved two scenarios as porosity change and permeability changes in the horizontal direction. We found that the maximum pressure loss occurs at a porosity value of 0.07, which later on, goes to pressure buildup due to reservoir saturation with the gas. Also we found that minimum pressure loss is encountered at porosity 0.46. Increases in both pressure and permeability in the horizontal direction result in corresponding increase in the production rate, and the pressure drop speeds up at the beginning of production as it increases. However, afterwards, this pressure drop results in an increase in pressure because of reservoir saturation. Besides, we determined the regression values, R, for the correlation between pressure and total production, as well as for the correlation between permeability and the total production, using neural network discipline.

1. Introduction

It is generally an accepted fact that at the present time, many of the major oilfields in Iran suffer from the incapable and undesirable conditions as they are in the second half of their life cycle process, and consequently, many of the oil reservoirs faced a shortage of gas injections for pressure maintenance. On the other hand, the need to continuously increase oil production requires high reservoir pressure which is to be maintained by gas or water injection. This discipline not only enhances the reservoir pressure to be adjusted at some needed levels, it also compromises the future production demands. Excluding gas injection as a consideration or any delay in implementing this unavoidable programme will lead to serious problems in meeting the future ever-increasing production rate needs. However, it is to be emphasized that only following-up gas injection at required volume may not lead to expected increase in recovery from the reservoirs. Thus, this implementation must be carried out in a more suitable period and situation; otherwise it could be unsuccessful to reach the desired results or demands [1-3].

In 2004, nitrogen injection into one of the gas caps of the southern Iranian fractured oil reservoir was simulated and it was claimed that matrix oil discharge by gas injection occurs effectively and rapidly when compared to the use of other miscible gases, which causes more oil contact that leads to an increase in sweeping efficiency [1]. In the present paper, we indicate that rock properties, specifically positive permeability change, gives high production rate by N2-injection into the selected oil reservoirs with the use of more accurate method of artificial neural network.

1.1. *Nitrogen behavior*
As an injectant under identical conditions, the isothermal compressibility coefficient,-(ϑV/ϑP)T, for nitrogen is greater than that of other commonly employed or promising gases as injection materials. For this reason, the injected nitrogen will occupy large volume of the reservoir and, therefore, relatively less amounts of nitrogen will be required for a specified injection[4]. Besides, nitrogen is non-flammable and less soluble in water. N2 injection is a more suitable option for deep reservoirs containing light oil [5]. Those reservoirs containing oils lighter than 34° API are good options for miscible injection of this gas where the minimum miscible pressure is usually high. For example, this pressure for a sample of light oil from North Sea under miscible situation with nitrogen has been reported as 4730 psi [5].

2. Modelling and Simulation

2.1. *Reservoir characteristics*
The present model involves 9600(32*25*12) blocks [6]. The initial reservoir pressure and the base depth are 5000 psi and 7055 feet, respectively, at a reservoir temperature of 150° F. The oil-water contact (OWC) and gas-oil contact (GOC) are 7130 and 7000 ft, respectively. The reservoir under this study includes two producing wells and one injection well. Figure 1 shows the start of the gas injection and the beginning of production from the reservoir at a pressure and rate of 5000 psi and 12000 Mscf (Thousand Standard Cubic Feet), respectively.

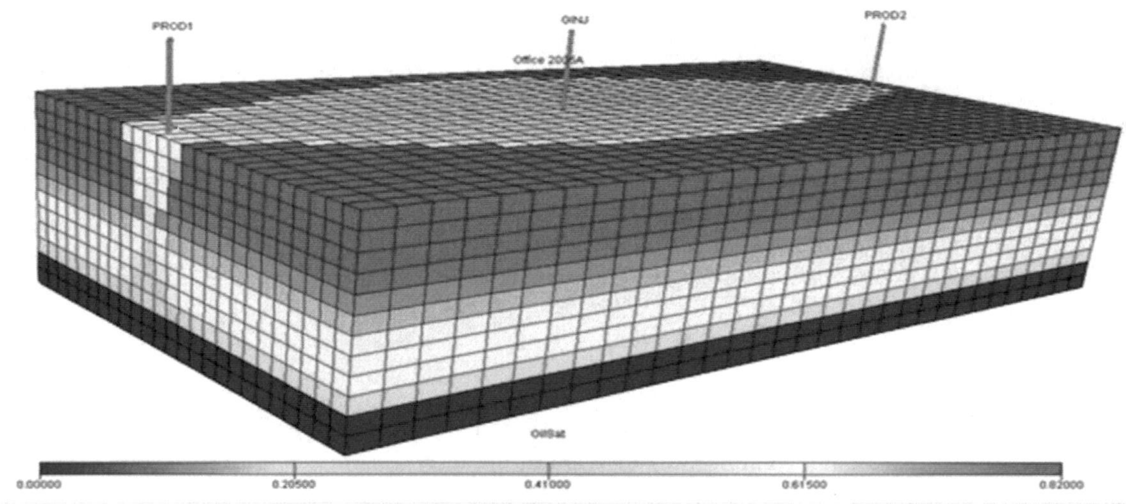

Figure 1. Reservoir Model

2.2. *Reservoir fluid properties and modeling*
The fluid undertaken in the present investigation is a 14-component hydrocarbon and non-hydrocarbon mixture as presented in the table 1. The Peng-Robinson equation of state (PR-EOS) with commonly used mixing rules have been used to evaluate the properties of the reservoir fluid [7]. The values of relative permeability and capillary pressure are given in table 2.

Table 1. Reservoir fluid composition

Components	N_2	H_2S	CO_2	C 1	C_2	C_3	I-C_4	N-C_4	I-C_5	N-C_5	C_6
ZI(percent)	0	0	0.24801	26.264	6.7921	5.9321	0.957	3.061	1.291	1.469	2.913

Table 2. Relative permeability and capillary pressure

S_w	K_{rw}	K_{row}	P_{cow}	S_{gas}	K_{rg}	K_{rog}	P_{cog}
0.18	0	0.9	50	0	0	0.9	0
0.2	0	0.808	43	0.05	0	0.555	0
0.32	0.001	0.412	22	0.1	0	0.371	0
0.42	0.082	0.26	11	0.12	0	0.212	0
0.5	0.165	0.068	508	0.18	0.002	0.104	0
0.59	0.249	0.015	2.36	0.25	0.005	0.04	0
0.68	0.38	0.002	1.412	0.3	0.013	0.011	0
0.73	0.482	0	1.09	0.33	0.036	0.001	0
0.82	0.82	0	0.8	0.36	0.06	0	0
1	1	0	0	0.8	0.9	0	0

3. Results

3.1. *Scenarios for reservoir simulation and results*

The scenarios undertaken in this study involve porosity change and permeability change in the X-direction as they are given in Tables 3 and 4, respectively. The variation of pressure (FPR) as well as total production rate with porosity and similarly with permeability at X-direction were determined. Figure 2 indicates the results obtained.

Table 3. Porosity Change

POROS1	0.07	**POROS21**	0.27
POROS2	0.08	**POROS22**	0.28
POROS3	0.09	**POROS23**	0.29
POROS4	0.1	**POROS24**	0.3
POROS5	0.11	**POROS25**	0.31
POROS6	0.12	**POROS26**	0.32
POROS7	0.13	**POROS27**	0.33
POROS8	0.14	**POROS28**	0.34
POROS9	0.15	**POROS29**	0.35
POROS10	0.16	**POROS30**	0.36
POROS11	0.17	**POROS31**	0.37
POROS12	0.18	**POROS32**	0.38
POROS13	0.19	**POROS33**	0.39
POROS14	0.2	**POROS34**	0.4
POROS15	0.21	**POROS35**	0.41
POROS16	0.22	**POROS36**	0.42
POROS17	0.23	**POROS37**	0.43
POROS18	0.24	**POROS38**	0.44
POROS19	0.25	**POROS39**	0.45
POROS20	0.26	**POROS40**	0.46

Table 4. Permeability Change in X-direction

K1	1	**K21**	360
K2	5	**K22**	380
K3	10	**K23**	400
K4	20	**K24**	420
K5	40	**K25**	440
K6	60	**K26**	460
K7	80	**K27**	480
K8	100	**K28**	500
K9	120	**K29**	520
K10	140	**K30**	540
K11	160	**K31**	560
K12	180	**K32**	580
K13	200	**K33**	600
K14	220	**K34**	650
K15	240	**K35**	700
K16	260	**K36**	750
K17	280	**K37**	800
K18	300	**K38**	850
K19	320	**K39**	900
K20	340	**K40**	950

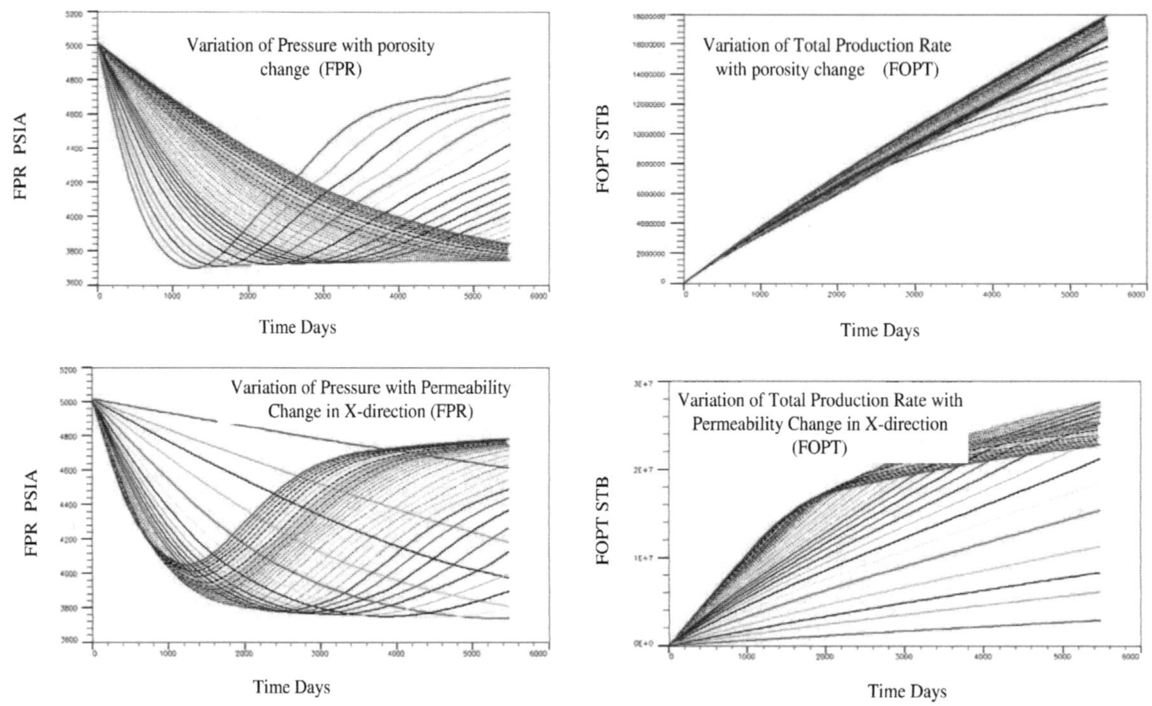

Figure 2. *Results of scenarios*

3.2. *Prediction through neural network*

Regression values, indicated by R2, for the above mentioned scenarios were also determined. These results are graphically presented in figures 3 to 6.

3.2.1. *Linear Regression for figure 2*

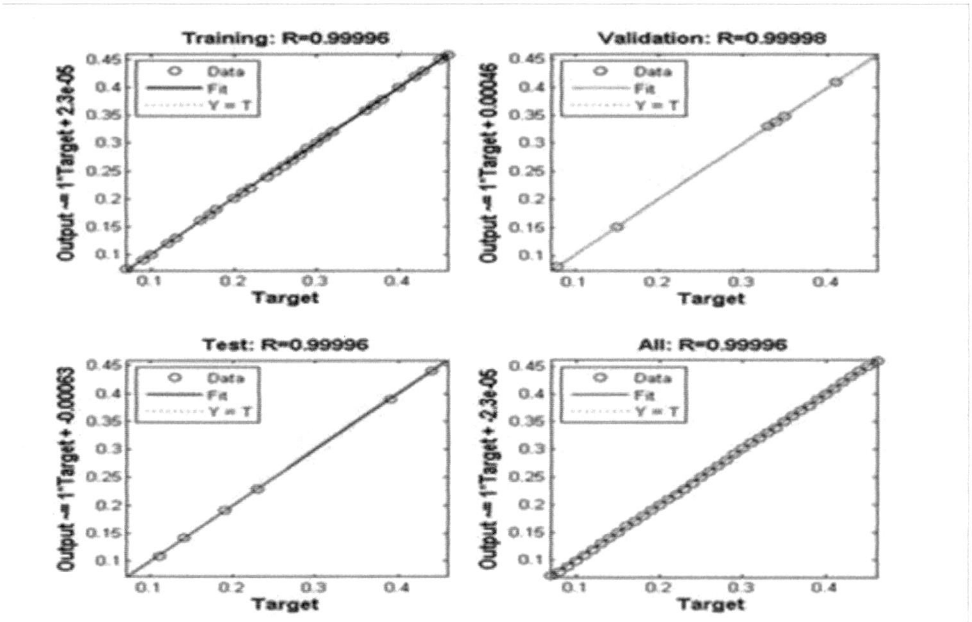

Figure 3. Linear Regression for Variation of Pressure with Porosity Change

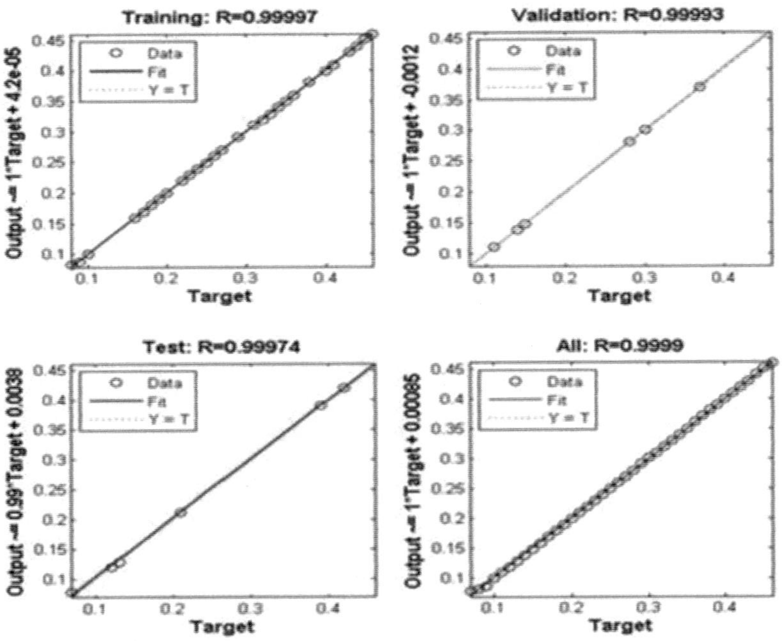

Figure 4. Linear Regression for Variation of Total Production Rate with Porosity Change

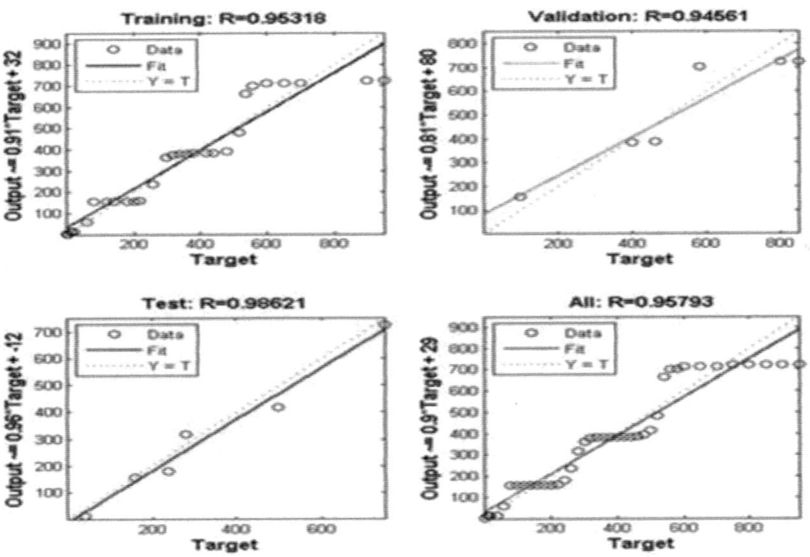

Figure 5. Linear Regression for Variation of Pressure with Permeability Change in X-direction

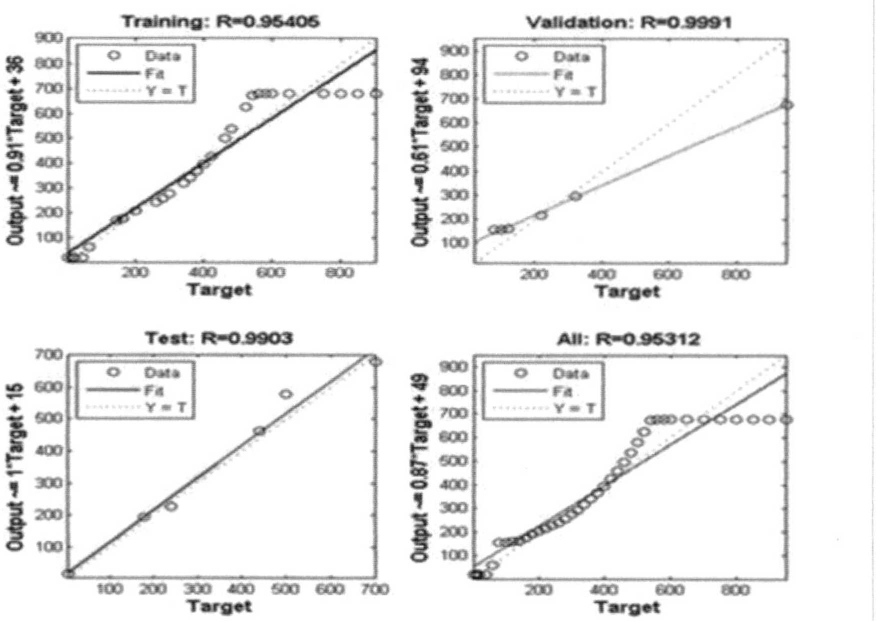

Figure 6. Linear Regression for Variation of Total Production Rate with Permeability Change in X-direction

4. Discussion

As it seems from the above figures indicating the pressure rate (FPR), most pressure loss occured at porosity 0.07. The pressure continued to fall for 1200 days, and when this period came to the end, there was no increase in pressure in the reservoir, because of occupation of pore spaces by fluids. Minimum pressure loss happened at a porosity of 0.46 which is considerably high. Also from the Figure (2) indicating the total oil production (FOPT), it is clearly observed that the most production and the least production are met at porosities 0.46 and 0.07, respectively. This obviously reflects the significant effect of high porosity on the increase in the total production rate. Certainly, at the lower porosity, the capillary pressure went up and the injected gas movement within the system was restricted. The most pressure loss situation is being indicated by the curves obtained at the permeability of 950 md (millidarcy), which continued for 1200 days. This is because of the reservoir containment by the gaseous Nitrogen. After this period, as the curve in the Figure (2) indicates, the pressure in the reservoir increased. This is because of the reservoir saturation by Nitrogen and its expansively. The least pressure loss is signified by the curve provided with a very low permeability of 1 md. In a similar manner, as it is seen from the curve presenting (FOPT), the highest production rate is met at a permeability of 950 md. So a high increase in production, from the beginning up to 2500 days, is reached continuously, and afterwards the pressure went up slowly. The least production becomes available at the permeability of 1 md. Through the predictions by neural networks for porosity changes, two profiles indicating changes in pressure rate (FPR) and total oil production (FOPT) have been produced. As it is observed from Figure 6, for changes in pressure rate within 20 hidden layers, R2 = 0.9999 is obtained. From Figure (4), for total oil production within 10 hidden layers, the value for R2 is 0.9995. Through additional predictions made by using neural networks, we produced another two profiles indicating changes in permeability in the X-direction and changes in pressure rate. These are presented in Figures 8 and 9. For the changes in pressure rate within 20 hidden layers and for total oil production within 12 hidden layers, the values of R2 were obtained as 0.9726 and 0.9807, respectively. These reliable data signify the accuracy of the results obtained which reflect, in turn, the extent and importance of parameters involved and the strong scenarios and method employed.

5. Conclusions

This investigation provides some guidelines as well as relevant limitations for oil production from an Iranian oil reservoir by N2 injection through variations occuring in rock properties. These are as follows:

(1) Increase in porosity results in pressure loss reduction and an increase in total production rate.

(2) The pressure loss raises up as the permeability in the X-direction increases. As the effect of injecting gas comes to action and reservoir get saturated by the gas, an increase in pressure as well as an increase in total production rate will be observed.

(3) For each couple of profiles for changes in pressure and total production which have been produced by neural network method, four values of regression R2 are over 0.9. This represents a good correlation between inputs and outputs and obviously indicates, in turn, the significant effects of variations in porosity and permeability on the pressure and total production.

Acknowledgements

The authors would like to thank National Iranian Oil Company-Southern Oil Fields Headquarter for providing data and facilities used in this study.

Nomenclature

ANN	: Artificial neural network
API	: American petroleum institute
FOPT	: Field oil production total
FPR	: Field pressure rate
N_2	: Nitrogen
T	: Temperature
ϑV	: Volume differential
ϑP	: Pressure differential

6. References

1. Fereidooni, A., et al., *Prediction of Nitrogen Injection Performance in Conventional Reservoirs Using the Correlation Developed by the Incorporation of Experimental Design Techniques and Reservoir Simulation.* Iranian Journal of Oil & Gas Science and Technology, 2012. **1**(1): p. 43-54.

2. Lambert, M., et al. *Implementing CO2 floods: no more delays!* in *Permian Basin Oil and Gas Recovery Conference*. 1996. Society of Petroleum Engineers.

3. Vicencio, O., K. Sepehrnoori, and M. Miller. *Simulation of nitrogen injection into naturally fractured reservoirs.* in *SPE International Petroleum Conference in Mexico*. 2004. Society of Petroleum Engineers.

4. Vicencio, O.A. and K. Sepehrnoori. *Simulation of nitrogen injection into naturally fractured reservoirs based on uncertain properties and proper matrix grid resolution.* in *International Oil Conference and Exhibition in Mexico*. 2006. Society of Petroleum Engineers.

5. Necmettin, M. *High pressure nitrogen injection for miscible/immiscible enhanced oil recovery.* in *SPE Latin American and Caribbean Petroleum Engineering Conference*. 2003. Society of Petroleum Engineers.

6. Momeni, M. and K.N. Teymourei, *Investigating and Modeling of the Effects of Rock Properties on CO_2, N_2 Injection Performance. 2013.*

7. Peng, D.-Y. and D.B. Robinson, *A new two-constant equation of state.* Industrial & Engineering Chemistry Fundamentals, 1976. 15(1): p. 59-64.

IOP Conf. Series: Materials Science and Engineering **328** (2018) 012022 doi:10.1088/1757-899X/328/1/012022

Application of Quantum Dot nanocrystal in Luminescent solar concentrators

Shokoufeh Bakhoda[1], Morteza Khalaji Assadi [1, *], Sohrab Ahmadi_Kandjani[3], Hussain H. Al_Kayiem[1] and Aamir Hussain Bhat[2]

[1]Department of Mechanical Engineering, Universiti Teknologi PETRONAS, 32610 Bandar Seri Iskandar, Malaysia.
[2]Applied physics and Astronomy (RIAPA), University of Tabriz-Iran, 51665_163
[3]Department of Fundamental & Applied Sciences, Universiti Teknologi PETRONAS, 32610 Bandar Seri Iskandar, Malaysia.

morteza.assadi@utp.edu.my, s_ahmadi@tabrizu.ac.ir

Abstract. The basic design of luminescent solar concentrator is a transparent plate doped with an appropriate luminescent material (organic dyes, quantum dots), which is able to absorb sunlight (direct and diffuse), and then guides photons produced by photoluminescence to its narrow edges where they are converted by photovoltaic cells. Unfortunately, LSCs have suffered from numerous efficiency losses. Therefore, new luminescent species and novel approaches are needed for its practical application. This paper deals with investigation of nonhazardous, environmental friendly luminescent species include CuInS2/ZnS core/shell QDs. The CuInS2/ZnS QDs possess advantages of Stocks shift as large as more than 130 nm and high photoluminescence quantum yield of 80%. The paper presents the effect of large stock shift CuInS2/ZnS QDs on reducing the reabsorption losses in LSC by using experimental investigation. The LSC sheets were fabricated by dispersing CuInS2/ZnS QDs particles in a polymethylmethacrylate waveguide. A series of LSCs (dimension 4.0 cm × 3.0 cm × 0.3cm) with different CuInS2/ZnS QDs particles concentration (0.015 and 0.03 wt.%) were fabricated and their optical properties (absorptions/emissions) were characterized. The results show that the CuInS2/ZnS QDs-LSC provides a promising way for the reduction of reabsorption losses in LSCs.

1. Introduction

In the 1950's, photovoltaic cells were invented to convert light into electricity(1) and in the 1970's one proposed method was to use fluorescent polymer waveguides to concentrate light onto traditional silicon solar cells (2-4). These devices were initially called fluorescent planar concentrators and later became known as luminescent solar concentrators (LSC). The general principle behind LSCs is illustrated in Figure 1. Incident photons from the sun are absorbed by luminophores in a waveguide, and the emitted photons are guided via total internal reflection (TIR) to an adjacent silicon solar cell where they are converted to electricity (5, 6).

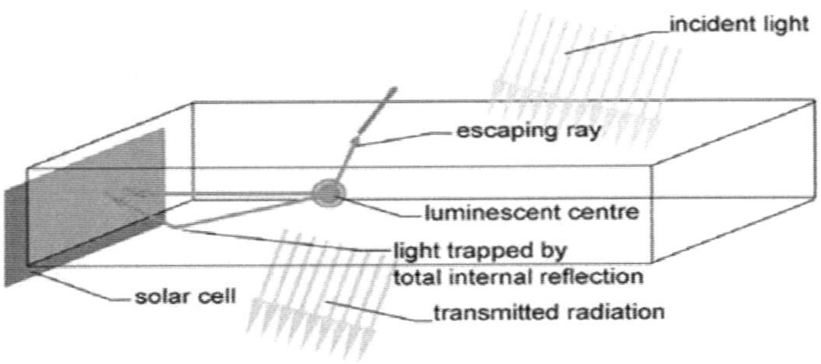

Figure 1. Operation principle of LSCs.(7)

The development of the LSC was initially limited by properties of organic dyes, such as narrow absorption band, large reabsorption and poor photo-stability. Recently, semiconductor QDs emerged as attractive candidates for development of stable and functional LSC devices. In 2015, the CuInS2/CdS nanocrystals (NCs) have been demonstrated to be attractive phosphors for high-gain full spectrum LSC applications. Their broad solar absorption and large photo luminescent quantum yield (PLQY) are combined to outweigh their moderate reabsorption losses(8).

LSCs could reduce the cost of solar energy by allowing replacement of expensive large-area PVs with cheaper solar-harvesting antennae coupled to small PVs (3, 6). LSCs are well suited for applications as semitransparent windows. They are used as exterior façade cladding, public transportation waiting area, shading device, street lightening, etc. However, they suffered from many losses. Self-absorption is an important loss factor and is thought to be a consequence of the spectral overlap between absorption and emission (3). Luminescent photons emitted in this overlap region, risk to absorbed again by another luminescent molecule and get lost instead of reaching the edge of the concentrator. By using experiments, this important loss mechanism will be studied. In this work, in order to minimize the reabsorption loss in luminescent solar concentrators, the LSC sheets in vary of quantum dots (QDs) particles concentration were investigated and their optical properties were characterized.

1.1. Quantum dots

Quantum dots are semiconductor nanocrystals that have an adjustable absorption and emission peak wavelength by varying the size of the quantum dot (Figure 2 and 3). Another advantage of quantum dots is the spectrally wide absorption band compared to organic dye molecules. QDs are more stable and less degradable than organic dyes. Several recent reports have recognized that the combination of large Stokes shifts and PL tunability displayed by copper-doped semiconductor nanocrystals makes them attractive luminophores for luminescent solar concentrators (LSCs) [6, 9-13].

By using nanotechnology and some special polymer components and quantum dots nanocrystals, the reabsorption loss of LSCs can be lessened. This fundamental study will investigate special quantum dot nanocrystals to reduce re-absorption in future work to design a new LSC with minimum re-absorption losses.

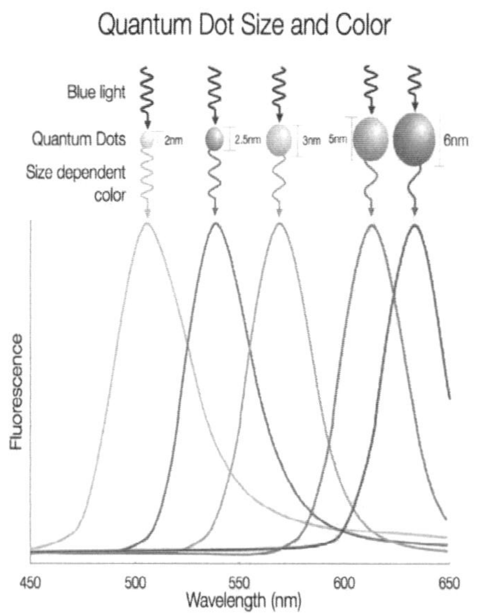

Figure 2. Emission spectra of different QDs. (9)

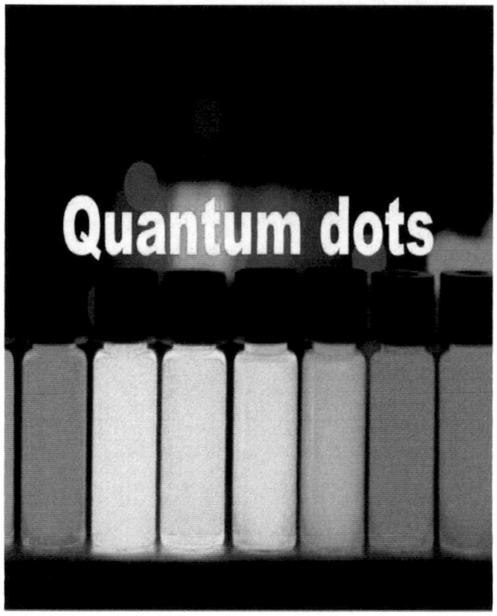

Figure 3. Different colure of QDs as increasing their size. (9)

1.2. The effect of stoke shift on reabsorption
A critical property of every fluorophore is its Stokes shift, which is the wavelength gap between the absorption and emission peaks; the bigger the Stokes shift, the smaller the overlap between the two spectra will be which will result in lower re-absorption as shown in Figure 4.

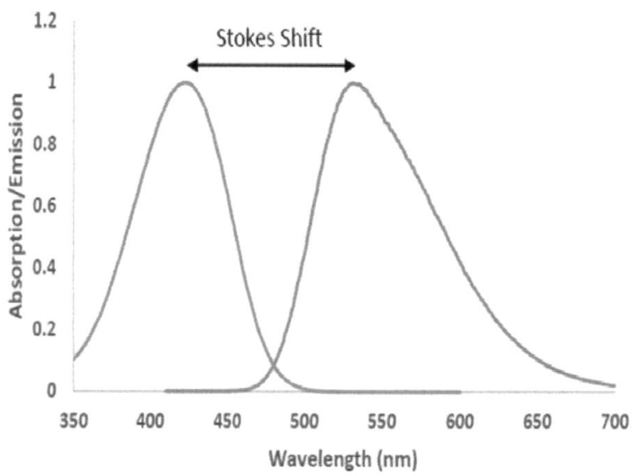

Figure 4. Absorption (orange) and emission (blues) spectra of Coumarin 153(10).

2. Experiment
LSCs were prepared with fixed dimensions of 4.0 cm × 3.0 cm × 0.3cm and varying concentration of CuInS2/ZnS QDs particles. These particles are purchased from a commercial supplier (UbiQD, USA) with a quantum efficiency of 80%. To fabricate the LSC, the ratio was set as 9:1 MMA: PMMA. The

CuInS2/ZnS QDs particles were first dispersed in a methyl methacrylate (MMA) monomer solution by ultrasonic stirring, then polymethylmethacrylate (PMMA) is added and stirred until all PMMA granules were dissolved completely. The mixture is cooled and azobisisobutyronitrile (AIBN) was added to the solution at an AIBN/MMA weight ratio of 0.1 wt. %. The mixture is heated to 60 °C by using a heater for more than 3hrs and then is cooled to room temperature. The solution is poured into molds (0.3 cm thick) and then placed in a water bath for longer than 18 hours to complete polymerization. Finally, the PMMA plate was cut with areas of 4 cm × 3 cm × 0.3 cm by laser cutting. By controlling the CuInS2/ZnS QDs particles concentration, a series of LSCs were fabricated at 0.015wt%, and 0.03 wt.%.

3. Result and discussion

Semiconductor crystalline CuInS2/ZnS core/shell QDs with QY 80% were fabricated in PMMA sheet at different concentration of 0.015 wt. % and 0.03 wt. %. Figure 5 and 6 show the absorption and emission spectra of CuInS2/ZnS QDs at different concentrations in PMMA sheet. The emitting peak of CuInS2/ZnS QDs observed at 625 nm with large Stocks shift 130 nm. These QDs are able to convert light with wavelength of less than 500 nm into around 630 nm effectively (Fig. 5), which could be absorbed by the c-Si PV cells more efficiently.

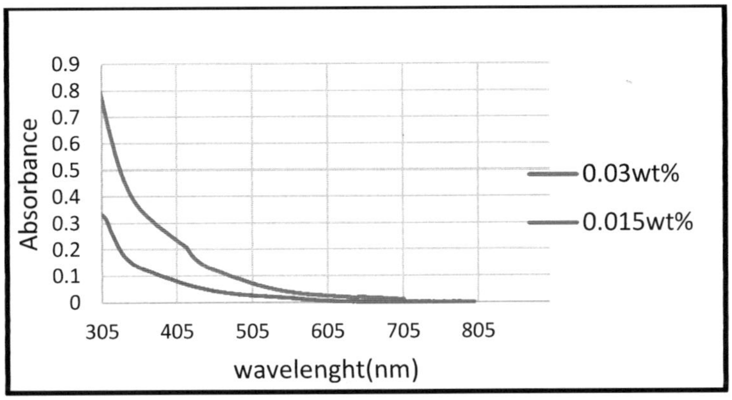

Figure 5. Absorption spectra of CuInS2/ZnS QDs at concentration of 0.015 wt. and % 0.03 wt. %.

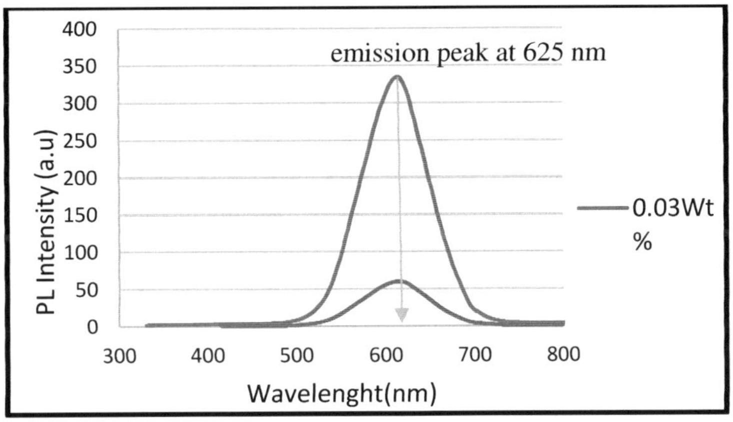

Figure 6. Emission spectra of CuInS2/ZnS QDs at concentration of 0.015 wt. and % 0.03 wt. %

Figure 7. shows spectra of photoluminescence measured at one of the slab edges (4 × 0.3 cm2) using optic fiber at 427 nm excitation positioned at different distances d (0-4cm) from the sample edge. By increasing distance (d), the PL intensity is observed to drop because of reabsorption by the quantum dots and by the polymer matrix and light escaping from the waveguide.

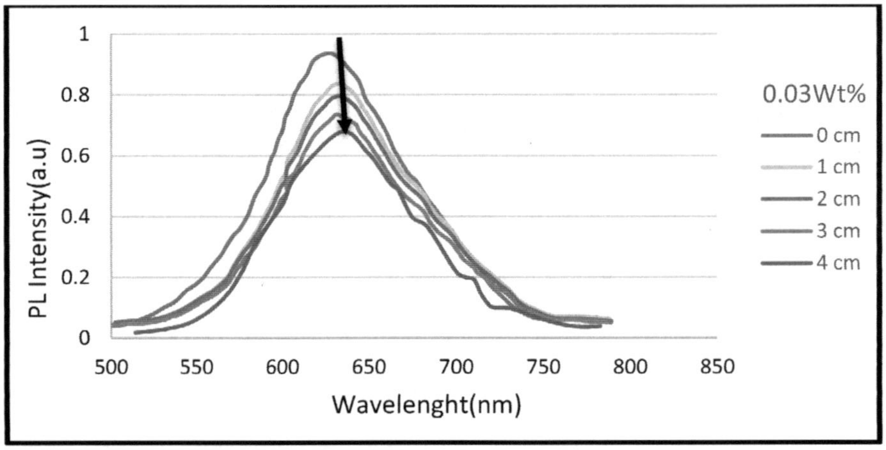

Figure 7. Photoluminescence spectra (excitation at 410 nm) as a function of distance (d) by increasing from 0 to 4 cm from the edge.

The shape of the photoluminescence spectra shows only a small change with d, suggesting that losses to reabsorption are not significant.

4. Conclusion

In this research, heavy metal free CuInS2/ZnS core/shell QDs with the large Stokes shift (130 nm) and high PLQY (80%) has been proposed. The QDs-doped PMMA LSCs were fabricated in 0.03 Wt. % and 0.015 Wt. % concentrations. Their large stock shift and broad solar absorption combined to reduce self-absorption losses in the LSCs sheet. Therefore, the CuInS2/ ZnS NCs can be attractive luminescent material to be used in LSCs applications and we can expect to see LSC-based products in the energy and built environment markets soon.

References

[1] Chapin, D.M., C. Fuller, and G. Pearson, A new silicon p- n junction photocell for converting solar radiation into electrical power. *Journal of Applied Physics,* 1954. **25(5):** p. 676-677.
[2] Goetzberger, A. and W. Greube, Solar energy conversion with fluorescent collectors. *Applied Physics A: Materials Science & Processing,* 1977. **14(2):** p. 123-139.
[3] Batchelder, J., A. Zewai, and T. Cole, Luminescent solar concentrators. 1: Theory of operation and techniques for performance evaluation. *Applied Optics,* 1979. **18(18):** p. 3090-3110.
[4] Batchelder, J., A. Zewail, and T. Cole, Luminescent solar concentrators. 2: Experimental and theoretical analysis of their possible efficiencies. *Applied Optics,* 1981. **20(21):** p. 3733-3754.
[5] Rowan, B.C., L.R. Wilson, and B.S. Richards, Advanced material concepts for luminescent solar concentrators. *IEEE Journal of selected topics in quantum electronics,* 2008. **14(5):** p. 1312-1322.
[6] Debije, M.G. and P.P. Verbunt, Thirty years of luminescent solar concentrator research: solar energy for the built environment. *Advanced Energy Materials,* 2012. **2(1):** p. 12-35.
[7] Krumer, Z., et al., Tackling self-absorption in luminescent solar concentrators with type-II colloidal quantum dots. *Solar Energy Materials and Solar Cells,* 2013. **111:** p. 57-65.
[8] Knowles, K.E., et al., Bright CuInS 2/CdS nanocrystal phosphors for high-gain full-spectrum luminescent solar concentrators. *Chemical Communications,* 2015. **51(44):** p. 9129-9132.
[9] Bradshaw, L.R., et al., Nanocrystals for luminescent solar concentrators. *Nano letters,* 2015. **15(2):** p. 1315-1323.

[10] Li, C., et al., Large Stokes shift and high efficiency luminescent solar concentrator incorporated with CuInS2/ZnS quantum dots. *Scientific reports,* 2015. **5:** p. 17777.
[11] Hermann, A., Luminescent solar concentrators—*a review. Solar Energy*, 1982. **29(4):** p. 323-329.
[12] Verbunt, P.P. and M.G. Debije. Progress in luminescent solar concentrator research: solar energy for the built environment. *in World Renewable Energy Congress-Sweden;* 8-13 May; 2011; Linköping; Sweden. 2011. Linköping University Electronic Press.
[13] Trupke, T., M. Green, and P. Würfel, Improving solar cell efficiencies by down-conversion of high-energy photons. *Journal of applied physics,* 2002. **92(3):** p. 1668-1674.
[14] https://commons.wikimedia.org/wiki/File%3AQuantum_Dots_with_emission_maxima_in_a_10-nm_step_are_being_produced_at_PlasmaChem_in_a_kg_scale.jpg.
[15] Erickson, C.S., Doped quantum dot luminescent solar concentrators. 2014

Acknowledgments
We gratefully acknowledge the support provided by Universiti Teknologi PETRONAS and Ministry of Higher Education Malaysia (MOHE) of Malaysia for this work under fundamental Research Grant (FRGS) with grant numbers 0153-AB-K88 and 0153-AB-K81.

IOP Conf. Series: Materials Science and Engineering **328** (2018) 012023 doi:10.1088/1757-899X/328/1/012023

A Preliminary Experimental Investigation of Wet Fine Erosion in Two-Phase Flow

H. H. Ya[1], Haziq luthfi[1], Nguyet-tran ngo[2], Suhaimi Hassan[1], William pao[1]*
[1]Mechanical Engineering Department, Universiti Teknologi PETRONAS,
32610 Seri Iskandar, Perak DR, Malaysia
[2]Worley Parsons Pte. Ltd., 111 Somerset Road, Singapore 238146
*Corresponding Author: william.pao@utp.edu.my

Abstract. Solid particles below 62 μm is classified as fine. In oil producing operation, the most commonly used downhole sand screen can only capture solid particles of 140 μm and above. Most predictive erosion model is limited to particle size of 100 μm with single phase flow assumption because it is commonly believed that erosion due to particles below 100 μm is insignificant and typically ignored by oil and gas consultants when proposing facilities design. The objective of this paper is to investigate the impact of fines particle on mild steel plate in two-phase flow at different collision angles. A two phase flow loop was set up. The average size of fine particle was 60 μm, mixed with water with sand to water ratio at 1:65 wt/wt. The mild steel plates were oriented at three different impact angles which are -30°, 30° and 90°, with respect to the horizon. Scanning electron microscopy (SEM), energy-dispersive X-ray spectroscopy (EDX), surface roughness and Vickers micro hardness techniques were used to quantify the effects of fine particle on the exposed surface.

Keywords: Two-phase flow, particle's impact, erosion, impact angle, mild steel.

1. Introduction

Solid particle erosion is one of the common problems in oil and gas industry. For many decades, it was believed that fine particles, with size below 62 μm, being very small in size; do not significantly damage the metallic flow-lines. However, the erosion studies on the microscopic scale have revealed that particles of even smaller sizes can be lethal during the production process [1]. They can escape through the most commonly used sand screens in the industry, sieve mesh 105. This made fines almost inevitable in oil and gas production. They can severely damage the installations at the location where coarse particles cannot reach in normal situations [2].

Many investigations have been conducted to develop empirical and numerical models of the solid particle erosion [2]. Upon careful study, it is found that many existing erosion correlations used in the industry has some serious drawback assumptions on single phase flow. It is also found that almost all correlations for solid particle erosion focused mainly on sand particle, which by definition is solid particles of size 62 – 2000 μm. In one of the most authoritative guidelines on sand controlled erosion for oil and gas industry, DNVGL-RP-O501 standard (2007) [3], the lower limit of the validity of model is also restricted to sand particles under single phase flow condition. The objective of this paper is to conduct a preliminary experimental investigation on the erosion caused by fine particle, i.e. solid particles of average size 60 μm; and their impact on mild steel.

IOP Conf. Series: Materials Science and Engineering **328** (2018) 012023 doi:10.1088/1757-899X/328/1/012023

The motivation is to determine if it is feasible to carry out longer term fine particle erosion research and to have a screening assessment of its impact on metallic oil and gas flow lines.

2. Literature Review

2.1. Mechanism of solid particle erosion

When a particle impacts a surface, it scars the surface. Shapes of these scars depend on many parameters including surface material, particle size, and impact angle. Researchers studied these scars to explain the mechanism of erosion and generally agree that the mechanism of erosion changes based on the ductility of the surface. [4] proposed a micro-geometry erosion model for ductile materials and suggested that erosion in ductile materials is the result of micro-cutting. When a particle impacts a surface at a low impact angle, it creates a crater. Other particle impacts make the crater larger and also pile up material around the crater. The piled up material is eventually removed by continued particle impacts. The micro-geometry model under-predicts erosion magnitude from the particles which impact the surface at higher angles compared to experimental data. [5] showed that initial erosion is lower than erosion from previously eroded surfaces and proposed a macroscopic erosion mechanism. He suggested that particles hitting the surface create shallow craters and platelet-like pieces. These platelets are easy to separate from the surface by subsequent particle impact as shown in Fig. 1. During the formation of platelets, adiabatic shear heating on the surface and work-hardening under the surface occur. The occurrence of these two processes helps platelet formation which explains the higher erosion rate for the steady-state condition compared to the initial erosion rate. Other solid particle erosion mechanisms for ductile materials are suggested by researchers and can be found in literature [6, 7, 8, 9]. Unlike the solid particle erosion mechanism for ductile materials, there is wide acceptance of the erosion mechanism for brittle material. It has been suggested that in brittle material, erosion is due to crack formation [10, 11, 12]. When a particle hits a brittle surface, it creates lateral and radial cracks. Other impacts cause these cracks to grow. These cracks divide the surface into smaller pieces which can be removed by other particles impacting the surface as shown in Fig. 2.

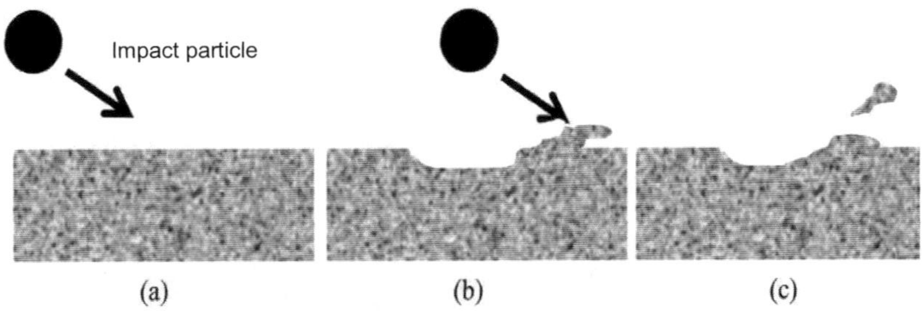

Fig. 1: Schematic of erosion mechanism in ductile material [12], (a): before the impact, (b): crater formation and piling material at one side of the crater, (c): separation of material from the surface

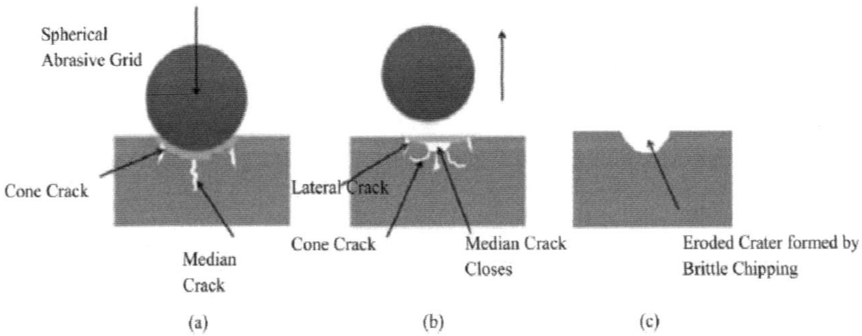

Fig. 2: Erosion mechanism by solid particle in brittle material [10, 11, 12]

IOP Conf. Series: Materials Science and Engineering **328** (2018) 012023 doi:10.1088/1757-899X/328/1/012023

2.2. Particle properties on erosion

Review from various researchers [13, 14, 15] yielded that the angularity has a great impact on erosion behaviours, whereby sharp particles are more aggressive than rounded ones. In particular, the study on erosion causing by angular solid particles has measured the influence of individual angular particles on lead and mild steel target surface [13]. [15] utilized sharp angular particles and spherical particles to investigate the effect of particles shape on erosion for steel material. They have found that angular particles imposed quadruple effect on the erosion as compared to spherical particles.

The size of particles impinged onto target surface defines the degree of erosion impact on the inner wall. The relationship between erosion rate, $\dot{E}$, and particle diameter, D_p, was revealed by Desale et al. [16] as

$$\dot{E} = f\left(D_p\right)^n , \tag{1.1}$$

where n is known as the particle size factor. According to [17], the value of this 'particle size factor' varied with particle size distribution, particle velocity, fluid properties, particle-particle interaction, material properties as well as experimental conditions. At $n = 1$, erosion rate has linear relations with sand size. Additionally, through the erosion analysis on aluminium alloy (AA 6063) using eight different sized quartz particles in between 37.5 μm and 655 μm at two different impact angles of 30º and 90º, Desale et al. [16] summed up that at the larger size of sand particles, the higher kinetic energy, thus greater erosion wear was observed. [17] studied the relationships amongst erosion rate and flow velocity and particle sizes. Particles size of 100 μm – 10 nm were studied, in a flow regime of 5-20 m/s with different volumetric concentration of particles. They concluded that there is a threshold velocity and particles size where erosion is significant and correlate the power law relationship between average erosion rate and particles concentration, velocity and particle size. In general, the rate of erosion is found to be proportional with the ratio of hardness of particles (H_p) to hardness of target surface (H_t), defined as

$$\dot{E} \propto \left(H_p / H_t\right). \tag{1.2}$$

Equation (1.2) was introduced by [18]. When the particle to targeted surface hardness ratio is greater than unity, erosion will occur. Also, the erosion rate will escalate corresponding to the increase of erodent to target hardness ratio. Through the investigation of the impact of particle and target wall materials on particles erosion characteristics, [19] also perceived that the increase in hardness of erodent particles will develop its erosivity. However, there is a critical hardness value; beyond which the degree of erosion does not considerably increase with an increase of its hardness. This shows that sand erosion is susceptible to harder sand particles as they shatter less when colliding with target wall. Besides, another important factor in erosion calculation is density. This is because denser particles contain more kinetic energy, producing greater impact force and generating higher erosion rates [15, 20]. Therefore, hard and dense particles are inherently more erosive than soft particles. Typically, density of sand particles in various studies is approximately 2650 kg/m^3 [21, 22]. Erosion rate is a power-law function of particle impact velocity, defined as

$$\dot{E} = f\left(v_p\right)^m \tag{1.3}$$

3. Methodology

3.1 Experimental setup

The setup and design of the experiment to conduct a single phase flow wet sand impact erosion test is shown below. The schematic diagram and experiment setup of the experiment is shown in Fig. 3.

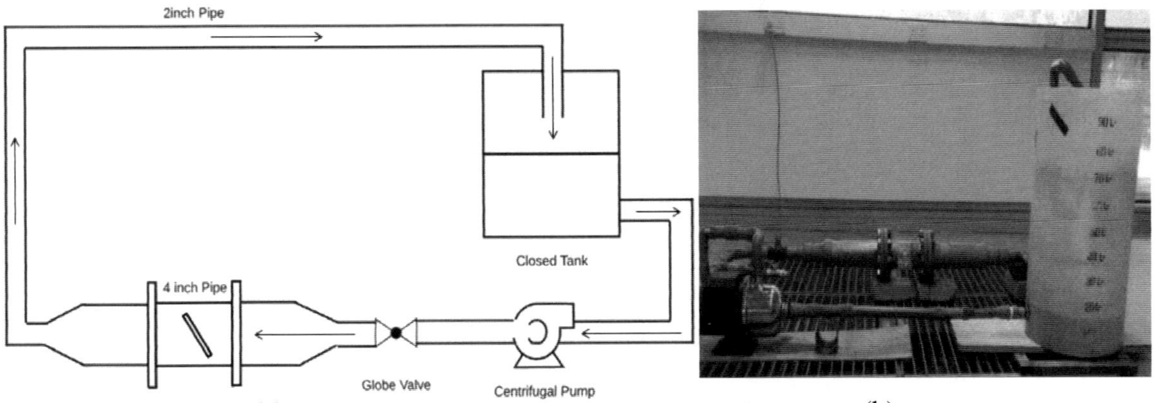

(a) (b)

Fig. 3: (a) Schematic diagram of the experiment and (b) Actual setup of the mixture flow loop

3.2 Sample preparation

The mild steel plates were cut into a square shape with a height of 50 mm and a length of 40 mm. The plates were cut by using an abrasive cutter machine. Then the mild steel plates were polished until they are free from scale, scratches and rough patches. The fine particles or silica samples average size of 60 μm were obtained from a ball mill factory. It is then mixed with water with fines to water ratio of 1:65 by weight.

Before and after the experiment, the specimen was subjected scanning electron microscopy to produce images for comparison. Energy Dispersive X-ray Spectroscopy (EDX) was also applied to the specimen before and after the experiments to analyse the differential chemical compositions of the exposed surface. The surface roughness of each mild steel plate were also recorded. The average surface roughness recorded before the experiment was found to be 0.17 μm. Finally, the specimens were subjected to Vickers micro hardness test to check the differential hardness of specimens before and after the experiments. The average hardness value recorded before the experiment was found to be 139.23 HV. This value is very close to the standard reference value 140 HV for mild steel.

The plates were placed inside the bombardment chamber at three different impact angles (-30°, 30°, 90° with respect to the horizon) and were bombarded with fluid containing fine particles with an average size of 60 μm for a fixed time interval of 5 hours. Due to the limitation of the test rigs capability, the flow rate and velocity of the fluid in the bombardment chamber were kept constant at 8.6 x 10-4 m3/s and 0.124 m/s respectively. The mixture filled up approximately 80% by volume when entering the bombardment chamber. The fluid hitting the mild steel plate inside the bombardment.

4. Results and Discussion

After 5 continuous hours of bombardment, the region of surface exposed to fine-water mixture showed observable and severe pitting, especially near the mixture free surface. Even though pitting is also observable at the bottom of the specimen, the distribution of the pits are scattered and the damage to the surface is relatively less. The level of surface damage was significantly influenced by the impact angles. From Fig. 4, the highest erosion rate was seen at an impact angle of 90° followed by -30 ° and finally 30 °. At 90°, the accelerated fluid containing fine particles does not exhibit any angular slip and imparted maximum energy to the surface of the mild steel plate. At inclined impact angles, the solid particles exhibited angular slip. The impact force by the particle can be split into axial and

IOP Conf. Series: Materials Science and Engineering **328** (2018) 012023 doi:10.1088/1757-899X/328/1/012023

vertical components. At inclined impact angles, the axial component is dominant and contributed to good amount of energy to the targeted surface. The energy transfer due to vertical component is almost negligible.

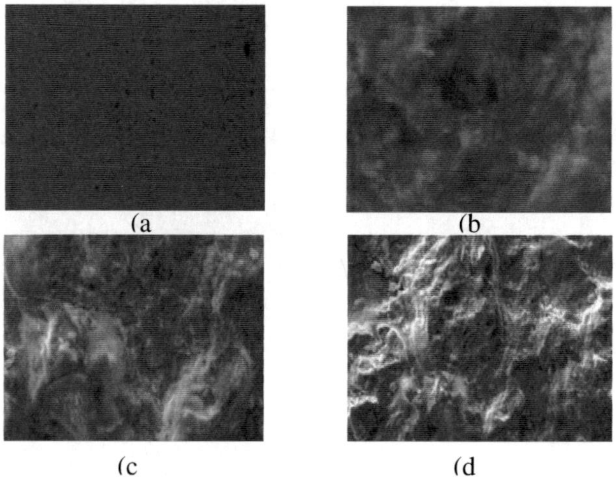

Fig. 4: (a) SEM micrograph of the untreated plate. (b) SEM micrograph of eroded plate at 30°. (c) SEM micrograph of eroded plate at -30°. (d) SEM micrograph of eroded plate at 90 °.

Figure 5: showed the EDX spectrum of the mild steel specimen before and after the experiment. It is clearly seen that before the experiment, the exposed surface showed composition of iron (Fe) and carbon (C) while after the experiment, the surface chemical composition of mild steel plates was significantly changed. It was noticed that some of the fine particles diffused onto the exposed surface of the mild steel plates. The new elements on the surface of the plates were believed to be introduced by the fines during the experiment. The EDX spectrum of the eroded specimen showed an additional two components, namely silicon and oxygen. This shows that erosion process does not only damage, the target surface it also changes its surface composition.

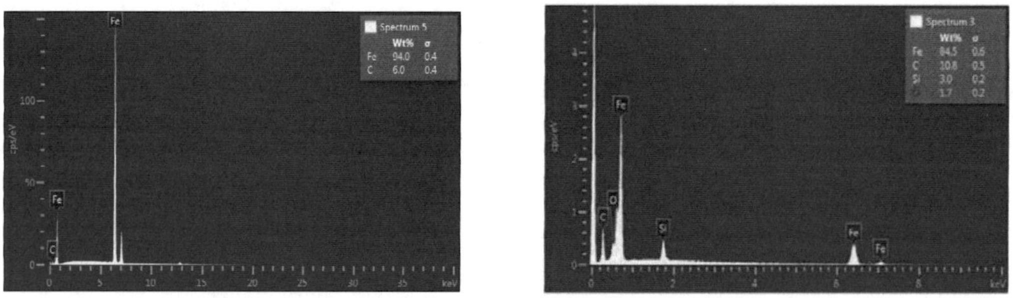

Fig. 5: EDX spectrum of mild steel specimen (a) before and (b) after experiment

Table 1 showed the surface roughness data for three specimens under investigation. The surface roughness of the specimens before experiment was determined to be 0.17 μm. After the experiments, the surface roughness of 30°, -30° and 90° were determined to be 0.56, 1.00, and 1.59 μm, respectively. The plate that was placed at an impact angle of 90° recorded the highest differential surface roughness, with change exceeded 800%.

Table 1. Surface roughness data

Impact angle (°)	Surface roughness (μm)			
	S1	**S2**	**S3**	**Average**
30	0.587	0.526	0.554	0.556
-30	1.032	1.043	0.929	1.001
90	1.678	1.505	1.572	1.585

The Vicker's micro hardness value of the treated samples as a function of impact angle is shown in Fig. 6. The non-eroded specimen was recorded with a hardness value of 139.23 HV. All three specimens recorded a reduction in their HV. The mild steel plate that was placed at an impact angle of 90° has the lowest hardness value followed by -30° and finally 30° which are 117.03 HV, 125.51 HV and 133.83 HV, respectively.

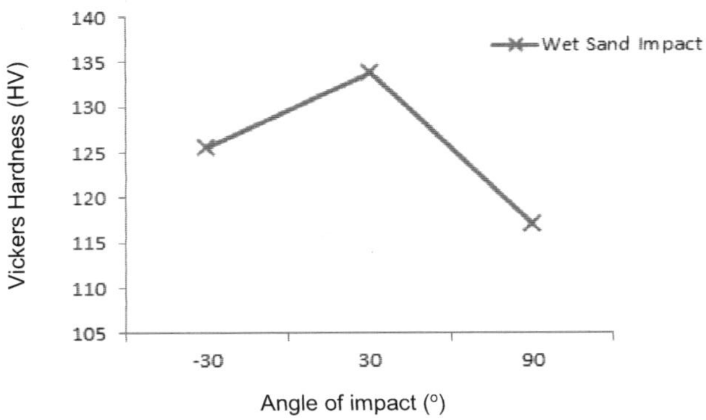

Fig. 6: Micro hardness of the eroded mild steel specimen

5. Conclusions

This study investigated the impact of wet erosion of mild steel, bombarded with water-fines mixture at three different impact angles which are -30°, 30° and 90°. The fines average particle is 60 μm and the water-fines mixture at a weight ratio of 1:65. It was found that the composition of the mild steel plates after the experiment changed significantly. Fine particles compositions were diffused onto the exposed metal surface. It could be concluded that fines changed mild steel surface roughness significantly, even for a short duration of 5 hours. A surprising 800% degradation of surface roughness was observed for 90° impingement. The hardness value of the specimen was observed to be reduced inversely proportional to the erosion.

References

[1] Kuncuro, B.; Ulummuddin, B.; Palar, S. (2011). Sand control for unconsolidated reservoirs. *ProceedingSimposium Nasional Yogyakarta*, 3.

[2] Carlson, J.; Gurley, D.; King, G. (2002). Sand control: why and how *Oil Field Review*, London.

[3] Naz, Y.; Ismail, N.I.; Sulaiman, S.A.; Shukrullah, S. (2015). Electrochemical and dry sand impact erosion studies on carbon steel, *Sci. Rep.*, 5, 16583.

[4] Finnie, I. (1960). Erosion of surface by solid particles, *Wear*, 3(2), 87–103

IOP Conf. Series: Materials Science and Engineering **328** (2018) 012023 doi:10.1088/1757-899X/328/1/012023

[5] Levy, A. (1995). *Solid particle erosion and erosion-corrosion of materials*. ASM International,Material Park, Ohio.
[6] Andrews, D.R.; Horsfield, N. (1983). Particle collisions in the vicinity of an eroding surface. *J. Phys. D: Appl. Phys*. 16(4).
[7] Chase, D.; Rybicki, E.; Shadley, J. (1992). A model for the effect of velocity on erosion of N80 steel tubing due to the normal impingement of solid particle. *J. Energy Resour. Technol.*, 114(1), 54-64.
[8] Hutchings, I. (1980). Some comments on the theoretical treatment of erosive particle impacts. In: Proceeding of the 5th International Conference on erosion by Liquid and Solid Impact, Newnham College, Cambridge.
[9] Jahanmir, S. (1980). The mechanics of subsurface damage in solid particle erosion. *Wear*, 61(2), 309-324.
[10] Kleis, I.; Kulu, P. (2008). *Solid Particle Erosion Occurrence, Prediction and Control*. Springer-Verlag London Limited. Library of Congress Control Number: 2007937988
[11] Srinivasan, S.; Scattergood, R.O. (1988). Effect of erodent hardness on erosion of brittle materials. *Wear* 128 (2), 139-152
[12] Sundararajan, G. (1991). A comprehensive model for the solid particle erosion of ductile materials. *Wear*, 149(1-2), 111-127.
[13] Winter, R. E.; Hutchings, I. M. (1974). Solid particle erosion studies using single angular particles. *Wear*, 29(2), 181-194.
[14] Levy, A.; Chik, P. (1983). The effect of erodent composition and shape on the erosion of steel. *Wear*, 89(2), 151-162.
[15] Parsi, M., (2014). *Sand erosion in vertical slug/churn flow*. Advisory Board Meeting, Erosion/Corrosion Research Centre, The University of Tulsa, May.
[16] Desale, G.R.; Gandhi, B.K.; Jain, S.C. (2009). Particle size effects on the slurry erosion of aluminum alloy (AA 6063). *Wear*, 266 (11-12), 1066-1071.
[17] Safaei M.R. et al., (2014). Investigation of micro- and nanosized particle erosion in a 90° pipe bend using a two-phase discrete phase model. *The Scientific World Journal*, Vol 2014, ID 740678.
[18] Wada, S.; Watanabe, N. (1987). Solid particle erosion of brittle materials (part 3), the interaction with material properties of target and that of impingement on erosive wear mechanism. *Joural of the Ceramic Association Japan*, 95(10), 573-578.
[19] Shipway, P.; Hutchings, I. (1996). The role of particle properties in the erosion of brittle materials. *Wear*, 193(1), 105-113.
[20] Nicolici, S.; Prisecaru, I.; Dupleac, D. (2013). Two-phase flow CFD modelling for evaluation of particulate erosion. *U.P.B. Sci. Bull., Series D*, 75(1), 187-198.
[21] Chen, J.; Wang, Y.; Li, X.; He, R.; Han, S.; Chen, Y. (2015). Erosion prediction of liquid-particle two-phase flow in pipeline elbows via CFD–DEM coupling method. *Powder Technology*, 282, 25–31
[22] Najmi, K.; Mac Laury, B. S.; Shirazi, S. A.; Cremachi, S. (2016). The effect of viscosity on low concentration particle transport in single-phase (liquid) horizontal pipes. *ASME J Energy Resour. Technol.*, 138(3), 032902.

IOP Conf. Series: Materials Science and Engineering **328** (2018) 012024 doi:10.1088/1757-899X/328/1/012024

Effects of Brass (Cu_3Zn_2) as High Thermal Expansion Material on Shrink Disc Performance During High Thermal Loading

MIS Mohd [1*], SA Mazlan[1], ND Bahar [1] and SAA Aziz [1]

[1] Universiti Teknologi Malaysia, Malaysia-Japan International Institute of Technology, Department of Mechanical Precision Engineering, Ikohza Vehicle System Engineering, Jalan Sultan Yahya Petra, 54100 Kuala Lumpur, Malaysia.

[*] idrulshah84@gmail.com

Abstract. This research work is focused on shrink disc operation at high temperature. Geometrical and material design selections have been done by taking into consideration the existing shrink disc operating at high temperature condition. The existing shrink disc confronted slip between shaft and shaft sleeve during thermal loading condition. The assessment has been obtained through virtual experiment by using Finite Element Analysis (FEA) -Thermal Transient Stress for 900 seconds with 300℃ of thermal loading. This investigation consists of the current and improved version of shrink disc, where identical geometries and material properties were utilized. High Thermal Expansion (HTE) material has been introduced to overcome the current design of the shrink disc. Brass (Cu_3Zn_2) has been selected as the HTE material in the improved shrink disc design due to its high thermal expansion properties. The HTE has shown a significant improvement on the total contact area and contact pressure on the shaft and the shaft sleeve. The improved shrink disc embedded with HTE during thermal loading exhibit a minimum of 1244.1 mm^2 of the total area on shaft and shaft sleeve which uninfluenced the total contact area at normal condition which is 1254.3 mm^2. Meanwhile, the total pressure of improved shrink disc had an increment of 108.1 MPa while existing shrink disc total pressure has lost 17.2 MPa during thermal loading.

Keywords: *Shrink disc, thermal expansion, thermal loading, contact pressure, contact area, shaft and shaft sleeve.*

1. Introduction

The interference fit has been used for torque transmitting from one component to other components since a long time ago [1]. It has been used in many industries such as automotive, plantation, medical, oil and gas and army application. It was designed due to drawbacks of other mechanical couplings such as pin, slots and key and screw which have less efficiency during operation compared to interference fit. On the other hand, interference fit has numerous capabilities such as simple structure, high load carrying capacity, excellent concentricity and etc. There are many researchers performing studies in the related field such as Sun et.al on his work, contributed on implementing 3D FEM modelling during shrink fit process of assembled crank shaft. Observation concluded that uniform expansion of crank shaft cannot be achieved by subjecting homogenous temperature distribution around the cylinder bore [2]. The crankshaft experienced distortion during assembly process which resulted the shrink fit interface undergoing inhomogeneous contact pressure. Meanwhile Bengeri et.al, has focused on the effect of temperature dependent yield stress on contact pressure during thermal assembly of shrink fit by analytical transient stress approach on plastic-elastic model [3]. The authors found that there was no significant effect of contact interface by introducing a temperature dependent and plastic–elastic model in transient stress analysis. Alexandre et.al in their study has widened observation of shrink fit interference pressure which are subjected to centrifugal force and temperature effect. The result revealed that the rotational speed and the existence of steady state thermal loading on the shrink fit has shown significant effect on the reduction of torque capacity [4]. Jianmei et.al investigated the interference stress on shrink fit by including several parameters as influence factors which consist of various range of interference, shaft diameter, wall thickness and mating length that might influence the interference stress on shrink fit. In his study, the various FEA contact analyses have been conducted by utilising orthogonal test approach. This approach used multiple step regression analysis by analysing data obtained from FEA in order to determine relationship toward contact stress. It was discovered that the interference parameter has strong sensitivity towards the interference stress [5]. Another study Strozzi et.al focuses upon determination of the significant contacting shape in a frictionless, keyless, shaft–hub press-fit profile which promotes uniform elastic contact pressure distribution with analytical study. They found that the infinite elastic stress build up at the edge of hub bore, otherwise the stress will reduce as the fillet shape is introduced on the profile. On the other hand, reduction of stress concentration is respected to various change of fillet radius [6]. Cerdán et.al investigated the stress distribution on the interference fit with having circumferential notched on the hub. They found that the stress concentration is higher on the hub notch. By modifying the notch shape, stress analysis was conducted which resulted in a reduction of stress concentration on that region [7].

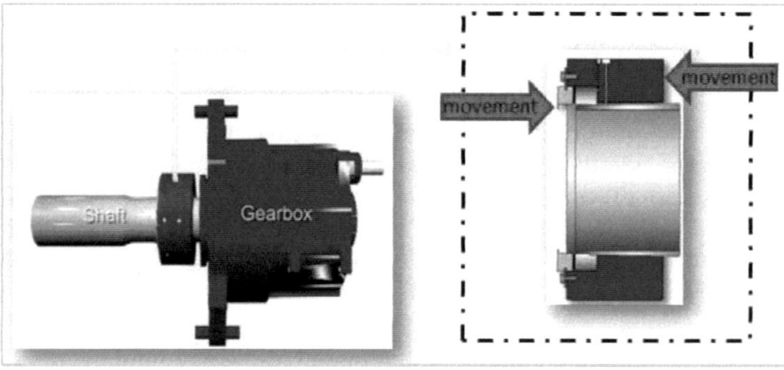

Figure 1 Shrink disc operating condition in pump

Figure 1 shows the current application of shrink disc embedded in pump. The obstacles are normally experienced by the system when operating at high temperature operating condition. The components encompass shaft and shrink disc tend to deform thus losing the grip of the shaft and the shaft sleeve whereby the torque will be unsuccessfully delivered. This study focused on the geometrical concept design and material properties modification of the existing shrink disc in order to enhance the contact area and contact pressure of the shaft and shaft sleeve thus simultaneously eliminating the occurrence of gap of shaft and shaft sleeve during thermal loading condition.

2. Simulation Methods

The simulation was performed by using Ansys Workbench version 17.2. The geometrical model has been designed using CAD Autodesk Inventor. The multilayer interference fit shrink disc consists of shaft, bolt, sleeve, conical ring – ferrule, clamping collar (1) and (2), thermal expansion components and pusher as shown in the **Figure 2**. Shrink disc improved design encompasses material and geometrical modification to ensure its performance even during high temperature condition. The thermal expansion components are built in shrink disc assembly as one of potential parameters in performance advancement studies. Brass (Cu_3Zn_2) has been selected due to its high thermal expansion properties compared to other materials [8].Geometrical simulations were conducted with the following parameters, boundaries and mechanical properties setting stipulates in **Figure 3**. The analysis were performed using thermal stress transient analysis to observe the change in contact pressure and contact area as time elapsed [9].

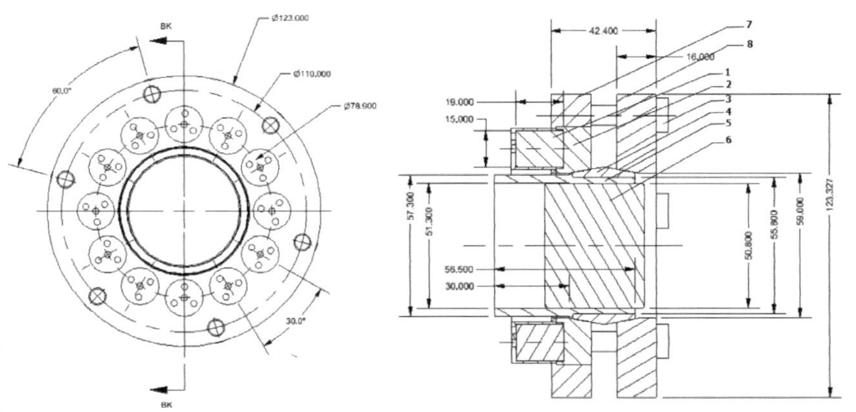

1. Thermal expansion components, 2. Pusher, 3. Bolt, 4. Conical ring (ferrule), 5. Shaft sleeve, 6. Shaft, 7. Clamping collar (1) and 8. Clamping collar (2)

Figure 2 Geometrical Concept Design

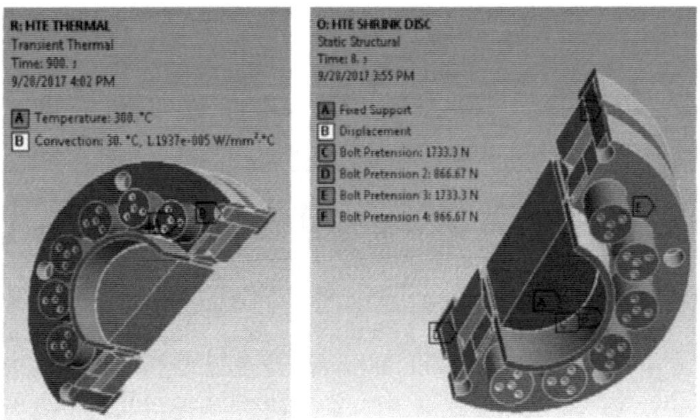

Parameters	*Components / Value*
Materials:	
Stainless steel (SS316)	Shaft
Stainless steel (SS17-4PH)	Clamping collar 1
Titanium Alloy (Ti6Al4V)	Clamping collar 2, Conical ring (ferrule), bolts, pusher and shaft sleeve.
Brass (Cu_3Zn_2)	Thermal expansion components
Mechanical Properties:	
Elastic modulus (GPa) (SS316, SS17-4PH, Ti6Al4V, Cu_3Zn_2)	193, 197, 113.8, and 97
Poison ratio	0.3
Boundary Conditions:	
Preload force (N)	1733.3
Thermal loading (°C)	300
Mesh	Hex Dominant
No. Elements/Nodes	142061/493679

Figure 3 Basic parameters and boundary conditions setting

Total preloading forces were computed by the following equation:

$$T = K.F_{sa}.d \qquad \textbf{Eqn. 1}$$

Where T is the tightening torque of the bolt (N mm); K is the torque coefficient; $F_{sa}$is the preloading force (N) and d is the nominal size of bolt diameter (mm) [10].

Contact pressure of the shaft sleeve and shaft is calculated by the following equation:

$$P_{min} = \frac{F_{Sa}(1-\mu_2 \tan\beta)}{\pi d_3 l_f(\tan\beta+\mu_2)} \qquad \textbf{Eqn. 2}$$

Where F_{Sa} is the total bolt preload force (N); μ_2 is the friction coefficient of conical ring - ferrule and clamping collar; β is the conical ring - ferrule angle (degree); d_3 is the external diameter of sleeve (mm) and l_f is the length of conical ring – ferrule [11].

3. Results and Discussion

Figure 4, **Figure 5** and **Figure 6** illustrate the results of improved and existing shrink disc design which are further explained in section 3.1 to 3.4 in detail.

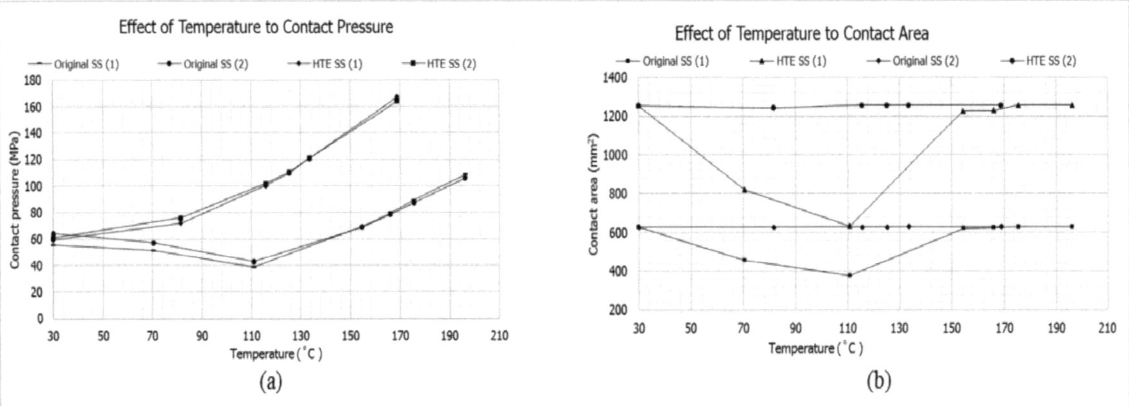

Figure 4(a) Effect of Temperature to Shaft Sleeve Contact Pressure and **4(b)** Effect of Temperature to Shaft Sleeve Contact Area

3.1 Effect of Temperature to Shaft Contact Pressure

The performance of contact area is analyzed by mating parts of shaft and shaft sleeve. This contact pressure is generated as torque is applied to the bolt which acts as an axial preload on the clamping collar. As the clamping collar travelled towards each other, the radial force was produced by conical ring which is located below clamping collar. Shrinkage of conical ring resulted in gripping between shaft and shaft sleeve. In **Figure 4(a)**, increase of contact pressure of 21.5% from 59.29 MPa to 72.04 MPa with High Thermal Expansion Shaft Sleeve 1 (HTE SS1) and 24% of increasedto75.84 MPa from 61.06 MPa are achieved by HTE SS (2) as the temperature of contact pressure between shaft and shaft sleeve elevated to 81.5℃. There was no pressure drop at HTE SS 1 and HTE SS 2as the temperature continuously increased. On the other side, the existing shrink disc with identical geometrical and material properties resulted in an inverse behaviour, where the contact pressure started to decrease as the temperature increased from 30℃ to 70℃. The original shaft sleeve (Ori. SS 1) shows reduction of approximately 7.7 % and 31% of contact pressure at 70℃ and 110℃ respectively as the temperature continuously increased [12].

3.2 Effect of Temperature to Shaft Contact Area

Contact area of HTE SS 1 in **Figure 4(b)** has shown tremendous improvement with minimal reduction of 0.8% while HTE SS 2 increased to 0.07% during high thermal loading. The Ori. SS 1 and 2 were severely reduced in contact area with approximately 1/2 and 2/3 from the total contact area before the application of thermal loading. The improved design of shrink disc with utilizing high thermal expansion materials have shown to successfully eliminate the occurrence of slip at high thermal loading condition.

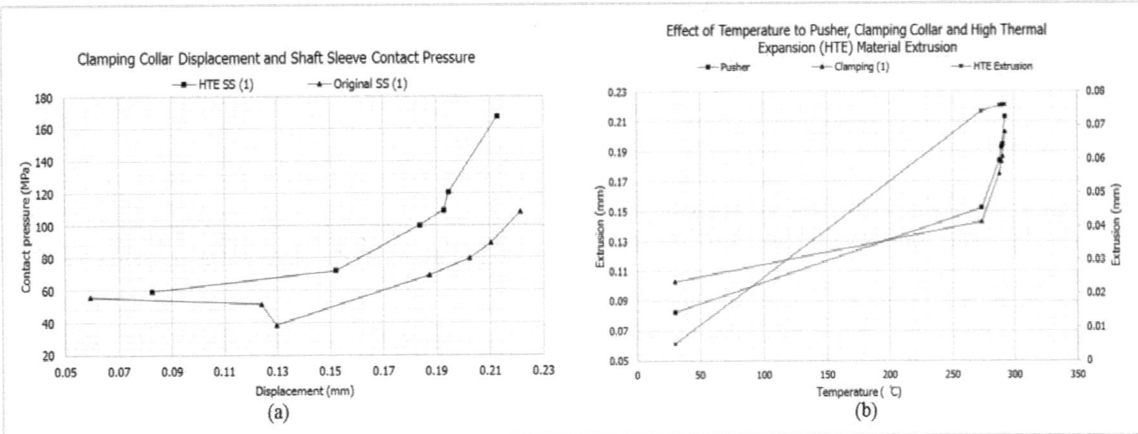

Figure 5(a) Clamping Collar Displacement vs. Shaft Sleeve Contact Pressure and **5(b)** Effect of Temperature to Pusher, Clamping Collar and HTE Material Extrusion

3.3 Clamping Collar Displacement and Shaft Sleeve Contact Pressure

Figure 5(a) shows clamping collar displacement of HTE SS (1) increased as the contact pressure increased and vice versa obtained by the original SS (1) where decline of 13 MPa has been obtained at 0.13 mm of clamping collar displacement, thus this would lead to slip of shaft and shaft sleeve.

3.4 Effect of Temperature to Pusher, Clamping Collar and HTE Material Extrusion

The initial extrusion of clamping collar, pusher and HTE component are due to bolt pretension at normal condition, 30℃ were 0.10 mm, 0.082 mm and 0.005 mm respectively. These components were consistently extruded due to thermal load applied. HTE expanded approximately 48% which is 0.069 mm from the normal condition, while pusher and clamping collar also showed the similar behaviour with 84%, 0.07 mm and 39%, 0.04 mm. This condition has successfully overcome the existing model drawbacks where slip of shaft and shaft sleeve occurred during thermal loading. It has been illustrated that with the High Thermal Expansion (HTE) design, the pusher managed to extrude to provide acting force on the shaft sleeve, whereby this was not achieved by the traditional shrink disc, thus this leads to the occurrence of slip between the shaft and the shaft sleeve.

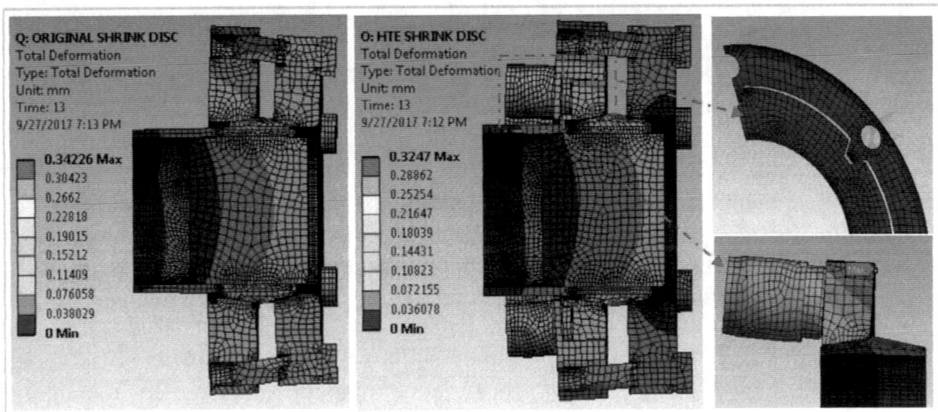

Figure 6 Simulation Results of Original and High Thermal Expansion Shrink Disc

Figure 6 further illustrates the extrusion condition encountered by the pusher, HTE and clamping collar at high temperature. The components experienced deformation and expansion during thermal loading. HTE components expanded and produced axial force which pressed the pusher forward simultaneously pushing the conical ring (ferrule) to shaft sleeve. This phenomenon has improved the contact pressure and area of shaft and shaft sleeve which are able to maintain the performance and deliverance of torque even at high thermal loading condition.

4. Conclusion

Compared to the existing shrink disc, the HTE concept model has improved the problem confronted by the existing shrink disc. The performance advancement on the contact pressure and contact area achieved by HTE model overcome the occurrence problems of slippage during thermal loading. Instead of producing gap the HTE holds the shaft and shaft sleeve tightly with contact pressure up to 167 MPa and 164 MPa at SS 1 and SS 2 respectively. The contact area also improved significantly with total contact area holds between shaft and shaft sleeve are at 629 mm^2 at shaft sleeve 1 and 1258 mm^2 at shaft sleeve 2. The reduction of 10.2 mm^2 at 81.5℃ deduced total contact area of 1244.1 mm^2, which is considered unaffected since the full contact area of shaft and shaft sleeve is 1254.3 mm^2at normal condition.

5. Acknowledgement

The author gratefully acknowledges ProEight Engineering and also Universiti Teknologi Malaysia for their technical advices and facilities.

References

[1] Bengeri, M. and W. Mack, *The influence of the temperature dependence of the yield stress on the stress distribution in a thermally assembled elastic-plastic shrink fit.* Acta Mechanica, 1994. **103**(1): p. 243-257.

[2] Sun, M.Y., et al., *Three-dimensional finite element method simulation and optimization of shrink fitting process for a large marine crankshaft.* Materials & Design, 2010. **31**(9): p. 4155-4164.

[3] Bengeri, M. and W. Mack, *The influence of the temperature dependence of the yield stress on the stress distribution in a thermally assembled elastic-plastic shrink fit.* Acta mechanica, 1994. **103**(1-4): p. 243-257.

[4] Malavolta, A.T., S.H. Evangelista, and M.E. Moreno, *The influence of thermo-mechanical stress state on the torque capacity of rotating interference fit assembly.* 2013.

[5] Jianmei, W., K. Jianfeng, and T. Liang, *Theoretical and experimental studies for wind turbine's shrink disk.* Proceedings of the Institution of Mechanical Engineers, Part C: Journal of Mechanical Engineering Science, 2014: p. 0954406214533529.

[6] Strozzi, A., et al., *Achievement of a uniform contact pressure in a shaft–hub press-fit.* Proceedings of the Institution of Mechanical Engineers, Part C: Journal of Mechanical Engineering Science, 2013. **227**(3): p. 405-419.

[7] Cerdán, J.P., et al. *Analysis of contact stresses in interference fit joints with circumferential round notch on the hub.* in *13th World Congress in Mechanism and Machine Science, Guanajuato, México.* 2011.

[8] He, G., et al., *Investigation of thermal expansion measurement of brass strip H62 after high current density electropulsing by the CCD technique.* Materials Science and Engineering: A, 2000. **292**(2): p. 183-188.

[9] Elhefny, A. and G. Liang, *Stress and deformation of rocket gas turbine disc under different loads using finite element modelling.* Propulsion and Power Research, 2013. **2**(1): p. 38-49.

[10] Croccolo, D., M. De Agostinis, and N. Vincenzi, *Failure analysis of bolted joints: Effect of friction coefficients in torque–preloading relationship.* Engineering Failure Analysis, 2011. **18**(1): p. 364-373.

[11] Jianmei, W., K. Jianfeng, and T. Liang, *Theoretical and experimental studies for wind turbine's shrink disk.* Proceedings of the Institution of Mechanical Engineers, Part C: Journal of Mechanical Engineering Science, 2014. **229**(2): p. 325-334.

[12] Mack, W. and M. Plöchl, *Transient heating of a rotating elastic–plastic shrink fit.* International Journal of Engineering Science, 2000. **38**(8): p. 921-938.

IOP Conf. Series: Materials Science and Engineering **328** (2018) 012025 doi:10.1088/1757-899X/328/1/012025

Validation of Analytical Damping Ratio by Fatigue Stress Limit

Faruq Muhammad Foong[1], Thein Chung Ket*[,1], Ooi Beng Lee[2], Abdul Rashid Abdul Aziz[3]

[1] School of Engineering and Physical Sciences, Heriot-Watt University, No. 1, Jalan Venna P5/2, Precinct 5, 62200 Malaysia.
[2] Intel PSG, PG 14, Plot 6, Bayan Lepas Technoplex, Medan Bayan Lepas, 11900 Penang, Malaysia.
[3] Center for Automotive Research and Electric Mobility (CAREM), Universiti Teknologi PETRONAS (UTP), 31750 Tronoh, Malaysia.

*c.thein@hw.ac.uk

Abstract. The optimisation process of a vibration energy harvester is usually restricted to experimental approaches due to the lack of an analytical equation to describe the damping of a system. This study derives an analytical equation, which describes the first mode damping ratio of a clamp-free cantilever beam under harmonic base excitation by combining the transverse equation of motion of the beam with the damping-stress equation. This equation, as opposed to other common damping determination methods, is independent of experimental inputs or finite element simulations and can be solved using a simple iterative convergence method. The derived equation was determined to be correct for cases when the maximum bending stress in the beam is below the fatigue limit stress of the beam. However, an increasing trend in the error between the experiment and the analytical results were observed at high stress levels. Hence, the fatigue limit stress was used as a parameter to define the validity of the analytical equation.

1. Introduction

The need for a sustainable energy source to power small electronics has caused an increase in research on energy harvesting. The idea of harnessing energy from ambient vibrations to power wireless sensors have been discussed in several literatures over the past decade [1–3]. The mechanism of a typical vibration energy harvester usually involves a cantilever beam with one end clamped to a vibrating structure; hence, subjecting the beam to base-excitation type motion. There are several approaches to convert the mechanical energy from the vibrating beam into electrical power, such as using piezoelectric transducers or through electromagnetic induction [4].

Optimisation of vibration energy harvesters have been the interest of most recent literatures [5–8]. The optimisation process generally involves design modifications in an attempt to produce more power or increase the frequency bandwidth of the device, while maintaining minimum size. Nevertheless, nearly all of the optimisation work involves experimental works or complex finite element algorithms due to the lack of analytical equations to fully describe the vibration system, particularly the damping parameter.

Often, the equations developed for a vibration system requires the damping parameter as an input. Previous work demonstrated a mathematical model equation describing the motion of a clamp-free cantilever beam under harmonic base excitation [9]. While this equation was proven to match closely with the experimental results, the equation was not fully analytical in terms that the damping parameter is expected to be obtained either from experiment or literature. Since damping varies according to the material and the dimensions of a cantilever beam, experimental input is preferred for better accuracy. This tells us that the mathematical model itself is not able to predict how the cantilever beam would behave prior to the experiment if an accurate damping ratio of the cantilever beam is unknown. Hence, the development of an analytical equation to accurately describe damping is highly necessary as this would allow for better hands-off predictions on the behaviour of the vibrating object and open up new paths to optimisation [10]. In an earlier work, an attempt was made to predict the damping of a material based on the stress distribution function of a cantilever beam during vibration [11]. Based on the study, an ideal damping equation was proposed, which relates the maximum bending stress and fatigue limit stress of a structure to its loss factor. However, this equation assumes the maximum bending stress as an experimental input.

In this work, an analytical equation describing the first mode damping ratio of a cantilever beam was derived by combining the equation of motion for the transverse vibration of a cantilever beam with a refined damping-stress equation [9,12]. The analytical equation can be used to predict the damping ratio of any clamp-free cantilever beam under harmonic base-excitation, provided that the beam's material properties and dimensions are known. After comparing the analytical results with the experimental findings, the importance of the fatigue limit stress on the quality of the analytical results was discussed and the validity of the equation was summarised.

2. Derivation of analytical equation

Figure 1. Cantilever beam under harmonic base excitation.

Figure 1 illustrates the movement of a vibrating clamp-free cantilever beam under harmonic base excitation. In general, the transverse motion of the clamp-free beam in Figure 1 at position x and time t can be described by equation (1).

$$a(x,t) = z(x,t) + y(t) \tag{1}$$

where $a(x,t)$ is the total amplitude of the vibrating beam, $z(x,t)$ is the amplitude of the beam relative to its clamped base and $y(t)$ is the harmonic base excitation amplitude. Based on the Euler-Bernoulli beam theory, the equation of motion for an undamped cantilever beam subjected to external force can be modelled using equation (2).

$$EI\frac{d^4z(x,t)}{dx^4} + \rho A\frac{d^2z(x,t)}{dt^2} = F(t) \tag{2}$$

IOP Conf. Series: Materials Science and Engineering **328** (2018) 012025 doi:10.1088/1757-899X/328/1/012025

where E is the Young's modulus of the cantilever beam, I is the second moment of area, ρ is the density, A is the beam's cross-sectional area and $F(t)$ is the forcing function distributed over the length of the beam. For cases of base excitation problem, the forcing function $F(t)$ signifies the driving force input of the base motion. Using the method of separation of variables, the term $z(x,t)$ can be separated into its spatial and temporal components.

$$z(x,t) = \sum_{n=1}^{\infty} \varphi_n(x)\eta_n(t) \tag{3}$$

where $\varphi_n(x)$ is the cantilever beam's modal-shape eigenfunction and $\eta_n(t)$ is the regular-response function. The subscript n in equation 3 denotes the correspondence to the nth mode of vibration. The cantilever beam's eigenfunction can be described as [9,13]

$$\varphi_n(x) = \left(\frac{1}{m}\right)^{1/2}\left[cosh\frac{\beta_n}{L}x - cos\frac{\beta_n}{L}x - \frac{sinh\beta_n - sin\beta_n}{cosh\beta_n + cos\beta_n}\left(sinh\frac{\beta_n}{L}x - sin\frac{\beta_n}{L}x\right)\right] \tag{4}$$

where m is the mass of the unclamped part of the beam, L is the length of the beam and β_n is a dimensionless constant. The regular response function is the solution to the following second-order differential equation. It is easy to notice that equation (5) can be solved by referring to a single degree of freedom vibration problem.

$$\ddot{\eta}_n(t) + 2\zeta_n\omega_n\dot{\eta}_n(t) + {\omega_n}^2\eta_n(t) = F(t) \tag{5}$$

where $\dot{\eta}_n$ and $\ddot{\eta}_n$ is the first and second derivative of the response function with respect to t, ζ_n, which is the cantilever beam modal damping ratio and ω_n is the nth mode natural frequency of the beam. The forcing function term can be expressed as [8].

$$F(t) = \frac{-m\omega^2 ye^{i\omega t}}{L}\int_0^L \varphi_n(x)dx \tag{6}$$

where ω is the frequency of the harmonic driving force input. Solving equation (7) and substituting the solution into equation (6), the solution for equation (6) for cases of harmonic motion can be obtained.

$$\eta_n(t) = \frac{2\omega^2 m^{1/2}(sinh\beta_n - sin\beta_n)}{\beta_n({\omega_n}^2 - \omega^2 + i2\zeta_n\omega_n\omega)(cosh\beta_n + cos\beta_n)} ye^{i\omega t} \tag{7}$$

Substituting Eqs. (7) and (4) into equation (3) and considering cases of first mode resonance where $\omega = \omega_n$ and $n = 1$, the real part of equation (3) becomes

$$z(x,t) = ye^{i\omega_1 t}\left[cosh\frac{\beta_1}{L}x - cos\frac{\beta_1}{L}x - \frac{sinh\beta_1 - sin\beta_1}{cosh\beta_1 + cos\beta_1}\left(sinh\frac{\beta_1}{L}x - sin\frac{\beta_1}{L}x\right)\right]\frac{sinh\beta_1 - sin\beta_1}{\beta_1\zeta_1(cosh\beta_1 + cos\beta_1)} \tag{8}$$

This accuracy of equation (8) has been validated by previous authors [9]. Applying the Euler-Bernoulli theory and Hooke's law, the stress acting on a cantilever beam can be described by the following equation

$$\sigma(x,t) = \frac{Eb}{2}\left(\frac{d^2z}{dx^2}\right) \tag{9}$$

where $\sigma(x,t)$ is the stress acting on the beam, b is the thickness of the beam and $\frac{d^2z}{dx^2}$ is the second derivative for the beam's vertical motion equation with respect to x. Substituting equation (8) into equation (9) and evaluating the beam at $x = 0$ and $ye^{i\omega_1 t} = y$, the expression for the maximum bending stress of a vibrating beam can be obtained.

$$\sigma_m = Eby\beta_1 \frac{sinh\beta_1 - sin\beta_1}{L^2\zeta_1(cosh\beta_1 + cos\beta_1)} \tag{10}$$

where σ_m is the maximum bending stress experienced by the beam. Equation (10) relates the maximum bending stress on a beam to its damping ratio. Equation (11) describes another equation relating these two variables. The equation is described as refined damping-stress equation that relates the maximum stress and fatigue limit stress of a cantilever beam to its first mode loss factor [12].

$$\sigma_m = Eby\beta_1\gamma_1 = E\left(2130.6\frac{{\sigma_m}^{0.3}}{{\sigma_f}^{2.3}} + 8176.7\frac{{\sigma_m}^{6}}{{\sigma_f}^{8}}\right) \tag{11}$$

where γ_1 is the first mode loss factor and σ_f is the fatigue limit stress of the beam. In general, the loss factor of a cantilever beam is twice the value of its damping ratio. Taking this into account and substituting equation (10) into equation (11), the analytical equation for the first mode damping ratio can be derived in the form of

$$\zeta_1 = \frac{E}{2}\left[2130.6\frac{\left(\frac{2Eby\beta_1K_1}{L^2\zeta_1}\right)^{0.3}}{{\sigma_f}^{2.3}} + 8176.7\frac{\left(\frac{2Eby\beta_1K_1}{L^2\zeta_1}\right)^{6}}{{\sigma_f}^{8}}\right] \tag{12}$$

where

$$K_1 = \frac{sinh\beta_1 - sin\beta_1}{cosh\beta_1 + cos\beta_1} \tag{13}$$

The dimensions and material properties of a cantilever beam are usually known. Hence, the only unknown parameter in equation (12) is the damping ratio of the beam. Since this unknown parameter appears on both sides of the equation, equation (12) can be solved using an iterative converging algorithm, such as the secant method or the Newton-Raphson method.

3. Results and Discussion

In this section, the analytical equation from equation (12) was compared to the experimental results. An aluminium and a steel cantilever beam were used in the experiments; their properties are tabulated in Table 1 below.

Table 1. Specifications of the specimen cantilever beams.

Material	E (GPa)	t (mm)	w (mm)	σ_f (Mpa)
Aluminium	60	1.20	18.14	95
Steel	190	0.50	12.75	310

IOP Conf. Series: Materials Science and Engineering **328** (2018) 012025 doi:10.1088/1757-899X/328/1/012025

The experiment involved subjecting a cantilever beam to a harmonic base-excitation vibration by clamping one end of the beam onto an electromagnetic shaker. The output response of the beam and the shaker was recorded using laser displacement sensors and the damping ratio was determined based on these responses. Figure 2 shows the schematic of the experimental setup.

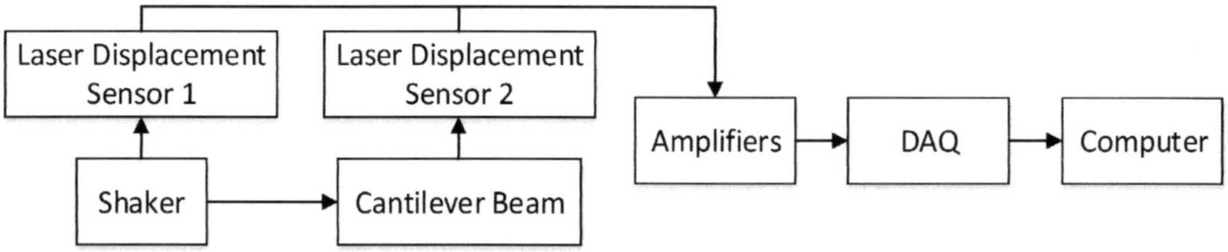

Figure 2. Schematic of the experimental setup used to determine the damping ratio of cantilever beams.

For each beam, the experiment was repeated several times to induce different damping ratios by changing the length of the beam or the base excitation amplitude. The thickness and the width of the beam remained fixed. Usually, the preferred method to determine the damping ratio of a beam specimen under base excitation cases is to use the half-power bandwidth method. However, this method can lead to inconsistent results as the damping ratio obtained is highly dependent on the quality of the frequency response curve. Hence, for a more substantial result, the experimental damping ratio was calculated by using equation (8), where the maximum relative amplitude at $x = L$ and the corresponding base excitation amplitude was obtained from the experiment. The maximum bending stress at the clamped end of the beam was calculated using equation (10) based on the determined damping ratio. Figure 3 and Figure 4 illustrate the comparison between the analytical and the experimental results.

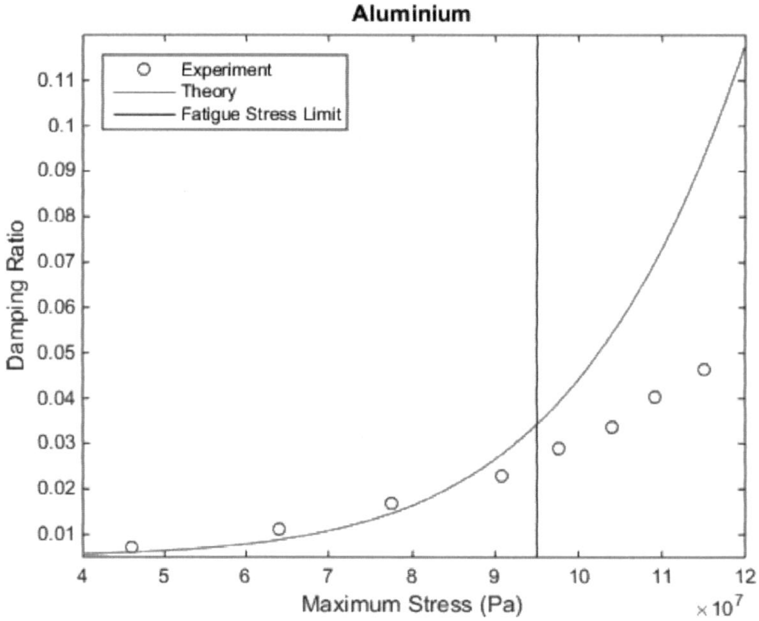

Figure 3. Comparison between experimental and analytical results for aluminium cantilever beam.

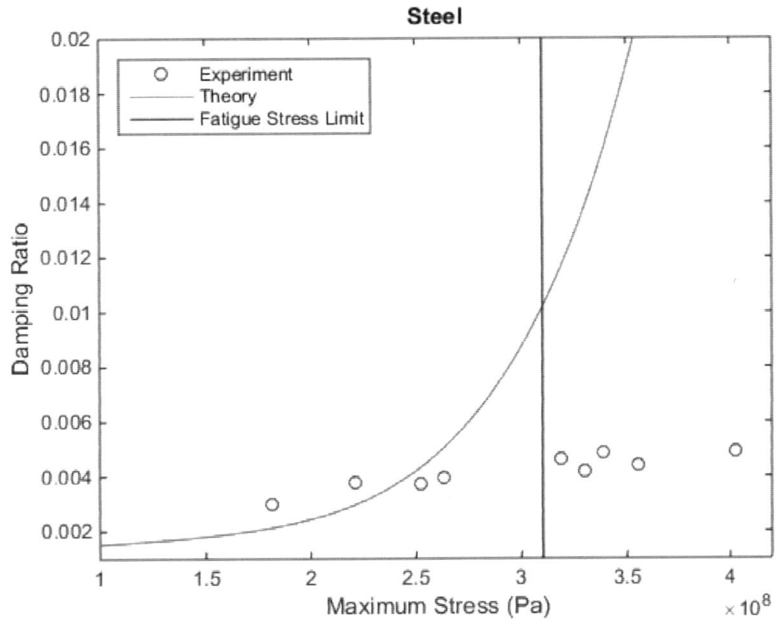

Figure 4. Comparison between experimental and analytical results for steel cantilever beam.

The results demonstrated that the experimental findings closely correlated to the analytical equation at low stress values, specifically below the fatigue limit stress of the beam. The maximum error recorded between the analytical and the experimental damping results at stresses below the fatigue limit stress for aluminium is approximately 21%, whereas for steel it is around 29%. The errors in both beams are seen to increase as the stresses exceed the fatigue limit stress, although steel recorded a more significant error with a maximum reaching 735%. It is easily noticed that although a large error was recorded for both beams at stresses above the fatigue stress limit, the damping ratio recorded at these stresses still displayed an increasing trend. This is in agreement to what was discussed in earlier studies in where the damping of a structure is highly dependent on the stress distribution of the structure [11]. The errors recorded in this work may be due to the fact that the damping-stress equation in equation (11) was a refinement of another damping-stress equation, which was described to be an idealised relation generalising high non-linear stress behaviour [11,12]. In reality, several other different types of damping-stress curve representations were also recorded in his work. Nevertheless, if the stress levels were constrained to values below the fatigue stress limit, the analytical equation can be concluded to provide a good estimation on the damping ratio of any cantilever beam. This, in turn, defines the validity of the analytical equation, where it is only valid within the mentioned constraints.

In terms of vibration energy harvesting application, current research focuses on design miniaturization as these devices are generally targeted to be installed in small electronics. Hence, the vibrating beam is not expected to vibrate at high stresses due to the volume constraint and would almost definitely not reach the fatigue stress limit. Therefore, it is possible to optimise these small devices and expect a fairly good prediction using the analytical equation. In general, the analytical equation can provide a good insight to designers or material scientists on the damping capacity of different materials, where instead of spending a large sum of money on material testing, one can apply the equation to have a broad idea on the average damping capacity of a material.

4. Conclusion

An analytical equation describing the first mode damping ratio of a clamp-free cantilever beam under base-excitation motion was proposed and the importance of the fatigue limit stress on the proposed equation was discussed. The analytical equation displayed a good agreement with experimental results when the cantilever beam experienced low stress levels. However, large errors were recorded when the maximum stress exceeded the fatigue limit stress of the beam, with the errors increasing further as the maximum stress in the beam increased. This limitation describes the validity of the analytical equation, where it was concluded that the equation would only be valid for cases where the maximum bending stress is below the fatigue limit stress. The work presented here does not account for cases where a lumped mass is placed at the free end tip of the cantilever beam. Hence, future plans include deriving the damping ratio equation for cases of cantilever beams with lumped mass and investigating the accuracy of this equation.

Acknowledgment
The authors would like to thank to Fundamental Research Grant Scheme (FRGS) from Ministry of Higher Education (MOHE) Malaysia, Grant No: FRGS/1/2016/TK03/HWUM/03/1, for funding this research.

References

[1] Ooi B L, Thein C K, Yew C K, Abdul Rashid A A 2016 *Biologically inspired energy harvesting through wireless sensor technologies* (Hershey: IGI Publishing) chapter 1 pp 1–22.
[2] Lu Y, Savvaris A, Tsourdos A, Bevilacqua M 2016 Vibration energy harvesters for wireless sensor networks for aircraft health monitoring *IEEE Metrology for Aerospace (MetroAeroSpace)* (Italy: Florence) 25–32.
[3] Wei C, Jing X 2017 A comprehensive review on vibration energy harvesting: Modelling and realization *Renew. Sustain. Energy Rev.* **74** 1–18.
[4] Kim M, Dugundji J, Wardle B L 2015 Efficiency of piezoelectric mechanical vibration energy harvesting *Smart Mater. Struct.* **24** L11.
[5] Thein C K, Liu J S 2017 Numerical modeling of shape and topology optimisation of a piezoelectric cantilever beam in an energy-harvesting sensor, *Eng. with Comput.* **33** 137–48.
[6] Thein C K Thein, Ooi B L, Liu J S, Gilbert J M 2016 Modelling and optimization of a bimorph piezoelectric cantilever beam in an energy harvesting application *J. Eng. Sci. Tech.* **11(2)** 212–27.
[7] Chen N, Bedekar V 2016 Modeling, simulation and optimization of piezoelectric bimorph transducer for broadband vibration energy harvesting *J. Mater. Sci. Research* **6(4)** 5–18.
[8] Dechant E, Fedulov F, Chashin D V, Fetisov L Y, Fetisov Y K, Shamonin M 2017 Low-frequency, broadband vibration energy harvester using coupled oscillators and frequency up-conversion by mechanical stoppers *Smart Mater. Struct.* **26** L11.
[9] Erturk A, Inman D J 2007 On mechanical modelling of cantilevered piezoelectric vibration energy harvesters *J. Intell. Mater. Syst. Struct.* **19** 1311–25.
[10] Danesh-Yazdi A H, Elvin N, Andreopoulos Y 2014 Green's function method for piezoelectric energy harvesting beams *J. Sound Vib.* **333** 3092–108.
[11] Lazan B J 1968 Damping of materials and members in structural mechanics (London: Pergamon Press).
[12] Kume Y, Hashimoto F, Maeda S 1982 Material damping of cantilever beams *J. Sound Vib.* **80(1)** 1–10.
[13] Sun W, Liu Y, Li H, Pan D 2013 Determination of the response distributions of cantilever beam under sinusoidal base excitation, *J. Physics: Conf. Ser.* **448** L11.

IOP Conf. Series: Materials Science and Engineering **328** (2018) 012026 doi:10.1088/1757-899X/328/1/012026

Hole Feature on Conical Face Recognition for Turning Part Model

A F Zubair[1] and M S Abu Mansor[2*]
[1,2] School of Mechanical Engineering, Engineering Campus, Universiti Sains Malaysia, 14300 Nibong Tebal, Pulau Pinang, Malaysia.
E-mail address: [1]afz15_mec008@student.usm.my, [2,*] Corresponding author: mesalman@usm.my

Abstract. Computer Aided Process Planning (CAPP) is the bridge between CAD and CAM and pre-processing of the CAD data in the CAPP system is essential. For CNC turning part, conical faces of part model is inevitable to be recognised beside cylindrical and planar faces. As the sinus cosines of the cone radius structure differ according to different models, face identification in automatic feature recognition of the part model need special intention. This paper intends to focus hole on feature on conical faces that can be detected by CAD solid modeller ACIS via .SAT file. Detection algorithm of face topology were generated and compared. The study shows different faces setup for similar conical part models with different hole type features. Three types of holes were compared and different between merge faces and unmerge faces were studied.

Keywords: *Automatic Feature Recognition, CAPP, lathe machining*

1. Introduction

Computer Aided Process Planning (CAPP) nowadays is the solution in bridging Computer Aided Design (CAD) and Computer Aided Manufacturing (CAM) system. Process planning encompasses the activities and functions to plan detailed set of procedures and instructions to produce a part. The first step on generating the process plan is feature recognition. Recent studies show automatic feature recognition (AFR) is among the best methods [1]. In AFR, there are several methods that had been used by researchers and feature recognition via topological data is among the best [2]. Other methods include attribute based [3], artificial neural network [4,5] and genetic algorithm (GA) [6]. Previous works on turning part model feature recognition had been done in [7] and more works on volume recognition of mill-turn part models had also been done[8,9].

Conical part model consist of unique characteristic and some considerations need to be made to recognise it. This paper provides useful information of part model with conical faces with two type of face setting (merges and unmerges) and the effect of hole feature on its surface. The topological of the part model will become distinctive as the characteristics change.

2. Process Overview

2.1 Part Model Preparation

CAD part models were modelled in commercial CAD software and being imported into .SAT file. The part model was then being added with three types of hole and this altogether generate four types of samples to be study. The part model shapes consist of two different types of conical faces. Samples are shown in figure 1. Besides having a separate two faces of each cone feature, one sample is being tested with merge cone faces. This will develop a single cone face for each cone. The merge faces sample is using the same sample of sample (a). The characteristics of conical features are calculated by sinus and cosines of the angle of the cone expressed in Eqn. (1) and Figure 2.

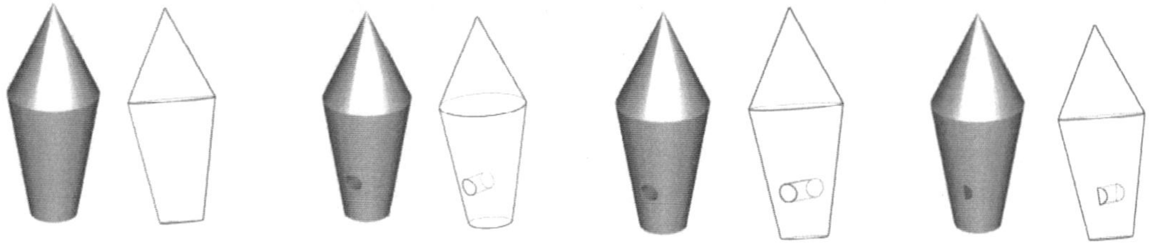

(a) (b) (c) (d)

Figure 1: Sample of conical faces and its wireframe model; (a) no hole (b) with blind hole (c) with through hole and (d) with half-circle hole

$$\boldsymbol{P(u,v)} = \boldsymbol{C_{base}} + \left(\mathbf{1} + \frac{\boldsymbol{s\,u\,s\,f_u}}{\boldsymbol{R}}\right)\boldsymbol{\widetilde{V}} + \boldsymbol{c\,u\,s\,f_u\hat{n}_{base}} \qquad (1)$$

Where $\boldsymbol{C_{base}}$ is ellipse's base centre, $\boldsymbol{sf_u}$ is u -parameter scaling factor negative if reverse when u is TRUE, $\boldsymbol{s}$ is sine angle if cosine angle is non-negative, or –sine angle if cosine angle is negative, $\boldsymbol{c}$ is cosine angle if cosine angle is non-negative or –cosine angle if cosine angle is negative, $\boldsymbol{\widetilde{V}}$ is vector from the base ellipse centre to a point on the base ellipse for a given $\mathcal{V}$-parameter value, $\boldsymbol{R}$ is length of the major axis of the base ellipse, and $\boldsymbol{\hat{n}_{base}}$ is base ellipse normal [10].

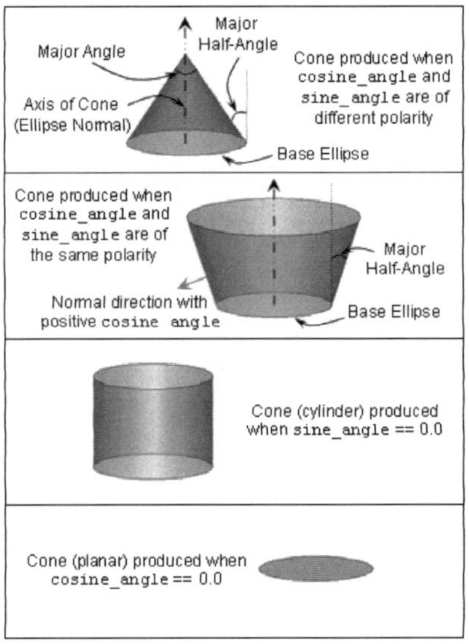

Figure 2: Cone definition [10]

2.2 Process framework

The setup of the algorithm will be based on the objective to detect the face orientation and topological structure. To carefully detect faces and topological behaviour, first of all, the topological structure of all body will be examine to know the overall lump, body, faces, edges and vertex of the 3D CAD model. The volume of the 3D CAD part model is then calculated for the comparison. Finally, the algorithm will detect every single face of the 3D CAD part model. By the detection, users will know each faces separately. In CAD modeller software, normally cone faces are treated by two faces. To make it one, user has to merge it. By merging the cone faces, the overall quantity of faces will be less. This will help the recognition of respective faces. Besides doing merging, faces also can be recognized by their loop and face type. By using loop type, one can recognize certain faces by separation loop, hole loop and periphery loop. Faces can be recognised by CAD solid modeller either as an analytic face or a spline face. Analytic face consists of planar face, cone face, sphere, torus face while spline face consists of B-Spline face. Figure 3 shows the flow of the algorithm.

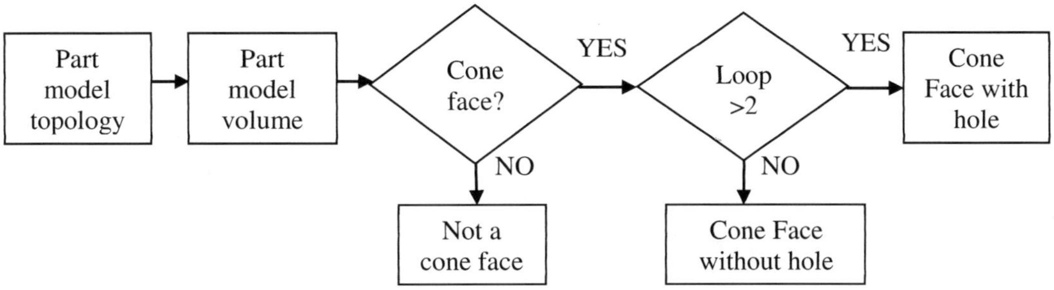

Figure 3: Process flow of algorithm

3. Results and Discussion

3.1 Topological data

Five samples are run through the algorithm. The volume of the part model sample was generated and faces were recognized. The calculation of the volume shows reduction if there is hole within the sample. As expected the volume of the through hole is greater than the blind hole. Table 1 shows the comparison of the volume and topological data of each sample. Sample 1 and Sample 6 shows the largest amount of volume as these samples had no hole. The differences of these two samples are the topological data of faces and edges. It is because the number of cone face had been reduced to half. Sample 5 shows three faces which are two cone faces and one planar face. Sample 1 shows five faces which are 4 cone faces and one planar face. Sample 2, sample 3 and Sample 4 show differences in volume as the types of hole are different. The samples with through hole (Sample 3 and Sample 4) show lesser volume compare to Sample 2 with blind hole. The number of faces of Sample 2 with blind hole is more than the through hole by the addition on one more face as the blind hole had another face at the bottom of the hole while through hole does not. This will reflect the number of edges and number of loop as well.

Table 1 Volume and topological data comparison

Sample	*Volume (mm3)*	*Lump*	*Faces*	*Edges*	*Loop*
Sample 1	60038.7	1	5	8	5
Sample 2	59360.4	1	8	15	8
Sample 3	58683.9	1	7	16	7
Sample 4	59360.4	1	7	16	7
Sample 5	60038.7	1	3	3	5

3.2 Face Recognition

The topological data and every face of the CAD model were being analysed. This individualised each face. By having the segregation, cone face, planar face and face with hole were detected. The behaviour of the detection was then concluded. Figure 4 shows the patent of face sequence of every sample. As algorithm detected the faces in loop sequence, face from F_{n-1} to F_n were displayed in the output window. By knowing the characteristic of each faces i.e : having hole, face type and number of loop, user will easily identify the faces sequence. By knowing this patent of sequence, future work can be done.

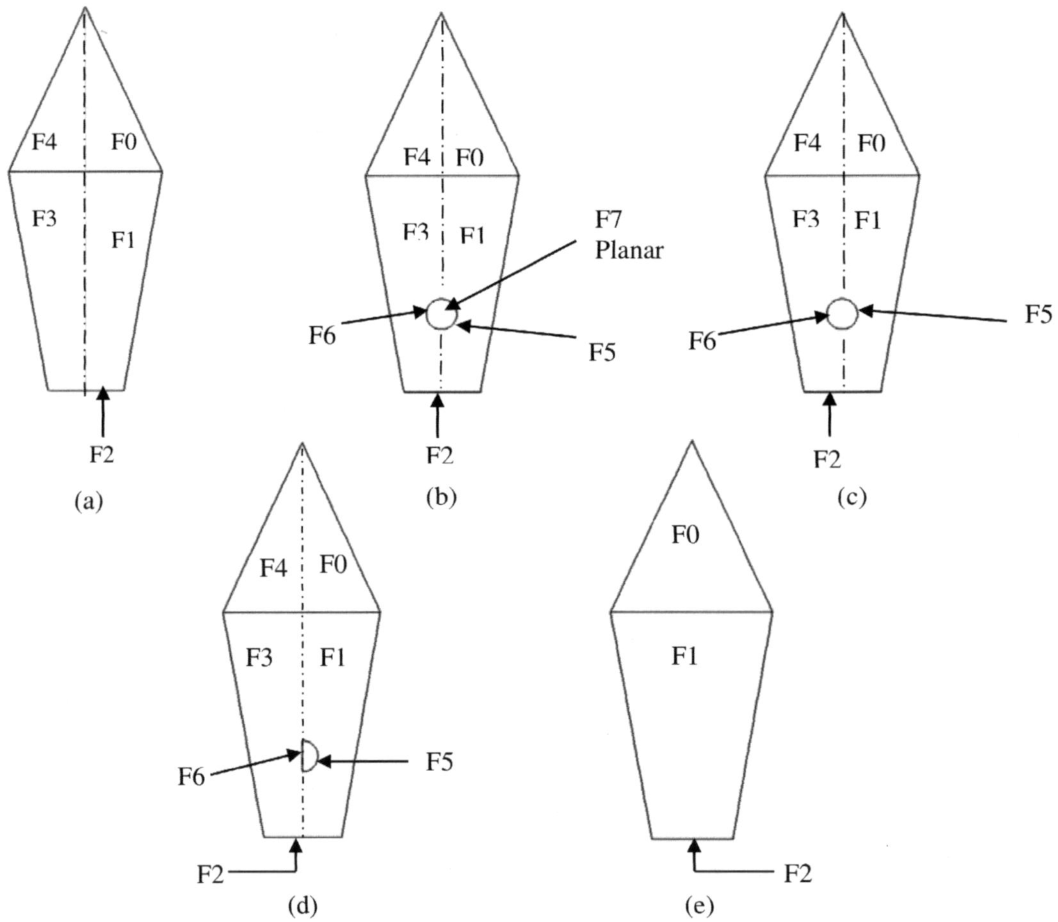

Figure 4: Face detection (a) Sample 1, (b) Sample 2, (c) Sample 3, (d) Sample 4, (e) Sample 5

Figure 4 shows the front view illustration of face configuration detected by the algorithm. Fig. 4 (a) shows face structure of Sample 1, it shows five faces starting from F0 to F4. The sequence is according to clockwise sequence. The middle line appears as the cone face is split by CAD modeller. F2 is planar face. Figure 4(b) acts the same as figure 4(a) with the addition of three faces from the blind hole: two faces of cylindrical faces and one more planar face. Sample 3 in figure 4 (c) shows some differences as the hole feature is through hole and two additional faces were generated. Next in figure 4(c) and 4(d), the configuration is in the same orientation which is clockwise. While in figure 4(e), there are only 3 faces; this is because F0 and F1 were generated from the merge face of F0 with F4 and F1 with F3.

4. Conclusions

The algorithm for cone face detection generation is proven to detect cone faces of CAD part model. The sequences of faces detected are shown in clockwise direction sequence. The volume differences of CAD part model with hole and without hole also as expected. Using rule of number of loop in each face, hole can be detected. Although more major works need to be done, this simple technique can be used in feature recognition method that will be useful for process planning.

Acknowledgment

This research is supported by Universiti Sains Malaysia under the Research University Grant (No: 814247). The first author also would like to thank the support of the Universiti Teknologi MARA for staff sponsorship.

References

[1] Xu X, Wang LH, Newman ST. *Computer-aided process planning: a critical review of recent developments and future trends.* Int J Comput Integr Manuf 2011;24:1–31. doi:10.1080/0951192x.2010.518632.

[2] Yusof Y, Latif K. *Survey on computer-aided process planning.* Int J Adv Manuf Technol 2014;75:77–89. doi:10.1007/s00170-014-6073-3.

[3] Abu R, Masine T. *Attribute based feature recognition for machining features.* J Teknol 2007;46:87–103.

[4] Deb S, Parra-castillo JR. *An Integrated and Intelligent Computer-Aided Process Planning Methodology for Machined Rotationally Symmetrical Parts.* Int J Adv Manuf Syst 2011;13.

[5] Younis M a., Abdel Wahab M a. *A CAPP expert system for rotational components.* Comput Ind Eng 1997;33:509–12. doi:10.1016/S0360-8352(97)00180-0.

[6] Su Y, Chu X, Zhang Z, Chen D. *Process planning optimization on turning machine tool using a hybrid genetic algorithm with local search approach.* Adv Mech Eng 2015;7. doi:10.1177/1687814015581241.

[7] Zubair AF, Abu Mansor MS. *Cylindrical axis detection and part model orientation for generating sub delta volume using feature based method.* ARPN J Eng Appl Sci 2016;11:13415–9. doi:ISSN 1819-6608.

[8] Kataraki PS, Abu Mansor MS. *Auto-recognition and generation of material removal volume for regular form surface and its volumetric features using volume decomposition method.* Int J Adv Manuf Technol 2016. doi:10.1007/s00170-016-9394-6.

[9] Bok AY, Abu Mansor MS. *Generative regular-freeform surface recognition for generating material removal volume from stock model.* Comput Ind Eng 2012;64:162–78. doi:10.1016/j.cie.2012.08.013.

[10] Spatial R25 Release Note. http: // spatial.com/Spatial/r25sp1_docs/articles/c/o/n/Cone.html 2015.

Auto-recognition of surfaces and auto-generation of material removal volume for finishing process

Pramod S Kataraki[1] and Mohd Salman Abu Mansor[2,*]

[1,2] School of Mechanical Engineering, Engineering Campus, Universiti Sains Malaysia,
14300 Nibong Tebal, Pulau Pinang, Malaysia.
E-mail address: [1] psk13_mec014@student.usm.my, [2,] * Corresponding author: mesalman@usm.my

Abstract. Auto-recognition of a surface and auto-generation of material removal volumes for the so recognised surfaces has become a need to achieve successful downstream manufacturing activities like automated process planning and scheduling. Few researchers have contributed to generation of material removal volume for a product but resulted in material removal volume discontinuity between two adjacent material removal volumes generated from two adjacent faces that form convex geometry. The need for limitation free material removal volume generation was attempted and an algorithm that automatically recognises computer aided design (CAD) model's surface and also auto-generate material removal volume for finishing process of the recognised surfaces was developed. The surfaces of CAD model are successfully recognised by the developed algorithm and required material removal volume is obtained. The material removal volume discontinuity limitation that occurred in fewer studies is eliminated.

Keywords: material removal volume, regional segmentation, lofting, finishing layer.

1. Introduction

The author defined the single setup machinable feature volumes as maximal features and the volume resulting from the subtraction of part model from stock model as delta volume. The delta volume was decomposed into large and simple maximal volumes and then maximal volumes were transformed into maximal features, by using a newly developed recursive maximal volume decomposition method. The recursive method consumes more time to decompose parts having feature intersections, and conversion of un-machinable maximal volumes into maximal volumes leads to generation of extra (unwanted) volume [1]. A hybrid feature recognition method (face pattern approach + volumetric decomposition method) was developed to recognise non-interacting parts and interacting features respectively. The developed algorithm was used only for cast-then-machined parts [2]. The author developed an algorithm to recognise parts of regular and freeform surfaces. The method recognised all forms of surfaces and also generated material removal volume for the finishing process and roughing process. However, the material removal volume of finishing process bore discontinuity between them and the percentage error obtained was greater than 1% [3]. An algorithm was developed for automatic recognition of only non intersecting machining features directly from 2D CAD input instead of converting 2D to 3D model for feature recognition. Methods were developed to determine the attributes of both isolated and non-isolated features. The developed algorithms are much simpler than the existing feature recognition algorithms that operate directly on 3D inputs [4]. An indirect extraction approach was applied to identify the lost design features and feature-related information of a CAD model from a data exchanged part model. A feature taxonomy was proposed based on the feature geometry and topological characteristics, from which design features were identified. The design features were classified into form features (convex, concave) and transitional features (generated by trimming and blending edges). The proposed approach was able to identify the lost design features and machining features from a data exchange part model. Simple and compound features can also be identified [5]. Although many researches have been carried out on automatic feature recognition, there is a need for a more optimal feature recognition and material removal volume generation method.

2. Methodology

The flow adopted by algorithm to recognise surfaces and their finishing layer generation is shown in **Figure 1**.

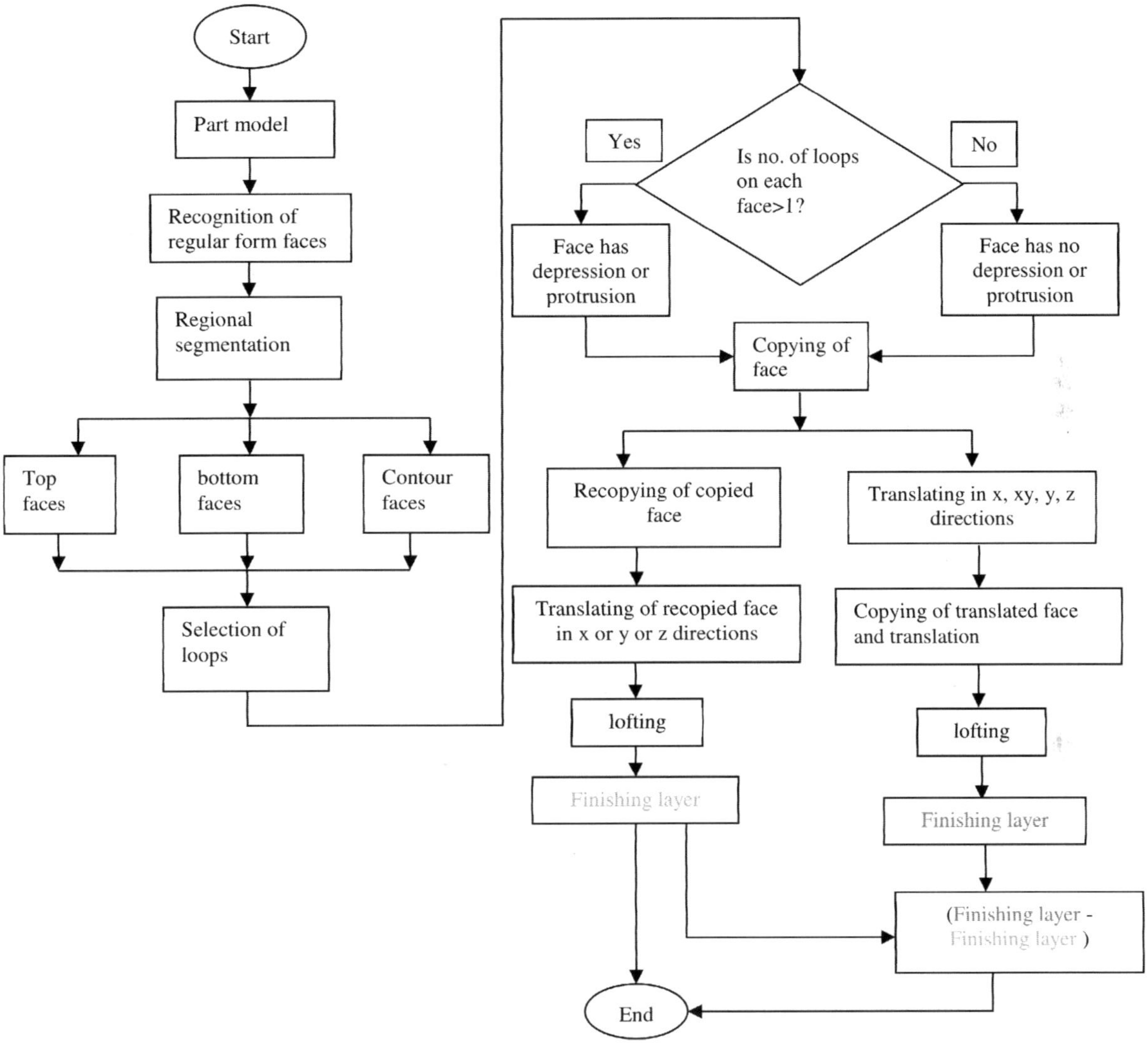

Figure 1 Flow chart of algorithm for recognition of surfaces and generation of finishing layer volume.

2.1 Part model

A part model, based on its geometrical shape can be of three forms: (i) regular form (ii) 2.5D form (iii) freeform. The part model having surfaces made of only straight and curve edges is said to be a regular form part model. The part model is said to be 2.5D form part model if its surfaces are made of straight, curve and spline edges, and part

IOP Conf. Series: Materials Science and Engineering **328** (2018) 012027 doi:10.1088/1757-899X/328/1/012027

model having surfaces made of only spline edges is said to be freeform part model. A solid body with 2-manifold boundary having CAD modeler file format extension must be input to algorithm.

2.2 Recognition of regular form faces

Surface is a superset of faces and is formed by joining of the set of faces. A component is formed by joining of these surfaces and geometrical evaluation application programming interface (API) functions. A surface recognition approach recognises the component face by face based on their geometrical shape. Faces having curve or NURBS patch or 3D B-spline surface are separated from the feces of cone, sphere, plane and torus geometrical shapes.

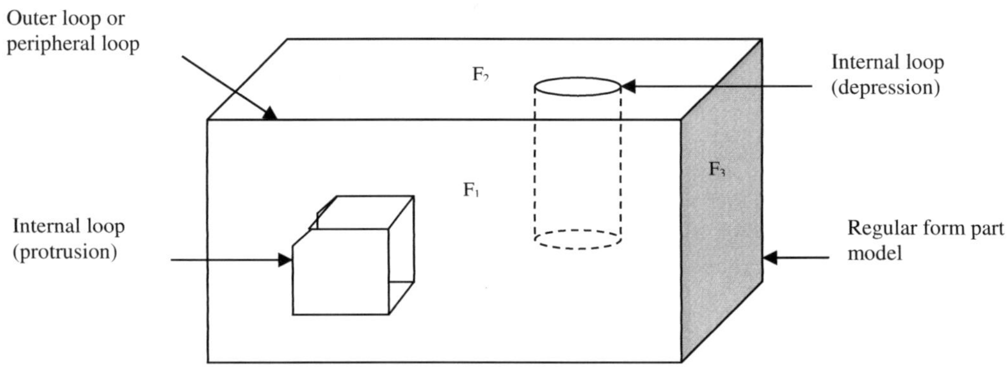

Figure 2 Geometric shapes of regular form part model

2.3 Regional segmentation

The segmentation of each face is achieved by considering its normal vector direction at midpoint. The three regions into which regular form faces are segmented: (i) top (ii) bottom (iii) contour. If a face's normal vector is greater than zero in z direction, then the face is segmented to top region. Similarly a face is segmented to bottom region if its normal vector is less than zero in z direction, and the face is segmented to contour region if its normal vector is equal to zero in the z direction and is less than or equal to 1 in x or y direction.

2.4 Selection of loops

The face is bound by loop and in turn loop is made of one or more co-edges. Each face is classified into face with depression or protrusion and face without depression or protrusion based on the following conditions (i) $2 > N > 0$; (ii) $\infty > N > 1$; 'N' is the number of loops on the face.

2.5 Regular form face with and without depression or protrusion

The surface recognition system recognises the faces of a CAD model that are joined together with a G0 continuity. The face with depression or protrusion will have a minimum of one loop within its peripheral loop, while a face without depression or protrusion consists of only a peripheral loop and no loops of void region, extruding volume exists within the peripheral loop. For example Face (F_3) contains no depression or protrusion (**Figure 2**) and satisfies the condition (i) as it contains only one loop. Faces (F_1 & F_2) contain protrusion, depression, respectively (**Figure** 2) and satisfy the condition (ii) as the faces contain more than one number of loops.

2.6 Copying, translation of face and lofting

The geometrical evaluation application programming interface (API) functions are used to copy segmented face. The copied face is recopied and translated in x or y directions for contour region, and in z direction for top and bottom regions. The table below shows the direction of translation for recopied faces. The copied face and translated face are lofted to generate finishing layer and the amount of thickness for finishing layer is given by the user as per the requirements.

Table 1 The face's translation direction based on its normal vector direction.

Face regions	Normal vector direction	Translation direction
Top	(Single direction) +z	+z
	(Multi direction) (+z) (+y)	(+z) (+y)
	(Multi direction) (+z) (-y)	(+z) (-y)
	(Multi direction) (+z) (-x)	(+z) (-x)
	(Multi direction) (+z) (+x)	(+z) (+x)
Bottom	(Single direction) -z	-z
	(Multi direction) (-Z) (+y)	(-Z) (+y)
	(Multi direction) (-Z) (-y)	(-Z) (-y)
	(Multi direction) (-Z) (-x)	(-Z) (-x)
	(Multi direction) (-Z) (+x)	(-Z) (+x)
Contour	(Single direction) +x	+x
	(Single direction) -x	-x
	(Single direction) +y	+y
	(Single direction) -y	-y
	(Multi direction) (+x) (+y)	(+x) (+y)
	(Multi direction) (-X) (-y)	(-X) (-y)
	(Multi direction) (+x) (-y)	(+x) (-y)
	(Multi direction) (-X) (+y)	(-X) (+y)

2.7 Boolean operation

The finishing layers of all region faces that are copied and translated only in single directions are Boolean subtracted from finishing layers of all region faces copied and translated into single and multi directions to obtain material removal volume for the discontinuity that exists between the two adjacent finishing layers generated from two adjacent faces that form convex geometry.

3. Results And Discussion

A regular form model with depression on one of its faces has been considered to validate the developed algorithm. **Figure 3(a)** shows the regional segmentation of part model and top region face contain depression, while contour and bottom region faces contain no depression and protrusion.

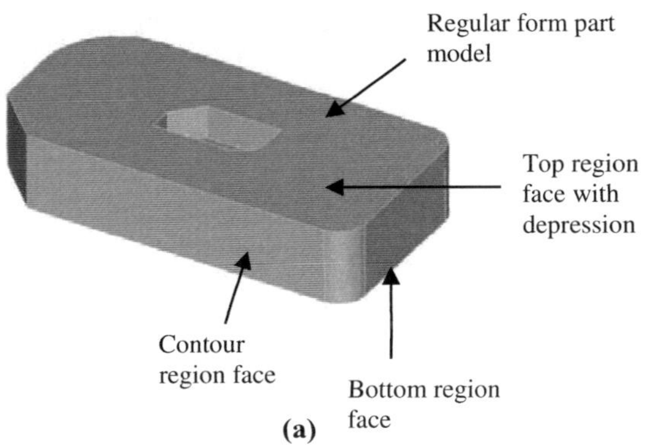

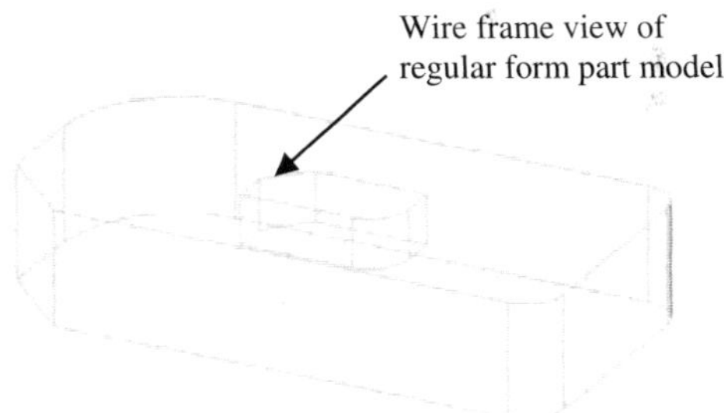

Figure 3. (a) Regular form part model with depression and regional segmentation. **(b)** Wire frame view of the regular form part model.

The generated material removal volume for finishing layer of all the faces of the regular form part model is shown in green and material removal volume that fills the discontinuity between the two adjacent finishing layers generated from two adjacent faces that form convex geometry is shown in red color. **Figure 4** shows finishing layers of all the faces and discontinuity filled material volume. The finishing layers of all region faces generated only in single directions are shown in **Figure 4(a)** and layers generated in both single and multiple directions are shown in **Figure 4(b).** The Boolean subtraction of layers of **Figure 4(a)** from layers of **Figure 4(b)** eliminates discontinuity problem, and a new layer of material volume is generated that fill all the discontinuity, as shown in **Figure 4(c)**.

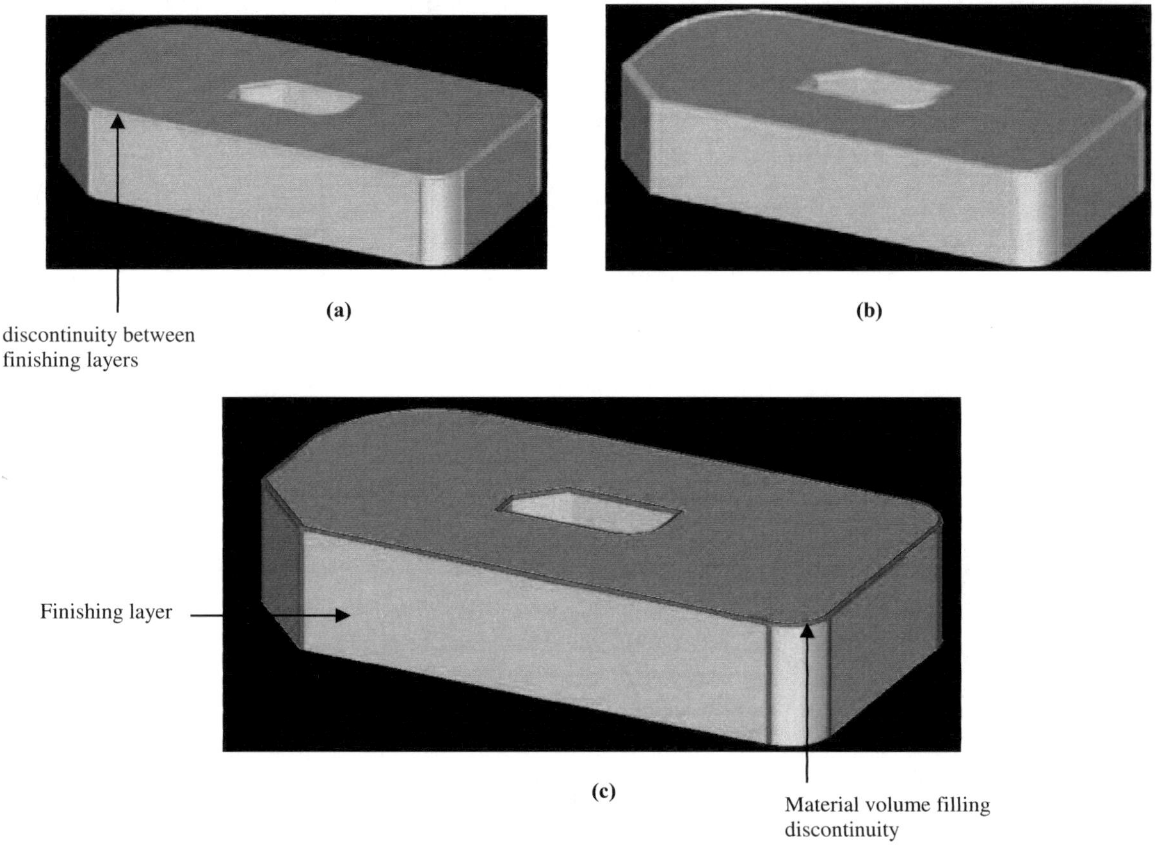

Figure 4. **(a)** Finishing layers (green color) of part model faces generated in single directions. **(b)** Finishing layers of part model faces are generated in both single and multiple directions. **(c)** Finishing layers and material removal volume (red color) filling the discontinuities.

Conclusion

The discontinuity that used to occur in previous research works is now eliminated and a new material volume is generated that fills the discontinuities between finishing layers. The current results show that a continuous tool path can now be maintained during machining process. The finishing layer thickness can be of any value as per users' requirements and with that, even a thickness of 1mm can be generated.

Acknowledgment

This research is supported by Universiti Sains Malaysia under the Research University Grant (No: 814247). The first author would also like to thank the support of National Overseas Scholarship by Ministry of Tribal Affairs, New-Delhi, India.

References

[1] Y. Woo and H. Sakurai, "Recognition of maximal features by volume decomposition," *Computer-Aided Design*, vol. 34, no. 3, pp. 195–207, 2002.
[2] Y. S. Kim and E. Wang, "Recognition of machining features for cast then machined parts," *Computer-Aided Design*, vol. 34, no. 1, pp. 71–87, 2002.
[3] A. Y. Bok and M. S. Abu Mansor, "Generative regular-freeform surface recognition for generating material removal volume from stock model," *Comput Ind Eng*, vol. 64, no. 1, pp. 162–178, 2013.
[4] L. Tyan and V. Devarajan, "Automatic identification of non-intersecting machining features from 2D CAD input," vol. 30, no. 5, pp. 357–366, 1998.
[5] M. W. Fu, S. K. Ong, W. F. Lu, I. B. H. Lee, and A. Y. C. Nee, "An approach to identify design and manufacturing features from a data exchanged part model," vol. 35, pp. 979–993, 2003.

Effect of Tube Diameter on The Design of Heat Exchanger in Solar Drying system

Shaymaa Husham Abdulmalek[1], Morteza Khalaji Assadi[1], Hussain H. Al-Kayiem[1], Ali Ahmed Gitan[2]

[1]Universiti Teknologi PETRONAS, Solar Thermal Advanced Research Center, Bandar Seri Iskandar, Perak, Malaysia.

[2]Tikrit University, Faculty of Engineering, Mechanical Engineering Department, Tikrit, Iraq.

morteza.assadi@utp.edu.my

Abstract. The drying of agriculture product consumes a huge fossil fuel rates that demand to find an alternative source of sustainable environmental friendly energy such as solar energy. This work presents the difference between using solar heat source and electrical heater in terms of design aspect. A circular-finned tube bank heat exchanger is considered against an electrical heater used as a heat generator to regenerate silica gel in solar assisted desiccant drying system. The impact of tube diameter on the heat transfer area was investigated for both the heat exchanger and the electrical heater. The fin performance was investigated by determining fin effectiveness and fin efficiency. A mathematical model was developed using MATLAB to describe the forced convection heat transfer between hot water supplied by evacuated solar collector with 70 °C and ambient air flow over heat exchanger finned tubes. The results revealed that the increasing of tube diameter augments the heat transfer area of both heat exchanger and electrical heater. The highest of fin efficiency was around 0.745 and the lowest was around 0.687 while the fin effectiveness was found to be around 0.998.
Keywords: (solar energy, air- water heat exchanger, silica gel, drying)

1. Introduction

High thermal energy storages are gained from evacuated tube solar collectors that use water to transfer heat to another system through an energy converter such as heat exchanger [1-4]. Several different types of heat exchangers have been developed for different applications being the shell-and-tube and plate-fin the most commonly utilized configurations. Because many heat exchanger arrangements involve multiple rows of tubes in a fluid cross flow, the heat transfer characteristics for tube banks are of important practical interest. The gas flowing on the outside impinges perpendicularly on these tubes [5]. With constant fluid velocity, this type of flow gives rise to an increase in turbulence compared with the case where the gas flows along the tubes and parallel to them. Heat exchangers are extensively used in diverse industrial processes nowadays. Many previous researches have presented different designs of heat exchanger used in different applications such as reactivation the silica gel as shown in Figure1 by using many sources of heat for example, solar energy [6, 7]. In an industrial application, the silica gel is regenerated by using fossil full like electrical heater at 100 °C. A desiccant system was used to dry crushed oil palm fronds [6, 8]. Solar energy was used to regenerate the silica gel through a heat exchanger. A theatrical analysis was carried out with an optimization of a gas-to-gas heat exchanger with a non-constant cross sectional area [9]. The results obtained from the modeling demonstrate the premise that it is possible to realize designs for heat exchangers that are highly exergy-efficient and very cheap. Some researchers studied the performance of heat exchanger starting from the CFD simulations at the micro-scale and was thus proposed. A set of control volumes that include the finned surfaces of both the cold and hot sides (separated by a wall) were detected [10-12]. This work aims to investigate the effect of diameter of tube bank heat exchanger on the heat transfer area of heat exchanger in comparison with an electrical heater for the purpose of desiccant regeneration in solar assisted desiccant drying system.

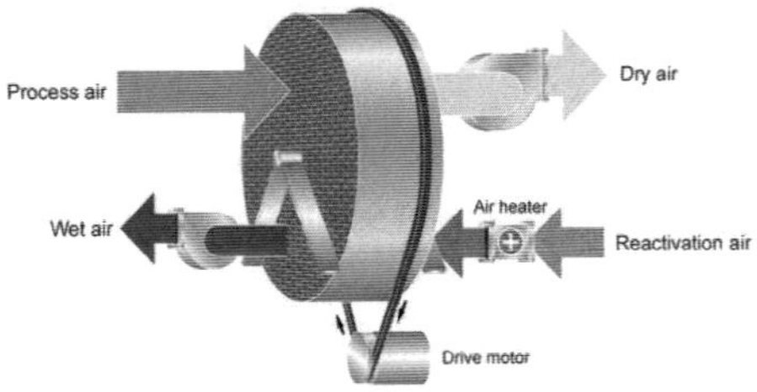

Figure1. Desiccant wheel air process

2. Heat exchanger input and mathematical model

The concept of the circular finned tub bank heat exchanger is shown in Figure 2 was adopted for building the mathematical model. The input parameters required for the mathematical model is defined in Table 1.

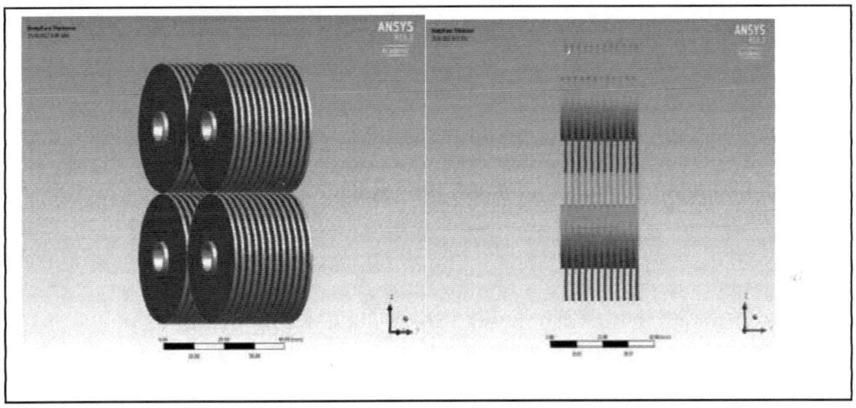

Figure 2. Tube bank water-air heat exchanger

Table 1. Input Parameters for Heat Exchanger Design

Parameter	T_{ai} (k)	D_o (inch)	T_{ao} (k)	T_{wi} (k)	S (W/m^2)	η_c (%)	F	C_n	Nt	N_f	A (m^2)	N_{row}	N_{col}
value	303	0.375	328	343	800	50	0.95	0.95	15	4	5	2	2

The heat gained from solar radiation (S) through solar collector is calculated by Eq. (1).

$$Q_{act} = \eta_c A_c S \tag{1}$$

The energy balance between air and water is modeled in equations (2-4).

$$Q_{act} = Q_w = Q_a \tag{2}$$

Heat absorbed by air, is

$$Q_a = m_a \, Cp_a \, (T_{ai} - T_{ao}) \tag{3}$$

Which as same as the heat released from water,

$$Q_w = m_w \, Cp_w \, (T_{wi} - T_{wo}) \tag{4}$$

To predict the heat transfer from the bank of tubes to air flow, heat exchanger roles may be applied. Hence, Reynolds number, as the key parameter in the internal forced convection heat transfer is a prominent factor. For the present geometry, Re for the tube water flow, Re_w is:

$$Re_w = \frac{\rho_w V_w D_i}{\mu_w} \tag{5}$$

The Nusselt number for internal flow inside circular tube is given by Equation (6) and (8) for laminar and turbulent flow, respectively [6].

$$Nu_w = 4.364\left(1+\left(\frac{Gz}{29.6}\right)^2\right)^{\frac{1}{6}}\left(1+\left(\frac{\frac{Gz}{19.04}}{\left(1+\left(\frac{Pr_w}{0.0207}\right)^{\frac{2}{3}}\right)^{0.5}\left(1+\left(\frac{Gz}{29.6}\right)^2\right)^{\frac{1}{3}}}\right)^{\frac{3}{2}}\right)^{\frac{1}{3}} \quad \text{For} \quad Re_w \le 2300 \tag{6}$$

Where:

$$Gz = \left(\frac{pi}{4}\right)\left(\frac{L}{DiRe_w Pr_w}\right)^{-1} \tag{7}$$

$$Nu_w = 0.023\, {Re_w}^{0.8} {Pr_w}^{0.3} \qquad \text{For} \quad Re_w > 2300 \tag{8}$$

The Nusselt number of external flow over hot tube bank array is given by equations (9-14) [6].

$$Nu_a = 0.9\, C_n {Re_a}^{0.4} {Pr_a}^{0.36}\left(\frac{Pr_a}{Pr_{a_s}}\right)^{0.25} \quad \text{For} \quad 1 < Re_a < 10^2 \tag{9}$$

$$Nu_a = 0.52\, C_n {Re_a}^{0.5} {Pr_a}^{0.36}\left(\frac{Pr_a}{Pr_{a_s}}\right)^{0.25} \quad \text{For} \quad 10^2 < Re_a < 10^3 \tag{10}$$

$$Nu_a = 0.27\, C_n\, {Re_a}^{0.63}\, {Pr_a}^{0.36}\left(\frac{Pr_a}{Pr_{a_s}}\right)^{0.25} \quad \text{For} \quad 10^3 < Re_a < 2\times10^5 \tag{11}$$

$$Nu_a = 0.033\, C_n\, {Re_a}^{0.8} {Pr_a}^{0.4}\left(\frac{Pr_a}{Pr_{a_s}}\right)^{0.25} \quad \text{For} \quad 2\times10^5 < Re_a < 2\times10^6 \tag{12}$$

$$\text{Where } Re_a = \frac{\rho_a V_a D_o}{\mu_a}.$$

All air properties are taken at film temperature defined as ($T_f = \frac{T_{ai}+T_s}{2}$) by considering surface temperature ($T_s = \frac{T_{wi}+T_{wo}}{2}$) except Pr_{a_s} which is at surface temperature. The overall heat transfer coefficient (U) of heat exchanger is determined as follows.

$$U = \frac{1}{P_i\, D_i L(R_w+R_s+R_a)} \tag{13}$$

Where R_w, R_s and R_a are thermal resistances of water side, solid wall and air side, respectively and they are given by

$$R_w = \frac{1}{A_w h_w},\ R_s = \frac{\ln(\frac{D_o}{D_i})}{2\pi L k_s} \text{ and } R_a = \frac{1}{A_a h_a} \tag{14}$$

Where $h_w = \frac{Nu_w k_w}{D_i}$, $h_a = \frac{Nu_a k_a}{D_o}$ and the thermal conductivity of copper (k_s = 400 W/m.K).

For unmixed flow heat exchanger, the temperature changes in logarithmic trend. The temperature difference can be calculated by using the logarithmic mean temperature difference (LMTD) as in the following equation.

$$\text{LMTD} = \frac{(T_{wi}-T_{ao})-(T_{wo}-T_{ai})}{\log\left(\frac{(T_{wi}-T_{ao})}{(T_{wo}-T_{ai})}\right)} \tag{15}$$

Equation 15 can be used for counter and parallel flow heat exchangers while for cross flow heat exchanger, it needs to use a correction factor (*F*). The total heat transfer area of heat exchanger is calculated as follows.

$$A = \frac{Q_{act}}{UF(LMTD)} \tag{16}$$

The fin performance can be determined by calculating fin efficiency and effectiveness.

$$\eta_f = \frac{\tanh(MLc)}{MLc} \tag{17}$$

Where $M = \left(\frac{h_f P_f}{K_s\, A_{fc}}\right)^{0.5}$ and the critical length ($Lc = L + \frac{t_f}{2}$).

$$\epsilon = 1 - \left(\left(\frac{A_u}{A}\right)(1-\eta_f)\right) \tag{18}$$

The MATLB algorithm is divided into two main parts. First part is related to heat exchanger calculations which aim to predict the length of heat exchanger from heat transfer area. Starting with initial value of heat exchanger length (L = 0.5 m) and set of input data (Table 1), the MATLAB code solves in closed iteration loop the value of heat transfer area as modeled in Eq. (3 – 16). A convergence criterion is adopted to predict the heat exchanger pipe length by comparing the old value with its new value. Second part is about the electrical heater calculations which performed separately out of the iteration loop to determine the heat transfer area of the heater.

3. Result and discussion

The water-air heat exchanger for regeneration purpose of silica gel material at 55°C was designed and analyzed. The aim of this work is to use renewable energy instead of fossil fuel in desiccant regeneration process. Usually the silica gel is regenerated by using electrical heater of 100 °C surface temperature. In the current work, a heat exchanger was designed to replace the electrical heater by solar energy. The results showed that the design of tube bank heat exchanger was almost the same in terms of heat transfer area with the electrical heater as illustrated in Figure 3 and Figure 4. The total and finned areas of heat exchanger were influenced by changing the diameter of the tube as shown in Figure 3. Generally, an increasing of tube diameter led to an increase in both areas. This gives an indication of that to minimize the size of heat exchanger a reduction in heat transfer area is required by selecting smaller diameter of 0.375 inch that provide the same outlet air temperature from the heater of around 55 °C. The finned area was slightly less than total area and the unfinned area represented that difference. The fin efficiency as a measurement of fin performance was investigated as well in this work. Figure 5 represents the variation of fin efficiency with the outlet diameter of heat exchanger tube as a design input parameter. The efficiency increases with the increasing of the diameter until reaches its maximum value of around 74.5% at a diameter value of around 20.3 mm. by further increasing of diameter, the efficiency degrades. The fin effectiveness has been considered to show the performance of fin and it was found to be around 0.998 which is almost unity. This means any additional increase in fin diameter will not enhance the heat transfer.

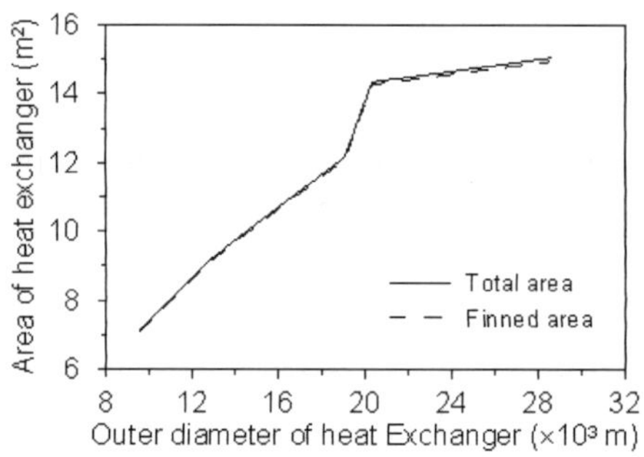

Figure 3. The effect of outer diameter on total area and finned area of heat exchanger

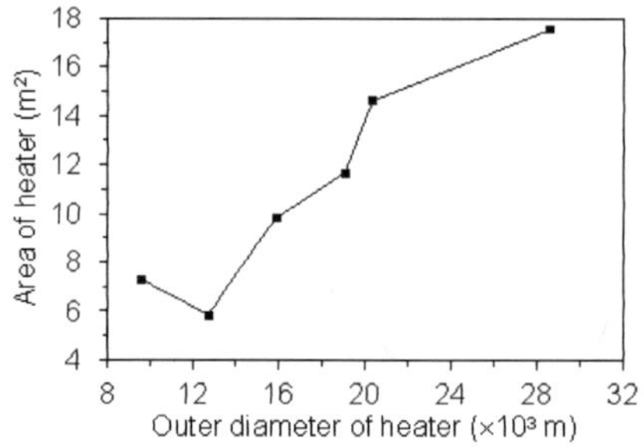

Figure 4. The effect of outer diameter on the area of the electrical heater

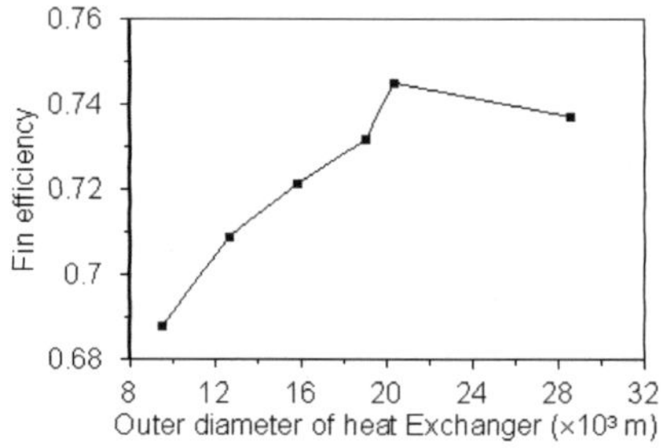

Figure 5. The effect of outer diameter on fin efficiency of heat exchanger

4. Conclusion

In conclusion, this work gives a clear indicate about the using of solar energy to regenerate the silica gel at a suitable temperature. The results illustrate the effect of tube diameter of tube bank water-air heat exchanger on different designs and operation conditions. It was founded that the heat exchanger at dimeter of 0.375 inch can provide hot air at 55°C to regenerate the silica gel. Also the fin efficiency was in the range of (0.745 - 0.687) while the fin effectiveness was found to be around 0.998.

Acknowledgment

This work was financially supported by Yayasan Universiti Teknologi PETRONAS (YUTP) grant number 0153-AA-A96

References

[1] A. Al-Abidi, S. Mat, K. Sopian, Y. Sulaiman, A. Mohammad 2016, *Heat Transfer Engineering* **37**(7-8) 705-712.
[2] S. Jakhar, M.S. Soni, N. Gakkhar, *Solar Energy* **157** (2017) 81-93.
[3] F. Zaversky, M. Sánchez, D. Astrain, *Energy* **65(Supplement C)** (2014) 647-664.
[4] Z.-Y. Liu, P.S. Varbanov, J.J. Klemeš, J.Y. Yong, *Applied Thermal Engineering* **105(Supplement C)** (2016) 755-762.
[5] A. Safwat Wilson, M. Khalil Bassiouny, *Process Intensification* **39**(1) (2000) 1-14.
[6] S. Misha, S. Mat, M.H. Ruslan, E. Salleh, K. Sopian, *Solar Energy* **112** (2015) 194-204.
[7] K.J. Chua, *Applied Thermal Engineering* **89** (2015) 703-716.
[8] S. Misha, S. Mat, M.H. Ruslan, E. Salleh, K. Sopian, *Solar Energy* **132** (2016) 415-429.
[9] B. Cárdenas, S.D. Garvey, B. Kantharaj, M.C. Simpson, *Journal of Energy Storage* (2017).
[10] O. Ipek, B. Kiliç, B. Gürel, 2017.
[11] N.H. Abu-Hamdeh, K.H. Almitani, *Solar Energy* **134** (2016) 202-210.
[12] Y. Jiang, T. Ge, R. Wang, *Energy and Buildings* **65**(Supplement C) (2013) 220-230.

IOP Conf. Series: Materials Science and Engineering **328** (2018) 012029 doi:10.1088/1757-899X/328/1/012029

Effect of inlet cone pipe angle in catalytic converter

Nurul Amira Zainal[1], Ezzatul Farhain Azmi[2] and Mohd Arifin Samad[3]

[1,2]Faculty of Engineering Technology,
Universiti Teknikal Malaysia Melaka,Hang Tuah Jaya, 76100 Durian Tunggal, Melaka, Malaysia.

Email:*[1]nurulamira@utem.edu.my

Abstract: The catalytic converter shows significant consequence to improve the performance of the vehicle start from it launched into production. Nowadays, the geometric design of the catalytic converter has become critical to avoid the behavior of backpressure in the exhaust system. The backpressure essentially reduced the performance of vehicles and increased the fuel consumption gradually. Consequently, this study aims to design various models of catalytic converter and optimize the volume of fluid flow inside the catalytic converter by changing the inlet cone pipe angles. Three different geometry angles of the inlet cone pipe of the catalytic converter were assessed. The model is simulated in Solidworks software to determine the optimum geometric design of the catalytic converter. The result showed that by decreasing the divergence angle of inlet cone pipe will upsurge the performance of the catalytic converter.

1. Introduction

Every vehicle production in automotive industries requires the main component to produce power and make the transport move. Vehicle engine acts as the core of the vehicle body that generate mechanical combustion power of the air-fuel mixture in its system. High pressure through the exhaust valve, the piston pushes the burning smoke out of the chamber as waste products is produced by the combustion of the air-fuel mixture in a combustion chamber and applies force to some part of the engine such as pistons. When the force is applied, piston move to top dead centre (TDC) and to bottom dead centre (BDC). Karuppusamy [1] stated that pressure rises when the piston moves from BDC to TDC and gases are pushed into the exhaust pipe afterwards or unwanted substances which contain chemical material such as Nitroxide (NOx), Carbon Monoxide (COx) and unburned Hydrocarbons (HC). This gas can cause air pollution and affect the environment. However, adding the catalytic converter in the middle of the exhaust pipe can diminish the waste product from emission to environments.

In recent years, much improvement has been made in computer modelling of catalytic converter to assist in design optimization with most of this work focused towards the automotive industry [2]. Major effort has been devoted into the design of a converter which lead to maximum use of catalyst volume. The maximal utilization of the catalyst volume would be attained by having a uniform flow distribution through the monolith substrate [3]. To achieve the uniform flow between sections of different cross-sectional areas, most modern catalytic converters have long tapered inlet and outlet headers. The flow distribution in the monolith catalytic converter have been widely investigated. Lemme [4] conducted an analytical and experimental treatment of flow through catalytic converters. Whereas Lai [5] simulated three-dimensional internal flow of the automotive catalytic converter. Jeong [6] implemented a CFD approach to investigate the three-dimensional unsteady flow in the catalytic converter. It is clearly known that the flow distribution in the monolith catalytic converter is influenced by many factors. Therefore, in order to improve the flow distribution, suggestions concerning baffle design and contoured substrates have been introduced. Howit [7] investigated the flow in round cross-section monolith converters with conical inlet and outlet headers. The study found that the monolith flow field was extremely dispersed. Additionally, Wollin [8] studied the flow performance of ceramic contoured substrates for automotive exhaust catalyst system. It is shown that a countoured substrate can provide improvements in flow uniformity. On another study, Wendland [9] found an insignificant effect of truncating the inlet and outlet diffusers of a monolith converter. Meanwhile, Lee [10] studied the effects of engine operating conditions in catalytic converter temperature in an SI engine. It was found that ignition retard and misfire can result in deactivation of the catalytic converter, which eventually leads the drastic thermal aging of the converter.

An experimental optimization of designing the catalytic converter is extremely expensive and time consuming since the design process involves building several prototypes with different geometries [11]. The flow inside the catalytic converter is very sensitive to geometric deviations thus these models must be exact and precise. Consequently, a computational approach to find an optimum design of the catalytic converter is established. Kumar [12] in his study employed a finite element approach to analyse a multi leaf spring using CAE tools. In the meantime, Vinkel [13] make comparisons between CAE approach and experimental results of leaf spring in the automotive industry. Wu [14] studied the effect of upstream flow distribution on the light of the performance of the catalytic converter by using numerical simulation. Also, Koltsakis[15] compared the cumulative emissions for the uniform and non-uniform distribution flow distributions by using numerical approach.

The design of the catalytic converter is very critical which require deep understanding. For the above reasons, the present work aims to study the effect of three different geometry angles of inlet cone pipe in the catalytic converter to get a better understanding of the catalytic converter processes. The purpose of this study is also to examine the relationship between the flow distribution and the performance of the catalytic converter inside three different models as proposed in order to optimize the catalytic converter design.

2. Design and Geometry

The catalytic converter is ideal based on a straight pipe. Size enlargement and additional angles were modified right after the inlet pipe hole gives some effect of the gas flow in the pipe. Figure 1 shows an example of a catalytic converter.

IOP Conf. Series: Materials Science and Engineering **328** (2018) 012029 doi:10.1088/1757-899X/328/1/012029

Figure 1: Basic Design of the Catalytic Converter

This study performed a complete drawing, including the inlet pipe, the substrate and the outlet of the catalytic converter. The inlet pipe diameter is set to be 47.6mm. The substrate length is 114.5mm where the diameter is 76.2mm. Total pipe length is measured as 31.5mm. As for the cone pipe angle, the model of the catalytic converter is set to be 30° degree (model 1) where the other two are 45° degree (model 2) and 60° degree (model 3). Figure 2,3 and 4 below show the different designs of the catalytic converter namely Model 1, Model 2 and Model 3.

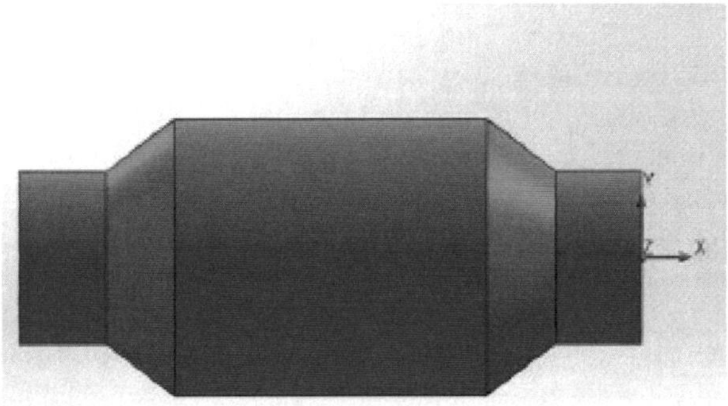

Figure 2: Inlet Cone of the Catalytic Converter 30° degree (Model 1)

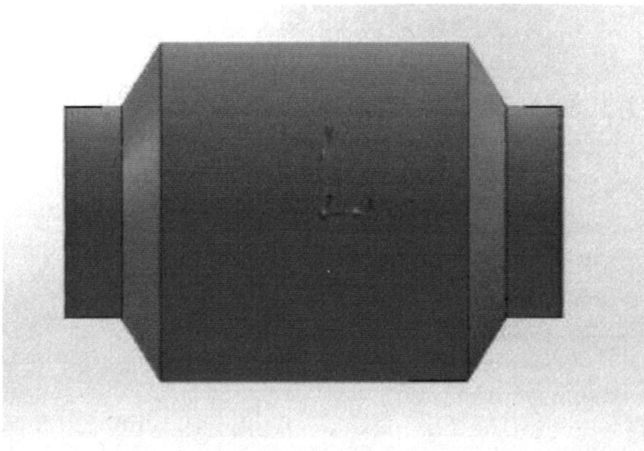

Figure 3: Inlet Cone of the Catalytic Converter 45° degree (Model 2)

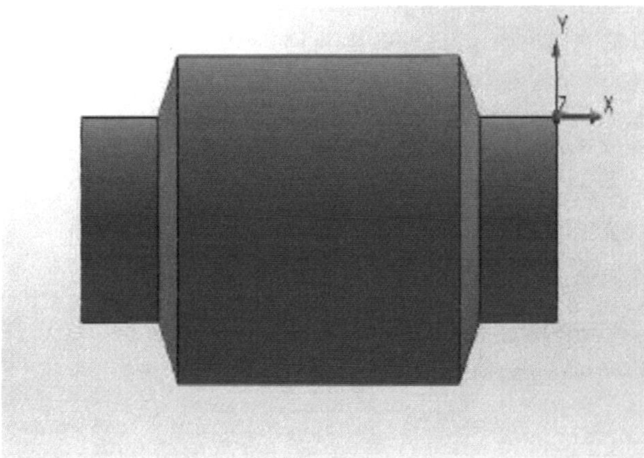

Figure 4: Inlet Cone of the Catalytic Converter 60° degree (Model 3)

IOP Conf. Series: Materials Science and Engineering **328** (2018) 012029 doi:10.1088/1757-899X/328/1/012029

3. Governing Equation and Boundary Conditions

Numerical study and algorithms are used to solve and analyze problem that involve fluid flows. The fundamental basis of almost all fluid problems is the Navier-Stokes equations. Principally, the equations are a set of coupled differential equation and could be solved for a given flow problem mathematically. However, in practice these equations are too difficult to solve analytically. Thus the computational approach was employed in order to solve approximations to the equations using a variety of techniques including finite element method. The Navier-Stokes equation consists of conservation of mass (Eq. 1), conservation of momentum (Eq. 2) and conservation of energy (Eq. 3).

Conservation of Mass:

$$\frac{\partial \rho}{\partial t} + \nabla \cdot (\rho \vec{V}) = 0 \qquad (1)$$

Conservation of Momentum:

$$\rho \frac{D\vec{V}}{Dt} = -\nabla P + \rho \vec{g} + \mu \nabla^2 \vec{V} \qquad (2)$$

Conservation of Energy:

$$\rho\left(\frac{\partial h}{\partial t} + \nabla \cdot (h\vec{V})\right) = -\frac{Dp}{Dt} + \nabla \cdot (k\nabla T) + \phi \qquad (3)$$

To solve the above equations, a turbulence model is needed. This study employed Spalart-Allmaras model. Spalart-Allmaras incorporates some of the recent advances in turbulence modelling. It regularly shows that it has equal or superior accuracy for nearly all classes of flows. In addition, the Spalart- Allmaras model is more computationally efficient compared to other model. Javaherchi [16] reviewed the Spalart- Allmaras model and its modification. The model is presented in Eq. (4).

Spalart- Allmaras model :

$$\frac{\partial}{\partial t}(\rho \vec{v}) + \frac{\partial}{\partial x_i}(\rho \vec{v} u_i) = G_\upsilon + \frac{1}{\sigma_{\vec{\upsilon}}}\left[\frac{\partial}{\partial x_j}\left\{(\mu + \rho \vec{\upsilon})\frac{\partial \vec{\upsilon}}{\partial x_j}\right\} + C_{b2\rho}\left(\frac{\partial \vec{\upsilon}}{\partial x_j}\right)^2\right] - Y_\upsilon + S_{\vec{\upsilon}} \qquad (4)$$

In this study, the fluid is set as air and the temperature is set to 873K assuming environment pressure at the outlet pipe of the catalytic converter. However, heat transfer from the fluid is not considered hence the fluid is reflected to be isothermal. The walls are assumed to have a smooth surface.

4. Results and Discussion

As mentioned earlier, the main objective of the present study is to investigate the effect of three different geometry angles of inlet cone pipe in the catalytic converter to get a better understanding of the catalytic converter processes, as well as to examine the relationship between the flow distribution and the performance of the catalytic converter inside three different models namely model 1, model 2 and model 3 in order to optimize the catalytic converter design.

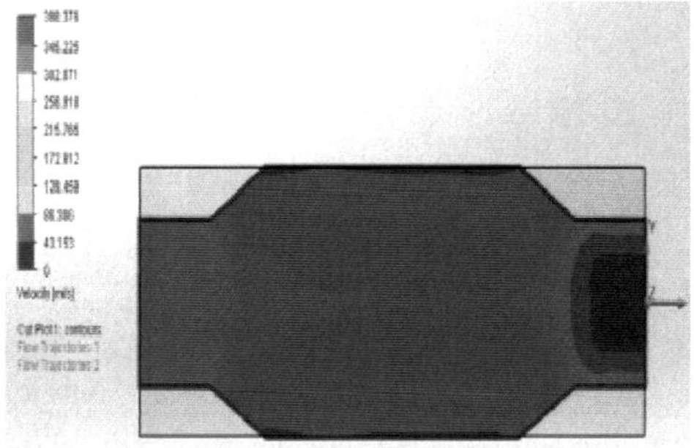

Figure 5: Velocity Distribution in Model 1

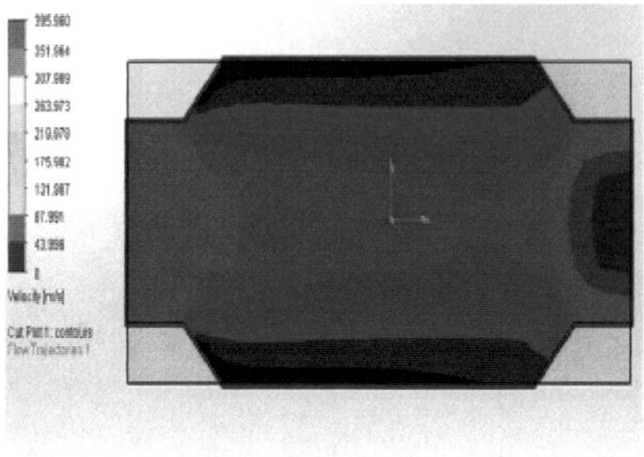

Figure 6: Velocity Distribution in Model 2

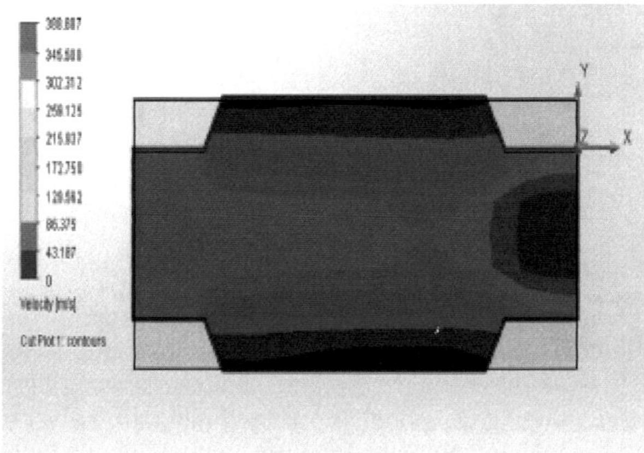

Figure 7: Velocity Distribution in Model 3

Figure 5,6 and 7 show the velocity distribution of the catalytic converter in model 1, model 2 and model 3 accordingly. In model 1, the velocity value of air start to decrease from 258.918m/s to 215.765m/s at the inlet cone of the catalytic converter before it reaches the end of the outlet substrate. However, the velocity increase back and reaches maximum at 388.378m/s after flowing out of the substrate region. In model 2, the mass flow moves roughly and some eddies are formed at the inlet cone of the catalytic converter. The velocity value of air start to decrease from 263.973m/s to 219.978m/s at the inlet pipe. Then the velocity increase back and reaches 395.96m/s. Comparing these two result observations, it is clearly shown that the velocity distribution is uniformly delivered in model 2 meanwhile in model 1, the velocity is highly non-uniform in the substrate inlet. Referring to Figure-6, model 3 shows differ result from model 1 and model 2. In model 3 the velocity flow only distributed in the middle of the catalytic converter mainly at the end of the substrate part. This may be caused from the bigger gradient (angle) of the catalytic converter. The non-uniformity increases in the substrate with the increase of the mass flow. Also, a formed of backpressure is detected in model 3 as the higher angle turns as a flow restriction. The velocity value of the inlet pipe is recorded to be 259.125 and the maximum value is 388.68m/s of the outlet substrate in model 3.

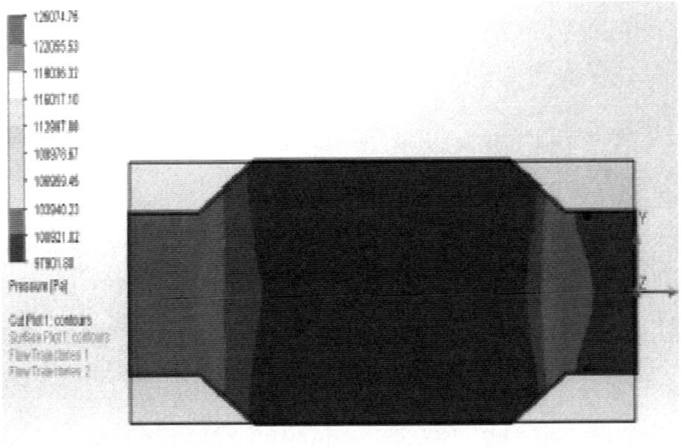

Figure 8: Pressure Distribution in Model 1

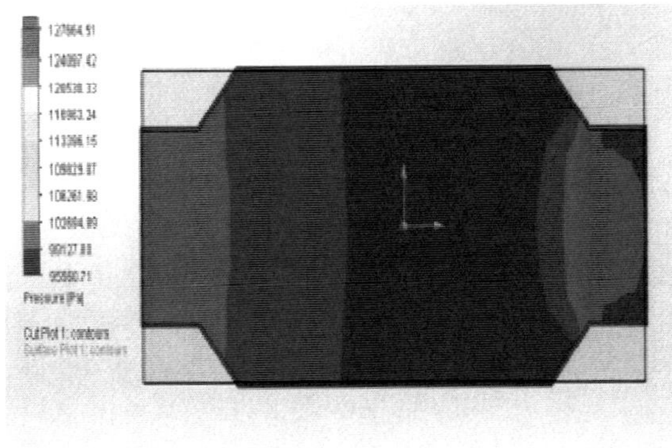

Figure 9: Pressure Distribution in Model 2

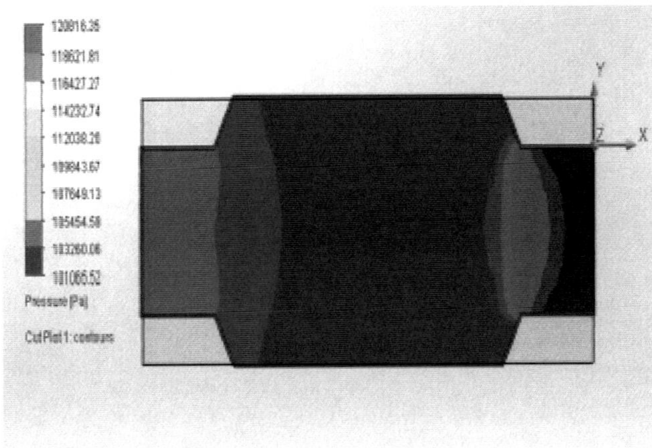

Figure 10: Pressure Distribution in Model 3

As for pressure distribution, the result in all models is shown in Figure 8,9 and 10. In model 1, the pressure starts to increase from 112997.88pa to 122055.53pa at the inlet cone of the catalytic converter and reach the maximum value of 125074.75pa at the substrate region. In model 2, the pressure start of 116963.24pa to 124097.42pa at the inlet cone and the value extent its maximum value of 127664.51pa. In the meantime, model 3 shows the maximum value of pressure distribution can be up to120816.35pa. Overall, the pressure distribution in model 2 is considerable compared to model 1 and model 3. The development of dispersion in pressure distribution in model 2 shows the best distribution flow between other cases.

5. Conclusion

From the present investigation, the results show that different geometries of the catalytic converter have a significant effect on the flow distribution in the monolith substrate. Furthermore, lower angles show less uniformity of flow distribution in the catalytic converter, whereas the flow tends to generate some additional backpressure in higher angles. Additionally, from these three cases it was revealed that if the angles are closer to the inlet cone pipe, the flow distribution is distributed uniformly. As a conclusion, cone inserted in model 2 is very effective and can increase the performance of the catalytic converter. The design of the catalytic converter is the key to provide a converter system to work proficiently as expected. The result is used to optimize and improve the development of new technologies for controlling regulated and non-regulated emissions.

The authors would like to acknowledge Universiti Teknikal Malaysia Melaka (UTeM) for financial support under UTeM short term grant PJP/2015/FTK(1C)/S01394.

References

[1] P. Karuppusamy and R. Senthil, "Design, analysis of flow characteristics of catalytic converter and effects of backpressure on engine performance," *Int. J. of Research in Engineering & Advanced Technology*, ISSN: 2330-8791, Vol 1, Issue 1, March 2013

[2] R.Wanker, H. Granter, G. Bachler, and G. Rabenstein, " New physical and chemical models for the CFD simulation of exhaust gas lines: a generic approach," *SAE Technical Paper*, 2002-01-0066, 2002.

[3] R.R.K. Thundil and R. Ramsai, "Numerical study of fluid flow and effect of inlet pipe angle in catalytic converter using CFD," *Res.J.RecentSci*, ISSN:2277-2502, Vol 1(7), pp39-44, July 2012.

[4] C. Lemme and W. Givens, "Flow through catalytic converters - an analytical and experimental treatment," *SAE Technical Paper*, 740243, 1974,

[5] M. Lai, J. Kim, C. Cheng, and P. Li, "Three-dimensional simulations of automotive catalytic converter internal flow," *SAE Technical Paper*, 910200, 1991

[6] S. Jeong and T. Kim, "CFD investigation of the 3-dimensional unsteady flow in the catalytic converter," *SAE Technical Paper*, 971025, 1997

[7] J. Howitt and T. Skella, "Flow effects in monolithic honeycomb automotive catalytic converters," *SAE Technical Paper*, 740244, 1974

[8] J. Wollin and S.F. Benjamin, "A study of the flow performance of ceramic contoured substrates for automotive exhaust catalyst systems" *SAE Technical Paper,* 1999-01-3626 , 1999

[9] D. Wendland, W. Matthes, and P. Sorrell, "Effect of header truncation on monolith converter emission-control performance," *SAE Technical Paper*, 922340, 1992

[10] S. Lee, C. Bae, Y. Lee and T. Han, "Effects of engine operating conditions on catalytic converter temperature in an SI engine," *Society of Automotive Engineers Inc.*, DOI: 10.4271/2002-01-1677

[11] B. Ramadan, P. Lundberg and R. Richmond, "Characterization of a catalytic converter internal flow," *SAE Technical Paper*, 2007-01-4024, 2007

[12] K. Kumar and M.L. Aggarwal, "A finite element approach for analysis of a multi leaf spring using CAE tools" *Res.J.Recent Sci.*, ISSN: 2277-2502,Vol. 1(2), 92-96, Feb. 2012

[13] A. Vinkel, M.L. Aggarwal and G. Bhusnan, "A comparative study of CAE and experimental results of leaf springs in automotive vehicles" *International Journal of Engineering Science and Technology*, ISSN:0975-5462 Vol. 3, No. 9,6856-6866, September 2011

[14] G. Wu and T. Song, "CFD simulation of the effect of upstream flow distribution on the light-off performance of a catalytic converter," *Energy Conversion and Management*, 46(13-14), 2010-2031, 2005

[15] G.C. Koltsakis, I.P. Kandylas and A.M. Stamatelos, "Three-way catalytic converter modelling and applications" *Chem. Eng. Comm*, Vol. 164:1 , pp. 153-189, 2007
[16] T. Javaherchi, University of Washington, pp 1–13, (2010)

IOP Conf. Series: Materials Science and Engineering **328** (2018) 012030 doi:10.1088/1757-899X/328/1/012030

Application of Electromagnetic Induction Technique to Measure the Void Fraction in Oil/Gas Two Phase Flow

H A Abdul Wahhab[1,2], A R A Aziz[1], H H Al-Kayiem[1], M S Nasif[1], and M N Reda[3]
[1]Mechanical Engineering Department, Universiti Teknologi PETRONAS, 32610 Bandar Seri Iskandar, Perak, Malaysia.
[2]Mechanical Engineering Department, University of Technology, Baghdad, Iraq. [3]Lehrstuhl für Thermodynamik, Technische Universität München, Boltzmannstr.15, 85748 Garching, Germany.

Email: Rashid@utp.edu.my

Abstract. In this work, electromagnetic induction technique of measuring void fraction in liquid/gas fuel flow was utilized. In order to improve the electric properties of liquid fuel, an iron oxide Fe_3O_4 nanoparticles at 3% was blended to enhance the liquid fuel magnetization. Experiments have been conducted for a wide range of liquid and gas superficial velocities. From the experimental results, it was realized that there is an existing linear relationship between the void fraction and the measured electromotive force, when induction coils were connected in series for excitation coils, regardless of increase or decrease CNG bubbles distribution in liquid fuel flow. Therefore, it was revealed that the utilized method yielded quite reasonable account for measuring the void fraction, showing good agreement with the other available measurement techniques in the two-phase flow, and also with the published literature of the bubbly flow pattern. From the results of the present investigation, it has been proven that the electromagnetic induction is a feasible technique for the actual measurement of void fraction in a Diesel/CNG fuel flow.

1. Introduction

The inference of bubbles and void fraction measurement is one of vast interest in two phase flow research and many technical applications, such as processes of petroleum, chemical, energy and other industries [1]. It is the main physical magnitude for determining almost other important parameters, such as the two phase flow properties (density and viscosity), for analysing the relative average velocity for two phases, and is of fundamental importance in two phase models for predicting flow pattern transitions, pressure drop and heat transfer [2]. Void fraction has been studied for many years on different types of two phase flows (but horizontal two phase flow orientation has received less attention compared to vertical flow), on different sized pipes, and several geometric definitions are used for characterizing the void fraction for two phase flows in channels and over tube bundles: cross-section, volumetric, local, and chordal [3]. Many measuring methods have been proposed: quick closing valve method, electrical method, ray absorption method, ultrasound method, image processing techniques, and

IOP Conf. Series: Materials Science and Engineering **328** (2018) 012030 doi:10.1088/1757-899X/328/1/012030

magnetic resonance method,...etc. [4-7], these techniques can be divided into intrusive and non-intrusive techniques according to whether flow fields are disturbed or not by the measuring equipment. Although many methods used to measurement this parameter, but it still one of difficulties of parameter detection in liquid-gas two phase flow [8]. One of the new methods used to measure void fraction in the bubbly flow supported the possibility of benefiting from the magnetization characteristic of liquid with gas bubbles under the magnetic field, uses the electromagnetic induction (EMI) [1, 9, 10]. The electrical conductivity of liquid in two phase flow is relatively higher than gas phase, which can be exploited with contactless methods based on electromagnetic induction [10]. Initially, the utilization of electromagnetic induction technique (EMI) to measure a void fraction in the liquid-gas fuel horizontal pipe has not been studied, due to the low electrical properties of liquid fuel and gas thereby creating a low magnetic field flux in the line. Lately, the nanotechnology technique was used to enhance characteristics of liquid fuel, reduce ignition delay time and emissions [11-14]. More so, nanotechnology helped improve the electrical properties of liquid fuel, and has opened the way for the application of EMI technique in liquid-gas fuel flows. From here, the EMI method is not considered a monopoly in the use of liquid metal but can be used in oil-gas flow [15, 16]. The chief purpose of this work is to establish the possibility of applying the electromagnetic induction (EMI) method of measuring a void fraction in a gas and oil flows such as diesel-CNG bubbly flow. The current article investigates from the diesel electric properties by its enhancement with iron oxide Fe_3O_4 nanoparticles. In addition, the use of electromagnetic induction system in measuring a void fraction of bubbly fuel flow is also studied. Table 1 shows the physical and chemical property of the diesel, diesel/CNG and Nano-additive diesel/CNG liquid in varying concentration.

Table 1. A Diesel fuel, CNG, and Diesel/Fe_3O_4 nanoparticles properties.

Parameter	**Diesel**	**Diesel & 3% Fe_3O_4**	**CNG**
Density (kg/m^3)	840	856	0.72
Kinematic Viscosity (cSt)	2.97	3.09	4.3*10-5
LCV (Mj/kg)	43.8	43.8	45.8
Octane Number	50	50	125
Carbon (%, w/w)	86.83	86.83	73.3
Hydrogen (%, w/w)	12.72	12.72	23.9
Oxygen (%, w/w)	1.19	1.19	0.4
Sulphur (%, w/w)	0.25	0.25	ppm < 5
Electrical conductivity (S/m)	$0.25*10^{-9}$	23.4*10-6	-
Relative permittivity	2.0	2.2	-

2. Experimental Implementation

Figure 1 (a) shows the schematic diagram of the experimental facility for the two-phase flow in a horizontal pipe. The facility test portion is made from transparent polycarbonate. The pipe dimension is 1.2 m long and has an inner diameter (Di) of 0.02 m. The CNG bubbles generator had an induction system that produce a liquid and gas mixture inside the test section at the inlet of the pipe. The diesel supply line and the compressed CNG line are directly linked to the bubble generator. The diesel is supplied by a fuel pump which has a flow rate and supply pressure measured by a flow meter and a pressure gauge respectively. The diesel flows through the bypass valve and is filtered by the fuel pump (90 watts, 5 l/min, and 1.5 m), then go into the flow meter, then, pumped along with the pressure gauge. After that, it goes into the side hole of the bubble generator and where it is combined with the CNG to produce the two-phase flow. Then, the Compressed Natural Gas CNG goes into the mixer situated at the first edge of the bubble generator. The CNG pressure is controlled using a pressure regulator while the flow rate is measured by a gas flow meter before it enters the bubble generator. A one way valve is used to control the CNG flow rate precisely. A thin pipe of 0.15 mm (inner diameter) was passed through the centre of the cross-sectional area of the pipe inside the bubble generator. The diesel flow meter was calibrated prior to its use. The electromagnet charger (excitation coils) is supplied by a DC power as described in Figure 1 (b), and is made up of two coils having 250 wraps of copper wire that was covered with enamel having a diameter of 2.1 mm and a U-type yoke. The yoke was made of steel; with the cross-sectional area for both poles was 22 mm, perpendicular to the direction of the flow (x-axis) and 60 mm that was parallel to the direction of flow (z-axis) to embrace the test pipe in order to reduce eddy current losses in the core. A pair of Tesla meter sensors was fixed between the yoke poles, and a test pipe was used to measure the magnetic intensity signing. The value of the magnetic field between the excitation coils had a variable magnetic field flux from 0 to 2000 Gauss. In capturing the induction voltage value from induction coils and the magnetic field intensity from excitation coils, the signal analyser (Oscilloscope) and AC/DC Magnetic Meter (Tesla-Meter: Model, MG-3003SD) were used to captured these values, as shown in Figure 1(a).

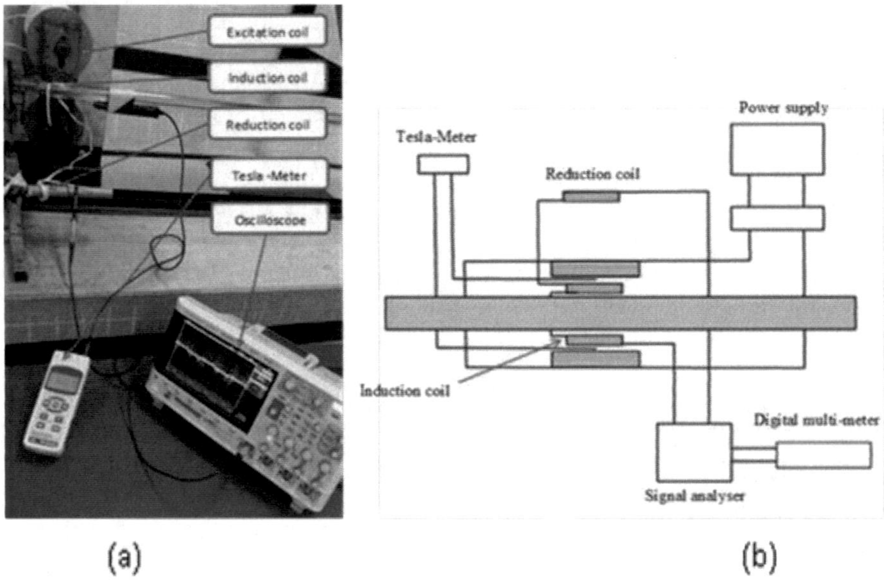

Figure 1. Components of an electromagnetic induction system.

3. Principle of EMI Measurement

The permeability of in a magnetic liquid differs from that of the gases because of the void fraction in the bubbly flow or Ferro-fluid dynamics [10, 17, 18]. When the magnetic field is excited, the coils generate eddy currents which serves as an insulating obstacle in the bubble created in the Ferro-fluid. The Ferro-fluid can detect the slight changes in the current distribution outside the fluid flow by measuring the amount of relative increase of the electromotive force between induction coils. It is assumed that there is no magnetisable material except for the magnetic fluid in the space affected by the magnetic field and no tangential components to the plane defined by either one of the induction coils in the space between the induction coils [18]. Thus, from the Gauss' low, the magnetic flux density becomes unique in the space between induction coils as well as in the test plug. In addition, in the induction coil, the induction current is small, (in the order of 10-9 amp) and the magnetic fluid is nonconductive, thus the mutual effect of magnetic field intensity by the excitation coils and induction coils is minimal.

Figure 2 shows a schematic sketch of the EMI system, (a) the field lines of the magnetic field generated by the excitation coil, while (b) field deformation due to the presence of a bubble at the magnetic field. The induced voltages in both induction coils appeared on a signal analyser; they have same amplitude but opposite signs so they cancel each other out. The challenge of this method is the reliable detection of these small variations which requires a sensitive measurement system. On the other hand, the deformation of the field lines results in a non-symmetrical magnetic field distribution in signal analyser which generates a net voltage signal at the terminals. If the bubble reaches the excitation field the induced voltage becomes zero again, because the secondary magnetic field is symmetric with respect of the signal analyser. When the bubbles move up further, again a voltage signal in the induction coil is generated, but now with an opposite sign [10]. The motion of bubbles in the horizontal pipe from the injection nozzle moved through the magnetic field thereby affecting the induction signal. Then, the difference in induction voltage values and magnetic field intensity traveling through diesel fuel recorded.

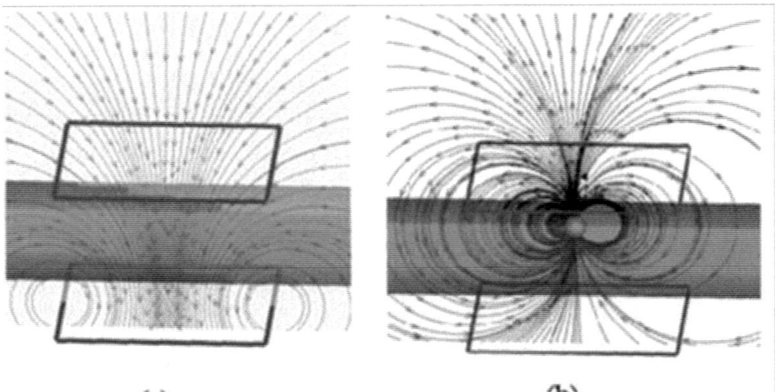
(a) (b)

Figure 2. Principle of electromagnetic induction measurement.

3.1. Measurement of Void Fraction

The diesel with 3% Fe_3O_4 nanoparticles fluid was electrically enhanced by improving the electromagnetic conductivity. The physical properties of diesel fuel and CNG are shown in Table 1. The magnetic intensity B is given by the following formula [9]:

$$B = \mu_o (H + M) \tag{1}$$

Where μ_o is the permeability (H/m), H is the magnetic field (A/m) and M is the magnetization (A/m). While, the magnetic intensity B_f passed through the magnetic fluid can be written by the following formula (this is based on a space distribution of CNG bubbles and a distribution of magnetization of a magnetic fluid is equivalent):

$$B_f = \mu_o (H + (1-\alpha) M_f) \tag{2}$$

Where M_f is the magnetization of magnetic fluid (A/m) and α is the void fraction. Thus, the CNG bubble behaves like a magnetized substance. The CNG bubble has demagnetization against the magnetization of magnetic fluid so that the value of magnetization in the CNG bubble can be expressed by using a demagnetization factor:

$$\Delta B = B_{\alpha=\alpha} - B_{\alpha=0} \tag{3}$$

With the assumption that the change in CNG bubble shape is ineffective, the detected magnetic intensity is simply written as follows:

$$\Delta B = \mu_o (1-\alpha) M_f \tag{4}$$

The differential value between the induced electromotive force in void fraction of 1.0 and the objective induced electromotive force was termed "detected voltage", as expressed:

$$\Delta V = RMS_{\alpha=\alpha} - RMS_{\alpha=1.0} \tag{5}$$

IOP Conf. Series: Materials Science and Engineering **328** (2018) 012030 doi:10.1088/1757-899X/328/1/012030

Where RMS represents the root mean square value of voltage.

4. Results and Dissuasion

Based on the measurement principle as described previously, the experiments were conducted at various conditions of two phase flow. The data obtained in these experiments depend on the master calibration line for actual flow experiments. Therefore, the void fraction measured by means of this electromagnetic induction method was compared with the void fraction data obtained by [9, 10]. In order to investigate how the measurement signal depends on higher gas flow rates. The first test carried out for measurement of detected voltage ($V_{\alpha=1}$) for gas flow where without liquid flow. Then, successive tests were carried out to find ($V_{\alpha=\alpha}$) for several conditions of a liquid superficial velocity 17.78, 20.79, and 28.9 cm/sec and for a wide range of a gas superficial velocity (5.5 to 63.4 cm/sec), as shown in Figure 3. From the figure it can be seen that the detected voltage for $\alpha = 1.0$ is 86 mV, while this voltage increase for liquid-gas flow with decreasing the gas superficial velocity until to 422 mV at $\alpha = 0.05$. To ensure the accuracy of the recorded reading of the detected voltage, the tests were repeated three times to detect and confirm the average value. More so, using a flat induction coil allows for a robust automatic detection of bubbles without any calibration, since the zero crossing of the induced voltage provides an exact time when the bubble is in the middle of this coil.

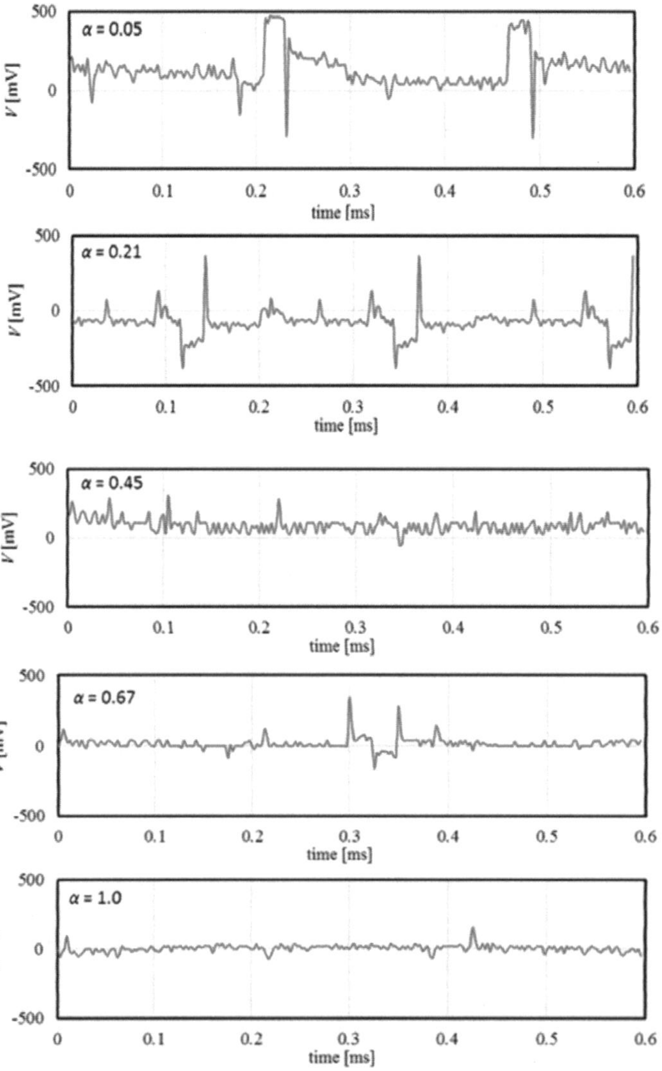

Figure 3. Induced voltage for void fraction analysis.

In Figure 4, the results obtained from the two phase flow experiments are displayed for the initial conditions of the liquid phase represented by the liquid superficial velocity 17.78, 20.79, and 28.9 cm/sec, respectively. In each diagram of (a), (b), and (c), the void fraction average was measured by recording the difference of magnetic field intensity ($B_{\alpha=\alpha} - B_{\alpha=0}$) indicated by data points, corresponding to the data obtained by the induction method, the trend line was plotted for each case. The low value of a liquid superficial velocity (17.78 cm/sec) tends to obtain higher void fraction value due to the intermittent flow mode of two-phase, because the flow mode tends to become the slug flow. From the Figure 4, increasing liquid superficial velocity from 17.78 cm/sec to 20.79 and 28.9 cm/sec decreases the trend line slope of void fraction value at 11.4 and 21.7 %, respectively. The obtained result are in line with the result obtained by Shuchi et al. [9]. It is concluded that the electromagnetic induction method is feasible for measuring the void fraction average in an actual two-phase flow of magnetic fluid.

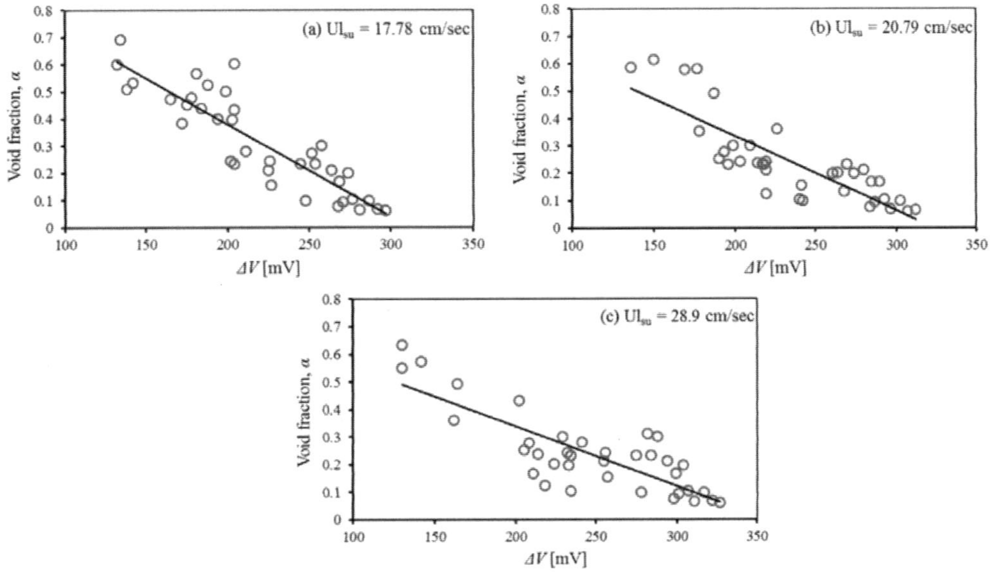

Figure 4. Void fraction versus ΔV in two-phase flow experiments.

The validation of results, the experiments were conducted at various conditions of two phase flow. The data obtained has been compared with the published literature for bubbly flow at same liquid and gas superficial velocity, as shown in Figure 5. The comparison of results for the time-averaged void fraction. Very good agreements between the user sensor and the results for Min et al. [19] were confirmed. For all flow-rate conditions, the maximum deviation between two sides was 8.3%. However, it was observed that the void fractions resulted from this technique were generally under compared.

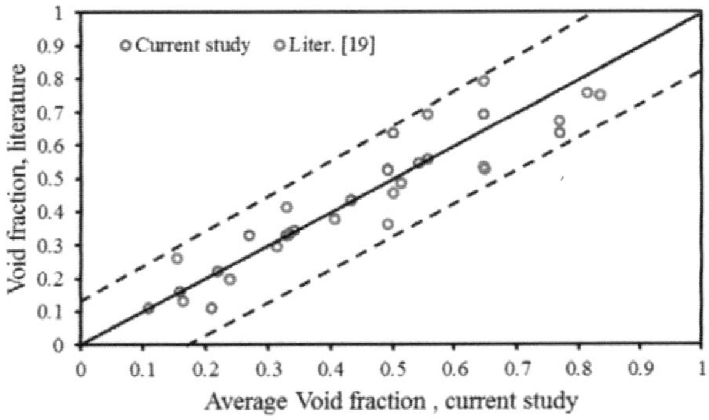

Figure 5. Comparison in results for average void fraction.

5. Conclusion

In this study, the electromagnetic conduction technique was adopted to determine void fraction in Diesel-CNG bubbly flow after enhancing Diesel by Fe_3O_4 nano particles, to improved electrical properties of a liquid fuel. This technique was applied to a horizontal pipe with a 20 mm inner diameter and a 1.2 m length. For several flow rate conditions covering stratified- and intermittent-flow regimes in bubbly flow. The chief purpose is to establish the possibility of applying the electromagnetic induction (EMI) method of measuring a void fraction in a gas and oil flows such as diesel-CNG bubbly flow. The experimental results was yielded quite reasonable account for measuring the void fraction, showing good agreement with the mechanical measured data in the two-phase flow, and with the published literature of the bubbly flow pattern.

References

[1] Kuwahara T, Yamaguchi H 2005 Void fraction measurement in magnetic fluid *Journal of magnetism and magnetic materials* vol 289 pp. 403-406

[2] Yeoh G, Cheung S C, Tu 2012 on the prediction of the phase distribution of bubbly flow in a horizontal pipe *Chemical engineering research and design* vol 90 pp. 40-51

[3] Sharaf S, Da Silva M, Hampel U, Zippe C, Beyer M, Azzopardi B 2011 Comparison between wire mesh sensor and gamma densitometry void measurements in two-phase flows *Measurement Science and Technology* vol 22 pp. 104

[4] Eckert S, Witke W, Gerbeth G 2000 A new mechano-optical technique to measure local velocities in opaque fluids *Flow Measurement and Instrumentation* vol 11 pp. 71-78

[5] Triplett K, Ghiaasiaan S, Abdel-Khalik S, LeMouel A, McCord B 1999 Gas–liquid two-phase flow in microchannels: part II: void fraction and pressure drop *International Journal of Multiphase Flow* vol 25 pp. 395-410

[6] Chen W, Twu M, Pan C 2002 Gas–liquid two-phase flow in micro-channels *International Journal of Multiphase Flow* vol 28 pp. 1235-1247

[7] Chaouki J, Larachi F, Dudukovic M P 1997 Noninvasive tomographic and velocimetric monitoring of multiphase flows *Industrial & engineering chemistry research* vol 36, pp. 4476-4503

[8] Ji H, Jiang B, Huang Z, Wang B, Li H 2013 Void fraction measurement of gas-liquid two-phase flow in mini-pipe based on image sequence in *2013 IEEE International Conference on Imaging Systems and Techniques (IST)* pp. 138-141

[9] Shuchi S, Yamaguchi H, Takemura M 2003 Measurement of void fraction in magnetic fluid using electromagnetic induction *Journal of fluids engineering* vol 125, pp. 479-485

[10] Gundrum T, Büttner P, Dekdouk B, Peyton A, Wondrak T, Galindo J 2016 V Contactless Inductive Bubble Detection in a Liquid Metal Flow *Sensors* vol 16 pp. 63

[11] Sarvestani N S, Rohani A, Farzad A, Aghkhani M H 2016 Modeling of specific fuel consumption and emission parameters of compression ignition engine using nanofluid combustion experimental data *Fuel Processing Technology* vol 154 pp. 37-43

[12] U-L S D E Envirofuels 2007 Operational Challenges Facing Engine Operators. Available: www.envirofuelslp.com/pdfs/TechnicalBulletin-070307

[13] Yu Q, Shi M, Cheng Y, Wang M, Chen H Z 2008 Fe_3O_4@ Au/polyaniline multifunctional Nanocomposites: their preparation and optical, electrical and magnetic properties," *Nanotechnology* vol 19 pp. 265

[14] S B R 2015 Ferrofluid-Diesel blend *International journal of scientific and Research Publications* vol 5 pp. 1-4

[15] Laurent S, Dutz S, Häfeli U O, Mahmoudi M 2011 Magnetic fluid hyperthermia: focus on superparamagnetic iron oxide nanoparticles," *Advances in colloid and interface science* vol 166 pp. 8-23

[16] Zhan J, Tian G, Jiang L, Wu Z, Wu D, Yang X 2008 Super paramagnetic polyimide/γ-Fe_2O_3 nanocomposite films: preparation and characterization *Thin Solid Films* vol 516 pp. 6315-6320
[17] Terzija N, Yin W, Gerbeth G, Stefani H, Timmel K, Wondrak T 2010 Use of electromagnetic induction tomography for monitoring liquid metal/gas flow regimes on a model of an industrial steel caster *Measurement Science and Technology* vol 22 p. 015501
[18] Yin W, Gerbeth G, Stefani F, Timmel K, Wondrak T, Terzija A P N 2011 Use of electromagnetic induction tomography for monitoring liquid metal/gas flow regimes on a model of an industrial steel caster
[19] Ko M S, Lee B A, Won W Y, Lee Y G, Jerng D W, Kim S 2015 An improved electrical-conductance sensor for void-fraction measurement in a horizontal pipe *Nuclear Engineering and Technology* vol 47 pp. 804-813

Acknowledgments
The authors are obliged to the Universiti Teknologi PETRONAS for providing the centre for automotive research and energy management (CAREM).

IOP Conf. Series: Materials Science and Engineering **328** (2018) 012031 doi:10.1088/1757-899X/328/1/012031

Early Shear Failure Prediction in Incremental Sheet Forming Process Using FEM and ANN

Majid Moayedfar[1], Hengameh Hanaei[2], Ahmad Majdi Rani[2], Mohd Azam Bin Musa[1] and MohammadSadegh Momeni[3]

[1]Faculty of Engineering and Technology, DRB-HICOM University of Automotive Malaysia, 20607 Pekan, Pahang Darul Makmur, Malaysia.
[2]Department of Mechanical Engineering, Universiti Teknologi PETRONAS, 32610 Bandar Seri Iskandar, Perak Darul Redzuan, Malaysia.
[3]Department of Petroleum Engineering, Universiti Teknologi PETRONAS, 32610 Bandar Seri Iskandar, Perak Darul Redzuan, Malaysia.

majid@dhu.edu.my

Abstract. The application of incremental sheet forming process as a rapid forming technique is rising in variety of industries such as aerospace, automotive and biomechanical purposes. However, the sheet failure is a big challenge in this process which leads wasting lots of materials. Hence, this study tried to propose a method to predict the early sheet failure in this process using mathematical solution. For the feasibility of the study, design of experiment with the respond surface method is employed to extract a set of experiments data for the simulation. The significant forming parameters were recognized and their integration was used for prediction system. Then, the results were inserted to the artificial neural network as input parameters to predict a vast range of applicable parameters avoiding sheet failure in ISF. The value of accuracy R2 ~0.93 was obtained and the maximum sheet stretch in the depth of 25mm were recorded. The figures generate from the trend of interaction between effective parameters were provided for future studies.

1. Introduction

The majority of systems considered are often too complex to yield simple analytical solutions and thus, numerical methods are generally used [1]. In many of complex systems, there are generally too many parameters to consider. Besides, experimentation may take a considerable period of time and is often costly. On the other hand, computational modelling, when combined with limited experimental tests, can provide important insights, parametric screening analyses and optimisation, and design guidance for further experimentation. The Finite Element simulation of ISF process requires a long computational time due to the large model deformation and non-linear conditions (large displacements, friction contact, and plastic deformation of the material) in the technique. Using ABAQUS® (Dassault systemes Corporation) analysing tool, there are two main approaches that can be used to reduce this computational time; time scaling and mass scaling. Mass scaling method is almost the first potential candidate for reducing the computational time. However, it still keeps the kinetic energy of deforming material below 4%-10% of the system total energy [2].

Mathematically, the Finite Element Method (FEM) is a numerical method to find an estimated solution for Boundary Value Problems (BVP). FEM employs a variation of methods effect on the

process parameters to minimize an error function and generate a stable solution. One of the most prevalent aims for new manufacturing processes is to optimise these process parameters to achieve an applicable part with a high geometrical accuracy together with an increase in productivity and minimizing production expenses [3]. There are some parameters, as per Gatea et al. (2016), which have a significant effect on the modelling and simulation of the ISF process. These parameters are the forming temperature, forming angle, incremental depth and feed rate, all of which have a strong influence on the formability and sheet failure of the ISF method [4].
A clear example of using numerical investigation in the ISF method is toolpath optimisation, which is determined by FEM, by applying a response surface method (RSM). This technique was first presented by Azaouzi and Lebaal (2012) to enhance the thickness distribution of parts with asymmetric designs [5]. It was found from their study that after three repetitions, the optimal solution proposed an enlargement of around 7% of the sheet thickness distribution. Moreover, another study using the FEM for toolpath optimisation is that of Filice et al. (2013). In their study, to solve the issues related to inhomogeneous sheet thickness distribution in an aluminium alloy part as an example, a model for optimisation is created to optimise the toolpath design by applying a trajectory related to a decrement of ±10° in slope differential [6].

2. Method

Simulation of ISF contains of multiple steps such as creating CAD models, generating toolpaths for monitoring the tool navigation, building FEM which consists applying boundary conditions, rendering material properties and contacting parameters and finally solving the FEM model [7]. In this study, a 3D thermo-viscoplastic FE model was used for the numerical investigation of ISF to obtain the thermo-mechanical responses of the procedure on Aluminium alloy 6061. The numerical computation was completed using the commercial code of the ABAQUS/CAE® software.
To find the optimum parameters, the Design-Expert® (State-Ease Corporation) software was employed to design the experiments for a better interaction between the parameters as the input data and the STD value as the output factor. Hence, the parameters were inserted in the software environment and the RSM with central composite design technique was applied to create 30 sets of data.
Table 1 presents the range of input data for the software for creating the set of data in this study. Based on the literature, the minimum and maximum values for each parameter were selected and inserted into the software. In regard to some inapplicable values suggested by the software, for example, -0.1 mm for depth step (or DOC), the Alpha function was activated to bring in real values and reduce the negative values since all the applied variables should be more than zero in these numerical study.

Table 1. RSM input variable range

Parameter	Unit	Range of values
Feed Rate	mm/min	10-200
Spindle Speed	rpm	50-200
Depth Step	mm	0.1-0.7
Width Step	mm	1-4

The toolpath generation is not a required step in simulation. Nevertheless, to simulate the ISF process, it is employed to move the tool along a predefined toolpath. For not complicated cases, toolpath coordinates are estimated using a Microsoft Excel® spreadsheet program, whereas in Pohlak et al. (2004), a Computer Automated Manufacturing (CAM) system was applied, as toolpath becomes much more complex in non-linear conditions [8].
The explicit method was applied due to the stable alteration of the contact boundary conditions and also small time steps in simulations. It should be noted that the explicit schema by increasing artificially the mass density will reduce the computational time of the numerical simulation.
In ISF process, a very small ratio between the thickness and other dimensions of the sheet call for the use of shell elements in FEM simulation. Therefore, to build up the model in this research work, shell elements were implemented. The tool is generally considered a rigid body, but the sheet is modelled as

an elastic-plastic material. Constraint conditions were determined by carrying out a preliminary static analysis based on a 3D model that includes the entire workbench [9]. This analysis proved the acceptability of constraining five degrees of freedom at the sheet border, leaving free the rotation about the axis parallel to the border only. The sheet was meshed with shell finite element with dimensions of 2×2 mm (in rectangular shape) allowing for reducing the integration. The final principal set up is provided at Figure 1.

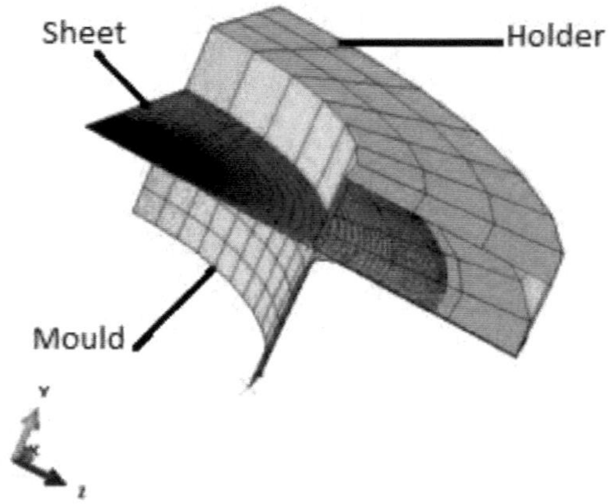

Figure 1. ISF process principle set up

The Johnson-Cook (J-C) constitutive model parameters of both sheet materials were considered for the numerical simulation of the material deformation in the ISF procedure. Therefore, the J-C flow stress was offered as the used material model for the sheet part. The parameter considered in simple forms of stress empirical relations with strain, strain rate, and temperature as per equation (1):

$$\sigma_y = [A + B(\varepsilon_P)^n]\left[1 + C\left[\frac{\dot{\varepsilon}_P}{\dot{\varepsilon}_0}\right]\right]\left[1 - \left[\frac{T - T_{room}}{T_{melt} - T_{room}}\right]^m\right] \tag{1}$$

Where A = 48 MPa which is the yield stress, B = 276 MPa (strain factor), C is the strain rate factor equal 0.015, T is (effective temperature), T_{melt} = 582°C (melting point), T_{room} = 20°C (ambient temperature), n = 0.042 (strain exponent) and m = 0.85 (temperature exponent). A, B, C, n, T_{melt}, T_{room} and m are material constants for the J-C strain rate dependent yield stress for Aluminium. The identified J-C constitutive model parameters that are relative to the applied material in this study are mentioned in Table 2.

Table 2. Material constants for the J-C constitutive model (aluminium)

A (MPa)	B (MPa)	C	n	m
48	276	0.08	0.042	0.85

The analytical tool will process all the points based on the STD (the units for STD would be logarithmic strain (LE) since the strain as the summation of numerous small differential segments is taken and thus it will express the final strain).
To check the accuracy of the calculated geometry, the highest normal distance d_{max} and the average normal distance d_{av} between the FE results and the experimental data were considered. The prediction of sheet thickness distribution was analysed using the highest deviation $d_{th,max}$ between the FEA and the experimental data.

To increase the accuracy of the model for prediction, in this study, back-propagation-type neural networks with 4 variables as inputs, an output, and a hidden layer were applied which in most cases, one hidden layer is satisfactory (Saljooghi and Hezarkhani, 2014; Van der Baan and Jutten, 2000) [10, 11]. The power of the network is increased through the increment of hidden layers or the number of neurons in the hidden layer. Hence, the error on the training data set will converge to a small value, however, when a new data is introduced, the error will be great. A complex network tends to memorize the data but it does not adapt to new data set. Therefore, a three-layer Artificial Neural Network (ANN) with a tangent sigmoid transfer function (tansig) at hidden layer and a linear transfer function (purelin) at output layer was implemented. To optimize ANN, 1 - 23 neurons in the hidden layer were trained at least five times using altered sets (which were randomly generated) of initial weights to find the lowest mean squared error. The data had been distributed randomly into three different groups (70% for training, 15% for the testing set, and 15% for cross validation). Overtraining was prevented by the validation data set, as the training continues; it lowers the training error function and the result of applying a validation set will improve. However, the likelihood of performance improvement on the validation data set with further iterations is not achievable. At this point, training should be stopped.

3. Results and discussion

The forming cycles from the corner to centre continued until the Ultimate Tensile Stress (UTS) appeared similar to Moayedfar et al. (2013) [7]. In this stage, as illustrated in Figure 2, the maximum STD for a particular set of experiment is accrued up to LE $8.452e^{-01}$, which showed a respectable agreement with achievements of Azaouzi et al. (2012) i.e. LE $8.227e^{-01}$ [5].

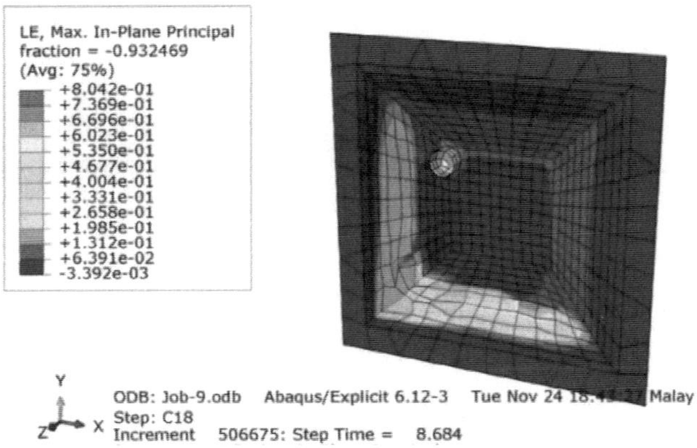

Figure 2. Maximum sheet thickness distribution

The large material deformation in ISF results in a several element entanglement and distortion. These may lead to reduce the size of stable time increment and also the accuracy of simulated results. Consequently, it is recommended to apply the adaptive meshing tool in the simulation of ISF process to enhance the meshing quality via a practical CPU cost for avoiding the elements distortion.

The result of the STD is inserted to the DE® to figure out the interaction between effective parameters with respect to the RSM. The influence of four parameters such as feed rate, spindle speed, depth of cut (depth of penetration in ISF), and width step versus the sheet stretching was examined via the statistical software. The behaviour of these four parameters versus sheet stretching during the simulation, where the (A) is the diagram of variation for feed rate, (B) spindle speed, (C) depth of cut, and (D) step width provided us a formula to predict the possible values of STD (Equation 2) which is not in presented 30 experiments.

$$STD = 0.42 + 0.081 * A + 0.032 * B + 0.020 * C + 0.072 * D + 0.091 * A2 + 0.056 * B2 + 0.044 * C2 + 0.066 * D2 - 0.012 * A * B + 4.919E\text{-}003 * A * C - 0.033 * A * D + 0.013 * B * C + 4.069E\text{-}003 * B * D - 4.244E\text{-}003 * C * D \quad (2)$$

The feed rate is the most influential parameter in this study. Moreover, step width also showed a considerable effect on the amount of STD in this analysis, while the spindle speed and depth of

penetration played a less important role in the forming process. High values of feed rate increased the friction and vibration of the process, which are the two major defects of ISF. In addition, a width step of more than 20% of the tool diameter increased the risk of sheet failure due to the huge pressure on the contact surface of the workpiece, which changes the process from stretching to chipping.
Respond values obtained from the RSM then, exported to the prediction software and the realizations were recorded. The result of ANN showed that the highest value obtained when R^2 was 0.931 (Figure 3) and the lowest value achieved in Means Square Error (MSE) that was 0.0073, with the application of 10 hidden neurons.

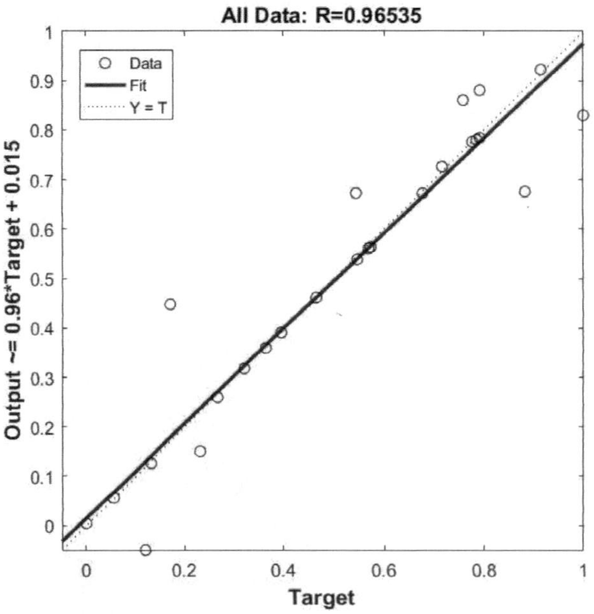

Figure 3. The predicted data *vs* the simulation data

The validation vectors are applied to stop the training early when the performance of network on the validation vectors remains the same or fails to enhance for the highest fail epochs in one row. Test vectors are applied as an additional check which shows the network is generalizing well, however, this test does not have any effect on the training. In order to prevent over training, a validation was performed. For this purpose, after eight times failing in finding the best network training was stopped (Figure 4).

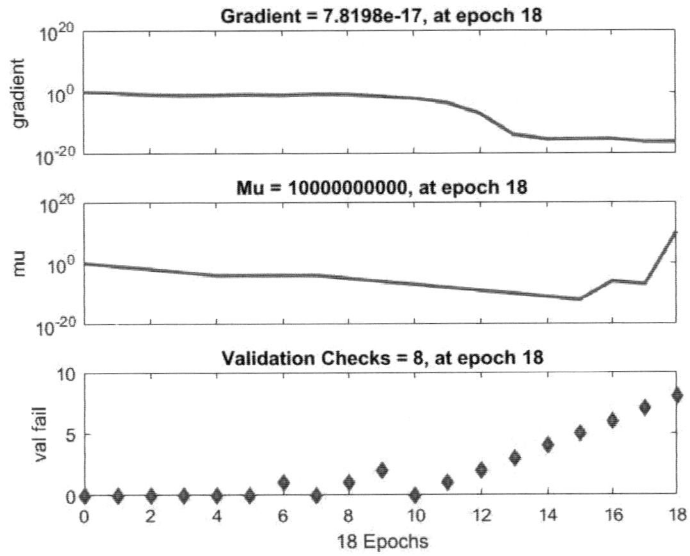

Figure 4. Normalized STD from ANN

Overfitting issue sometimes occur when a very low MSE is obtained, then, it will be a cause of confusion with a high accuracy. The final MSE created is due to the isolated test data that is random 15% of the samples. When the generalization is very poor, a low MSE may be possible tough. The validation MSE curve and follows the test MSE curve showing enhanced generalization (Figure 5).

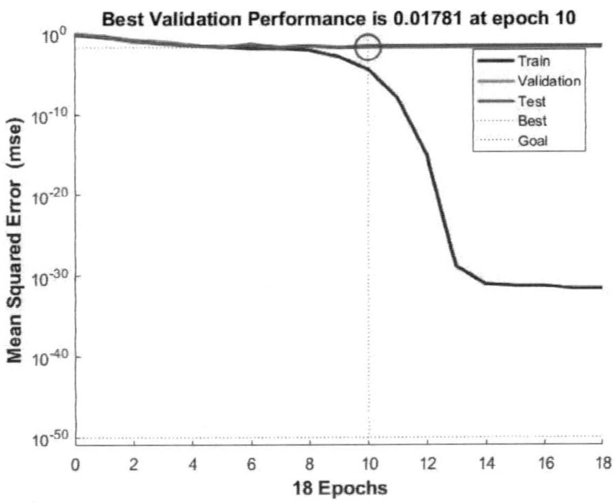

Figure 5. MSE *vs* the number of epochs

The results of this study illuminated that the ANN predicted model for the sheet thickness distribution based on the RSM analysing having a good correlation with the numerical simulation input data when the MSE = 0.0011

4. Conclusion

Last but not least, the Design of Experiment method was applied for 30 tests using numerical modelling together with a proper experimental study that would lead to the lowest and the highest STD when the maximum sheet stretch considered 25mm (in Z direction) in ISF for AISI 316. The significant parameters of feed rate, spindle speed, depth step, and width step are demonstrated and interactions between these effective factors are formulated for future studies. This formula is exclusive to this study but can also be applied in a variety of ISF procedures on AISI 316 material. With respect to the results of the design analytical software, a good agreement was achieved between the available numerical simulation and the literature. However, the number of parameters, which was used in this work, was more than the current studies in which used an explicit technique similar to this study. The respond of each set of parameters inserted as an input data to ANN for predicting a wide range of values for these four effective parameters with the R^2 equal 0.931. The results of this study will help other researchers to find the amount of sheet thickness distribution to predict the sheet failure in manufacturing process.

Acknowledgment

The Authors are grateful to DRB-HICOM University of Automotive Malaysia for providing financial aids via grant IRGS/012017/003.

Reference

[1] Z. Jin, 2014 1 - Fundamentals of computational modelling of biomechanics in the musculoskeletal system in Computational Modelling of Biomechanics and Biotribology in the Musculoskeletal System ed: Woodhead Publishing pp. 3-11.
[2] Le Van Sy, 2009 Modelling of Single Point Incremental Forming Process For Metal and Polymeric Sheet PhD thesis University Of Pauda
[3] I. Bagudanch, M. L. Garcia-Romeu, and M. Sabater 2016 Incremental forming of polymers: process parameters selection from the perspective of electric energy consumption and cost, Journal of Cleaner Production vol. 112 Part 1 pp 1013-1024
[4] S. Gatea, H. Ou, and G. McCartney 2016 Review on the influence of process parameters in incremental sheet forming," The International Journal of Advanced Manufacturing Technology pp 1-21.
[5] M. Azaouzi and N. Lebaal 2012 Tool path optimization for single point incremental sheet forming using response surface method, Simulation Modelling Practice and Theory vol 24 pp 49-58
[6] L. Filice, G. Ambrogio, and M. Gaudioso 2013 Optimised tool-path design to reduce thinning in incremental sheet forming process, International journal of material forming vol 6 pp 173-178
[7] M. Moayedfar et al 2013 The Effect of Forming Parameters on the Sheet Stretch in Incremental Sheet Forming (ISF) Process on CNC Lathe Machine, Advanced Materials Research Vols 634-638 pp 2894-2898
[8] M. Pohlak, R. Küttner, J. Majak, K. Karjust, and A. Sutt 2004 Simulation of incremental forming of sheet metal products, in Proc of the fourth International DAAAM Conference Tallinn Estonia.
[9] V. Fontanari, M. Benedetti, S. Bruschi, and A. Fuganti 2012 Numerical and experimental analysis of the single point sheets incremental forming process, in 15th International Conference on Experimental Mechanics Porto/Portugal pp 22-27
[10] Saljooghi, B. S., & Hezarkhani 2014 A. Comparison of WAVENET and ANN for predicting the porosity obtained from well log data, Journal of Petroleum Science and Engineering pp 172-182
[11] Van der Baan, M., & Jutten, C 2000 Neural networks in geophysical applications, Geophysics 65(4) pp 1032-1047

A Comparative Experimental Study on the Use of Machine Learning Approaches for Automated Valve Monitoring Based on Acoustic Emission Parameters

Salah M. Ali[1*], K.H. Hui[1], L.M. Hee[1], M. Salman Leong[1], M.A. Al-Obaidi[2], Y.H. Ali[3] and Ahmed M. Abdelrhman[4]

[1]Institute of Noise and Vibration, Universiti Teknologi Malaysia, 54100 Kuala Lumpur, Malaysia.

[2]Energy and Renewable Energies Technology Centre, University of Technology, Baghdad, Iraq

[3]Department of Refrigeration and Air-conditioning, Technical College of Mosul, Northern Technical University, Mosul, Iraq.

[4]School of Engineering, Bahrain Polytechnic, 33349 Isa Town, Kingdom of Bahrain.

[*]E-mail adress: salah.obaidi@pioneers-group.com

Abstract. Acoustic emission (AE) analysis has become a vital tool for initiating the maintenance tasks in many industries. However, the analysis process and interpretation has been found to be highly dependent on the experts. Therefore, an automated monitoring method would be required to reduce the cost and time consumed in the interpretation of AE signal. This paper investigates the application of two of the most common machine learning approaches namely artificial neural network (ANN) and support vector machine (SVM) to automate the diagnosis of valve faults in reciprocating compressor based on AE signal parameters. Since the accuracy is an essential factor in any automated diagnostic system, this paper also provides a comparative study based on predictive performance of ANN and SVM. AE parameters data was acquired from single stage reciprocating air compressor with different operational and valve conditions. ANN and SVM diagnosis models were subsequently devised by combining AE parameters of different conditions. Results demonstrate that ANN and SVM models have the same results in term of prediction accuracy. However, SVM model is recommended to automate diagnose the valve condition in due to the ability of handling a high number of input features with low sampling data sets.

Keywords: Faults diagnosis; Acoustic emission; Signal analysis; Machine learning; SVM; ANN.

1. Introduction

Reciprocating compressors are one of the most essential machines in many industries, for instance in power plants, gas transmission and storage, petrochemical industry and many other applications. The ongoing growth in these sectors, have increased the demand of reciprocating compressors worldwide and the demand is expected to grow more in the next couple of years [1]. The compressor reliability is an essential factor required by customers who are expecting a reduction of unscheduled shutdowns and extending the critical components lifetime. This drives the challenge among rival firms in term of design development and employing advanced materials in the compressors manufacturing. However, fatigue failure in some compressor components cannot be avoided. Compressors valves are one of the critical components that play a significant role on both efficiency and reliability of reciprocating compressor. Due to repetitive pressure loads, valves failures are accounted by 29% of unscheduled shutdowns in reciprocating compressor [2]. This issue drives the consideration of effective valves monitoring methods to enable scheduling maintenance shutdown according to the valve condition.

Many methods have been proposed for valve fault detection in reciprocating compressor. Ren et al. [3] detected valve faults using extracted features of vibration signal and SVM. Tran et al. [4] attempted to use three signals including pressure, vibration and current of the induction motor to diagnose valve condition in reciprocating compressor. They extracted the Teager-Kaiser energy from the acquired signals then they employed the deep belief networks (DBN) to diagnose the valve condition. Manepatil et al. [5, 6] investigated the fault detection in reciprocating compressor based on modelling and simulation study of pressure signal. Tiwari and Yadav [7] proposed a method to quantify the valve leakage percentage by analyzing the acquired pressure pulsation using artificial neural network (ANN). Some of the previous studies have analyzed the acquired signal using the same machine learning approaches that are proposed in this paper, for instance [3, 7]. However, they did not consider the variation of speeds and loads during compressor operation which is necessary for any automated system to cope with this fact. In addition, vibration is not the most sensitive method to detect the fault at incipient stage and the pressure measurements are not preferred for issues related to intrusiveness into the compressor operation.

The feasibility of using AE for reciprocating compressor valve condition monitoring has been investigated by some researchers. Gill et al. [8] revealed the advantage of using the AE technique for valve faults detection in a reciprocating compressor. They further concluded that vibration analysis is less sensitive to the higher-frequency noise emitted by fluid-mechanical motion. El-Ghamry et al. [9] suggested a method based on AE statistical feature isolation to diagnose several reciprocating machinery faults. Wang et al. [10, 11] proposed a diagnosis method for reciprocating compressor valve condition by comparing the AE waveforms for normal and faulty valves in simulated valve motion. Unfortunately, limited operational conditions have been used and some faults could not be identified. Sim et al. [12] proposed a valve fault detection method by analysing the AE signal. The authors employed wavelet packet transform (WPT) to decompose the acquired AE signals to different frequency ranges then they used statistical analysis to detect the valve fault based on RMS value. Although the AE could detect the valve's faults but the analysis was complicated and not practical to use in the industry.

AE signal analysis based on artificial intelligence (AI) has been reviewed in [13, 14]. The research revealed a potential for use AI in automated machinery fault detection. Moreover, AE features extractions have been investigated in our previous research using statistical analysis [15]. This paper aims to automate valve fault detection for reciprocating compressor based on the most common machine learning approaches ANN and SVM with selected AE signal parameters. Moreover, this paper investigates the performance of selected AI methods. The paper structure presented as follows. Section 1 reviews the state of the art the techniques used in valve fault detection and the shortcomings of the existing AE analysis methods. Section 2 briefly describes the theoretical background, including AE parameters, ANN and SVM. Section 3 explains the research methodology, including the test rig, instrumentation and experimental procedure. Section 4 illustrates the modelling results and models validation. Section 5 concludes the paper.

IOP Conf. Series: Materials Science and Engineering **328** (2018) 012032 doi:10.1088/1757-899X/328/1/012032

2. Theoretical Background

2.1. AE Signal Parameters

Acoustic emission refers to the generation of transient elastic waves produced by a rapid release of energy from a localized source within the surface of material, as reported by the American Society for Testing and Materials (ASTM) [16]. In this paper, AE is defined as transient elastic waves produced by the impact of one surface on another in a reciprocating motion. In other words, the transient elastic waves produced by the impingement of the plates inside the valve with the upper and lower plate housing during the reciprocating compressor operation. AE hit has specific parameters related to the signal event. The interpretations of AE parameters often related to the machine condition [17]. In this study, AE parameters have been extracted from the acquired AE hits include amplitude, counts, duration, energy, absolute energy, ASL and signal strength. See Fig. 1 and Table 1.

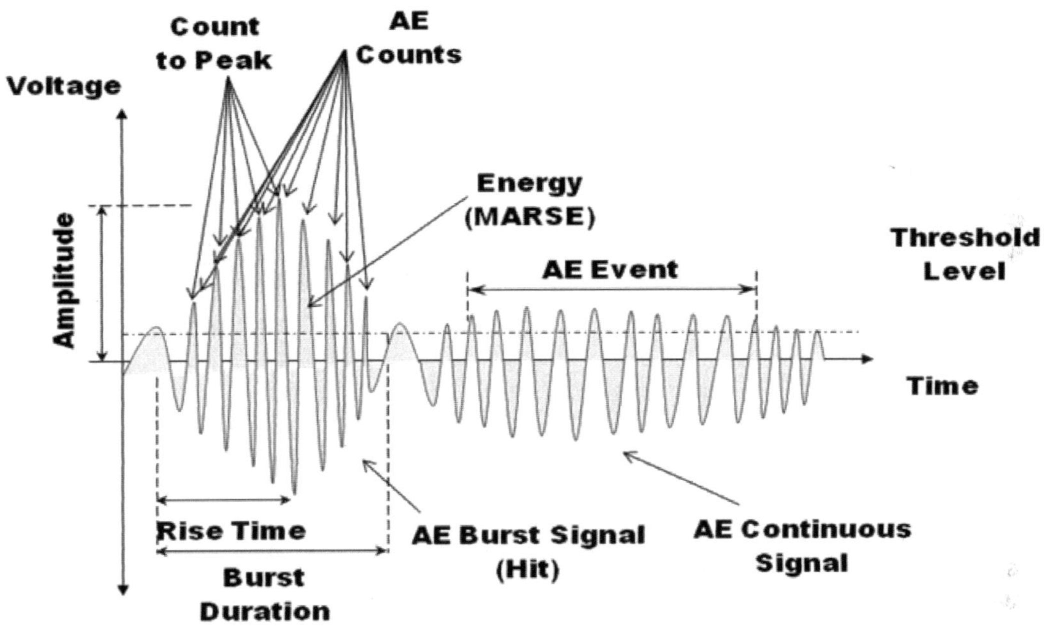

Fig. 1: AE signal parameters.

Table 1: AE main parameters according to ASTM E1316-05 standard.

AE Signal Parameters	Description	Units
Amplitude	The greatest measured voltage in a waveform.	Volt
Counts	The number of times the AE signal exceeds a preset threshold during an event.	Counts
Duration	The time between AE signal start and AE signal end.	μsec
Energy	The mean area under the rectified signal envelope.	MARSE
Absolute energy	The real amounts of AE signal energy.	Attojoule (aJ)
ASL	The average signal level of the AE amplitude.	db
Signal strength	The integral of the rectified voltage signal over the duration of the AE waveform packet	V.sec

2.2. Support Vector Machine

Support vector machine is a supervised machine learning method that relies on statistical learning theory with an ability to handle high input features with low sampling data sets [18]. This learning technique uses input vectors for pattern classification. During the training process, SVM creates a hyperplane that allocates the majority points of the same class in the same side, while maximizing the distance between the two classes to this hyperplane [2]. See Fig. 2.

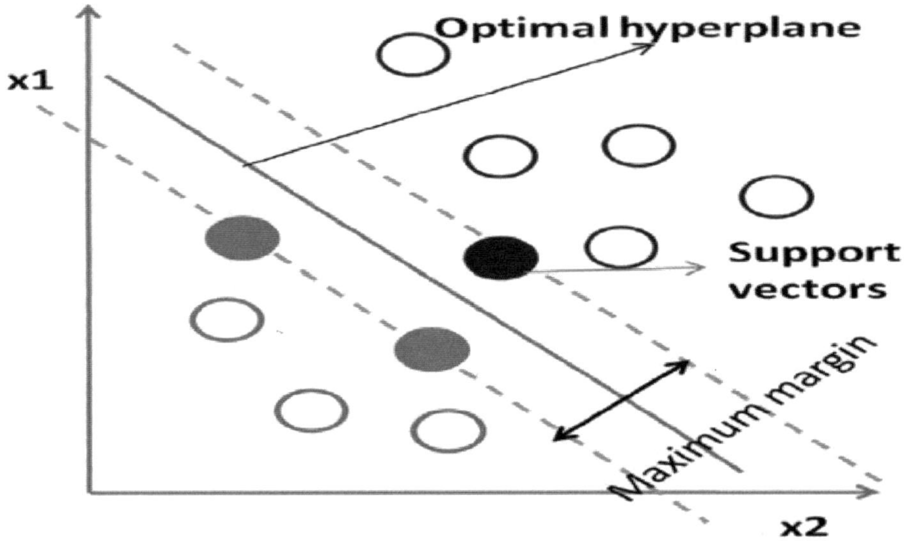

Fig. 2: SVM's decision boundary.

This hyperplane could be either linear or non-linear, which is also relevant to the kernel function [19]. SVM training seeks a globally optimized solution and avoids over-fitting so that it can deal with a large number of features. A comprehensive description, limitations and drawbacks of SVM method are available in [20, 21]. In the linearly separable case, there exists a separating hyperplane whose function:

$$\mathbf{w} \cdot \mathbf{x} + b = 0 \tag{1}$$

Where:

w: weight
x: input factor
b: bias

which implies:

$$y_i(\mathbf{w} \cdot \mathbf{x} + b = 0) \geq 1, i = 1, \ldots, N \tag{2}$$

Where:

y_i: the labels of the training samples
N: number of samples

IOP Conf. Series: Materials Science and Engineering **328** (2018) 012032 doi:10.1088/1757-899X/328/1/012032

The SVM algorithm tries to determine a distinctive separating hyperplane with minimizing $\| \mathbf{w} \|$ which represents the Euclidean norm of $\mathbf{w}$; the distance between the hyperplane, by adjusting the data points of each category using $2/\| \mathbf{w} \|$. When Lagrange multipliers $\boldsymbol{\alpha}_\mathbf{i}$ introduced, the SVM training process is to solve a convex quadratic problem (QP). The solution employs the following equation:

$$w = \sum_{i}^{N} \alpha_i y_i \mathbf{x_i} \tag{3}$$

Where:

$\boldsymbol{\alpha}_\mathbf{i}$: Lagrange multipliers

Only if corresponding $\alpha_i > 0$, this $\mathbf{x}_i$ is known as support vectors. During the model training process, the decision function is represented by:

$$f(\mathbf{x}) = \operatorname{sign}\left(\sum_{i=1}^{N} \alpha_i y_i (\mathbf{x} \cdot \mathbf{x}_i) + b\right) \tag{4}$$

In this study, the SVM tries to place a margin between the faulty-healthy data and adjusts it in a way to keep the distance between the data points and the margin as maximal in each group. The nearest data points are used to define the margin and are known as support vectors. However, in most cases the patterns are not linearly separable; therefore, a kernel function is used to perform the transformation. Hsu et al. [22] proposed RBF kernel function to be the first try kernel function for an SVM model. Chen et al. [23] found that RBF kernel gives a better test accuracy compared to the polynomial kernel. Therefore, SVM with RBF kernel function was deployed in this study.

2.3. Artificial Neural Network

Artificial neural networks (ANNs) are a family of statistical learning models inspired by biological neural networks of human brain. The ANN may don't have exact precision and of the traditional computing approach; yet, they are sufficient to get close approximations to the system that we have inadequate information to make a suitable solution [24]. ANNs are generally presented as systems of interconnected "Neurons" which exchange messages between each other. A neuron often received several inputs x_i. Each input should multiply by a value of weight w_i and then compile with a bias b. Thus the neuron's activation z will calculate as a result. The activation function to calculate as following:

$$z = \sum_{i=1}^{p} w_i x_i + b \tag{5}$$

When the activation z is obtained by the neurons, it is then ready to calculate the activation function $f(z)$. There are many types of activations functions and the selection depends on the network application and ANN architect choice. However, the most used activation functions include linear, threshold and sigmoid function. Thus the output of an activation function is the output of certain neuron. Moreover, the ANN consists of interconnection for all neurons that they are arranged in form of layers. The most common ANN includes three layers (input layer, hidden or interactive layer and output layer). In this study, feed forward network was employed. The neurons in each layer are connected to the next layer in one direction. See Fig.3.

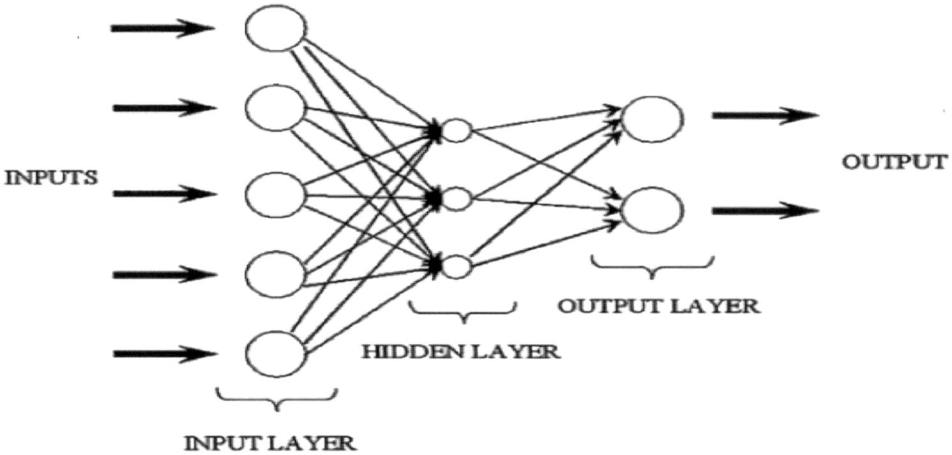

Fig. 3: A simple three layered ANN.

In this study, the network was trained to receive 9 inputs (duration, count, amplitude, ASL, signal strength, energy, absolute energy, speed and flow rate) and 2 output (healthy and faulty). The data were divided randomly into three groups: 70% as the training set, 15% as the validation set and 15% as the testing set [25]. Training and validation samples were used to develop the model, whilst the testing samples were held out and then applied to the developed model to evaluate the model performance. Once training is complete, a relationship between input and output data can be established. In training stage, the node weighting is adjusted till the value get close to the real value of all available inputs. However, if over fitting is identified, the computational processing will stop. Over fitting takes place when the model is performed well during training; then it starts to decline when tested with hidden data.

Two error values: mean squared error (MSE) and the percent error (%E) were used to check the network during training, validation and testing process. MSE is defined as the average squared difference among outputs and targets. The lower values of MSE are better (when MSE = 0, means no error). The percent error defined as a fraction of samples which are misclassified by ANN (when %E = 0, indicates that no misclassifications happened by the network. While a value of 100 indicates maximum misclassifications). Feed forward supervised network was used to classify the inputs according to targeted classes. Scaled conjugate gradient back propagation (TRAINSCG) was utilized as training algorithm. See Fig. 4. Thus, training will stop when validation data set MSE has increased more than maximum fail times; set to be 6 times, to avoid over fitting [26, 27]. Further, hyperbolic tangent sigmoid (TANSIG) has been employed as a transfer function. Thus, the network process input of both negative and positive values in fast time [28]. Equation 6 presented the TANSIG transfer function:

$$f(z) = \frac{e^{2x} - 1}{e^{2x} + 1} \qquad (6)$$

The number of neurons in the hidden layer was optimized through changing the number of neurons and trains the network until the highest accuracy and MSE of the network was achieved.

IOP Conf. Series: Materials Science and Engineering **328** (2018) 012032 doi:10.1088/1757-899X/328/1/012032

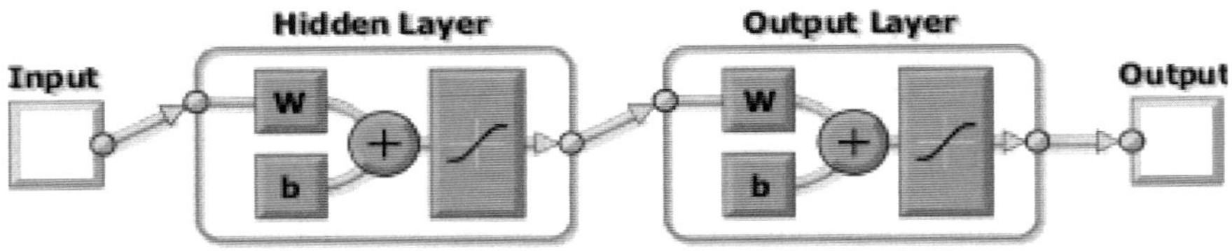

Fig. 4: ANN structure and transfer function of feed-forward back propagation.

3. Experimental Study

3.1. Test Rig and Instrumentation

The test rig employed in this study consists of a single-stage, two-cylinder air-cooled reciprocating compressor with a 1.5 kW/2 hp motor that can provide a maximum speed of 820 rpm. The compressor consists of two plate valves mounted over each cylinder. The valve consists of two parts, suction and discharge. Each part includes one plate, and both plates are move up and down opposite to each other during the compressor cycle for the suction and discharge process. During the opposite movement of the valve plates (up and down), the plates will impact the upper and lower valve housing. This impact is a rapid release of energy that generates a transient elastic wave, which moves through the valve up into the valve/cylinder cover and is detected by the AE sensor. See Fig.5.

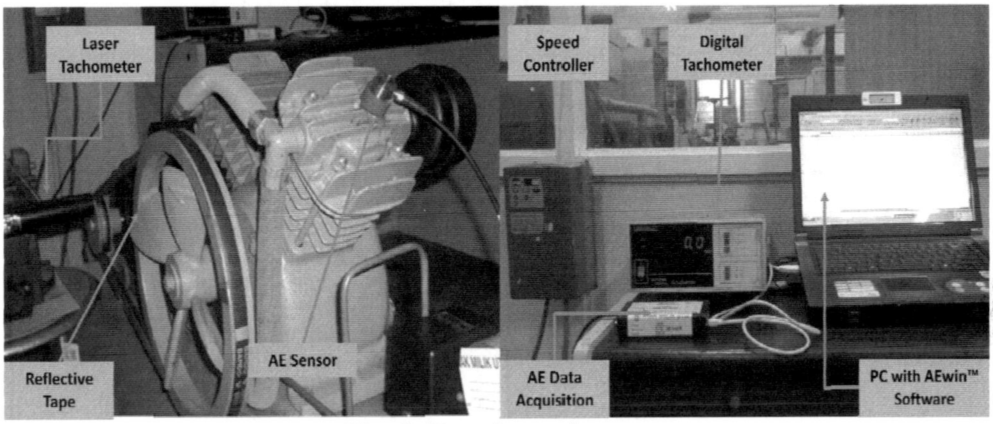

Fig. 5: Experimental setup.

A digital laser tachometer was used to show the compressor speed and to record the compressor cycle. The tachometer was installed near to the compressor flywheel to receive a pulse from a reflective tape attached to the flywheel. An AE sensor (model: PKWDI) was used to acquire the signal in this research. The sensor was placed at the centre of the valve/cylinder cover (the left cylinder of the reciprocating compressor) and fixed firmly to the surface by super glue. A single channel AE data acquisition (DAQ) system (model: USB AE Node) used for AE signal collection. AEwin™ software was used for recording AE hits and extracting AE parameters. The AE signal for valve faults diagnosis is less than 250 kHz [29], the AE signals were acquired at a sampling rate of 500 kHz.The signal was recognized perfectly at a threshold level of 55 db. The AE signals was digitized and conditioned by the DAQ device before transmission to a computer for further analysis.

IOP Conf. Series: Materials Science and Engineering **328** (2018) 012032 doi:10.1088/1757-899X/328/1/012032

3.2. Experimental Procedure

The experiment began by acquiring the AE signal (baseline signals) from the compressor with the valve in a healthy condition. The experiments were conducted in various operational conditions in terms of speed and airflow rate. Thirteen operational speeds ranging from 200-800 rpm (with incremental increasing by 50 rpm) and three flow rates (0%, 50% and 100%) were employed. Speeds were controlled by the speed controller, whilst the flow rates were controlled using a flow metre at the compressor outlet. Next, the experiment was repeated with the same operational conditions but emulated two types of actual faults, corrosion and clogged, individually at the compressor valve (including both the suction and discharge parts). Corrosion was introduced into the valve plates, whilst clogs were introduced into the valve body. Each fault was simulated with different severity levels to simulate progressive fault deterioration. Table 2 illustrates the types of defects with their severities. Fig.6 and Fig.7 illustrates the defects simulation.

Table 2: Types of defects and defects severities.

Valve Condition	Defect Type	Defect Severity	Defect Symbol	Defect Size
Healthy Condition	No defect	No defect	ND	No defect
Faulty Condition	Corrosion Defect	Very small corrosion	VSC	37.07 mm^2
		Small corrosion	SC	56.57 mm^2
		Medium corrosion	MC	79.63 mm^2
		Large corrosion	LC	106.27 mm^2
		Very large corrosion	VLC	136.48 mm^2
	Clogged Defect	Moderate clogged	MCL	40 %
		Intense clogged	ICL	80 %

All defects in the experimental specimens (spare valves) were simulated in advance. Next, the first defective valve was configured inside the reciprocating compressor. The first AE signal was acquired when the test rig was operated at the first speed and flow rate. The test was repeated for the other speeds and flow rate conditions until the signal was acquired for all the operational conditions. Then, the test rig was shut down, and the valve was replaced with the second specimen with another fault severity. The procedure was repeated, and another set of AE signals was recorded.

IOP Conf. Series: Materials Science and Engineering **328** (2018) 012032 doi:10.1088/1757-899X/328/1/012032

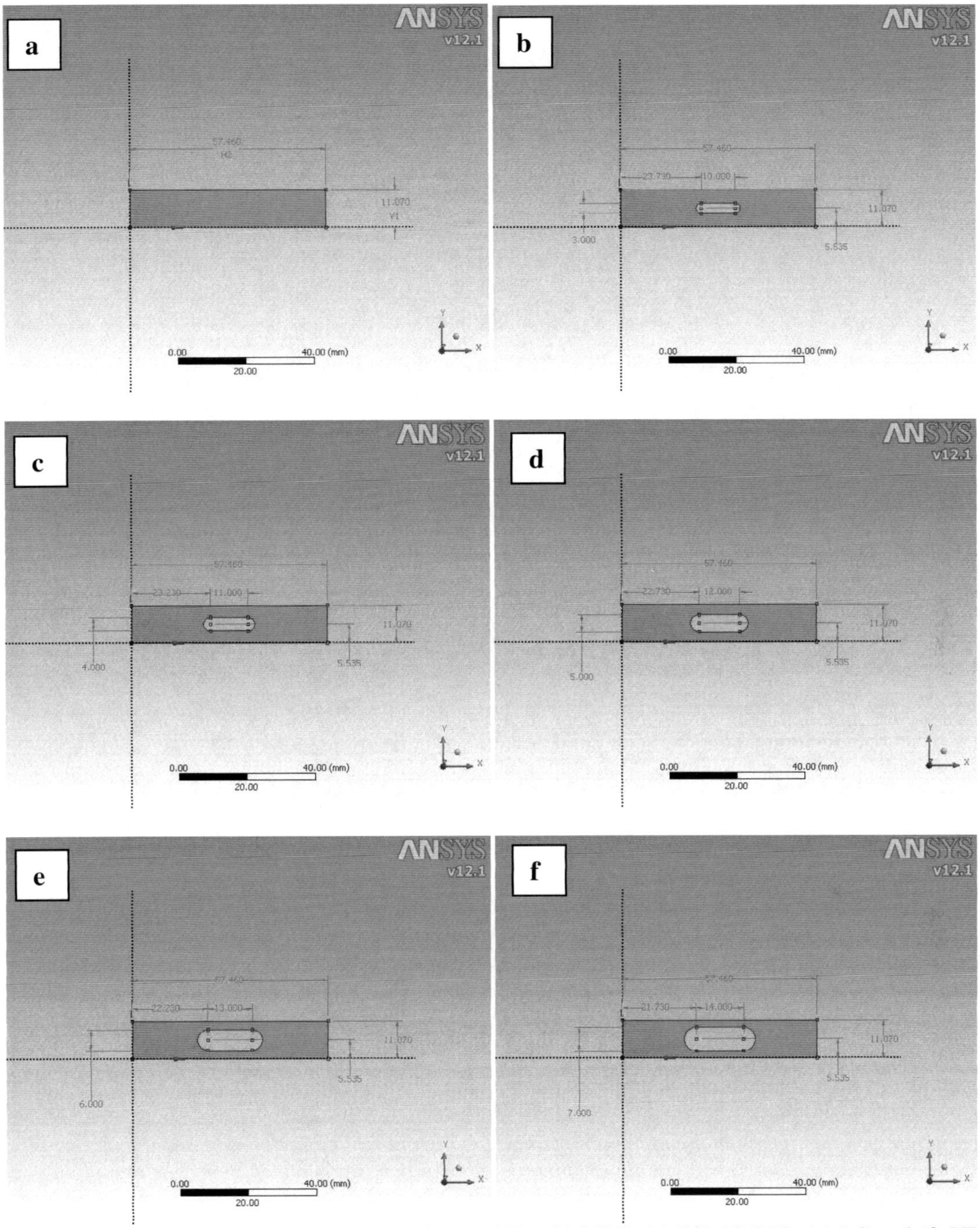

Fig. 6: The valves corrosion severity simulation (a) ND, (b) VSC, (c) SC, (d) MC, (e) LC and (f) VLC.

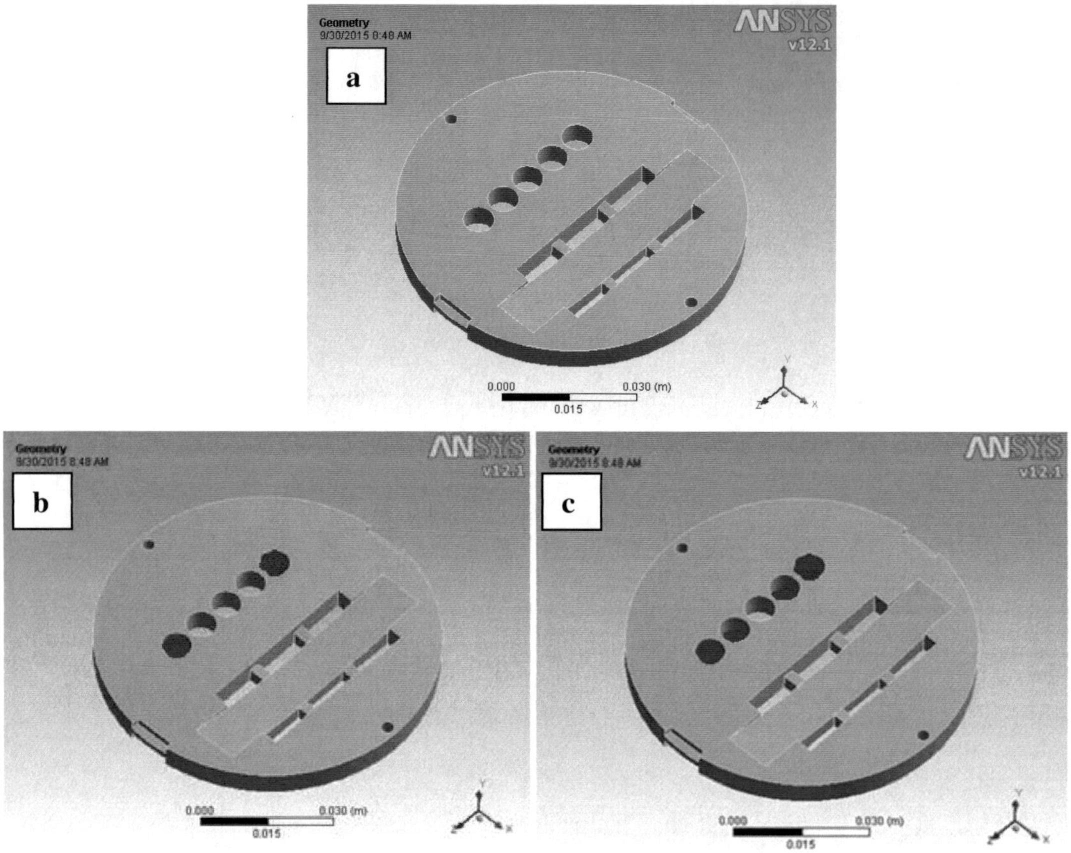

Fig. 7: The valves clogged severity simulation (a) ND, (b) MCL and (c) ICL.

To complete the experimental procedure, the test-rig was operated for 39 different operational conditions (13 speeds × 3 flow rates = 39) and sixteen valve conditions (8 valve conditions ×2 fault locations = 16) for a total of 624 tests. Each test was conducted for 30 sec and repeated 3 times, and the average was calculated. All experiments were conducted at laboratory temperature range between 25-30°C and standard atmospheric pressure. According to hold and train method [25], the data were divided randomly into two groups: 85% as the training set and 15% as the validation set. Training samples were used to develop the model, whilst the validation samples were held out and then applied to the developed model to evaluate the model performance.

4. Results and Discussions

4.1. Artificial Neural Network Model

The best network was selected according to the network highest classification accuracy and minimum MSE through changing the number of neurons in the hidden layer and trains the network until the highest accuracy and minimum MSE was achieved. Twelve attempts of changing the number of neurons have been investigated to select the network. The MSE value shows a significant decrease when the number of neurons increases until it has achieved a minimum value < 0.01 at the network with 40 neurons in the hidden layer. Next, any try to increase the number of neurons for more than 40 was resulted in increasing the value of MSE. Likewise, the network observed to achieve the highest accuracy when the proposed number of neurons in the hidden layer was 40 with dramatic decrease after that. Accordingly, the ANN with 9-40-2 configuration (Input layer: 9 neurons, Hidden layer: 40 neurons and output layer: 2 neurons) has been selected as the best network to classify the valve condition into healthy/faulty with an accuracy of 99.4% and minimum MSE of 0.0053. See Fig.8 and Fig. 9.

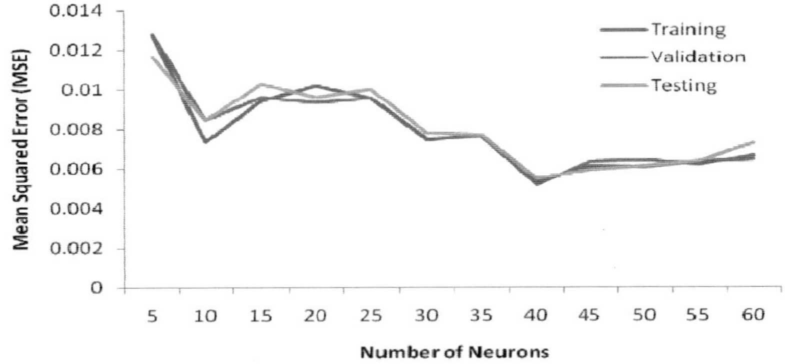

Fig. 8: Mean Squared Error versus Number of Neurons.

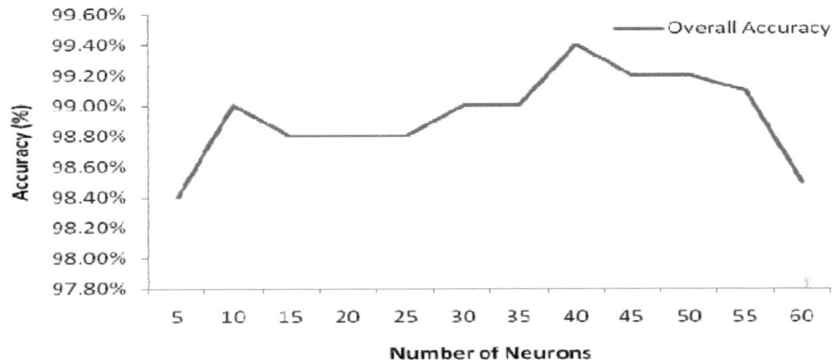

Fig. 9: The Overall ANN Accuracy versus Number of Neurons.

4.2. Support Vector Machine Model

SVM algorithms (svmtrain) and (svmclassify) were used in to train and classify the AE data. In this method, the SVM model was generated as follows: first, map the inputs vectors nonlinearly into one features space. Then, within the feature space from the first step, seek an optimized boundary division, that is, construct a hyperplane which separates the two classes of faulty/healthy. Thus, SVM training seeks a global optimized solution and avoid over-fitting. Table 3 illustrates the summary of SVM model based on 85% of training samples.

Table 3: SVM model structure based on training samples.

Output Arguments	**Value**
Support vectors	range: -7.69 to 7.47 for 3511 samples
Alpha	range: -0.74 to 1.53
Bias	0.0829
Kernel function	RBF kernel
Group names	120823 samples
Support vector indices	range: 5 to 120492
Scale shift	range: -4.07 to -0.22
Scale factor	range: 1.76 to 6.42

Table 3 shows the output arguments for SVM model, the support vectors are the range of data points with each row after normalization has been applied. Alpha are the weights values for the support vectors. The sign of the weight is positive for support vectors belonging to the first group (healthy) while negative for the second group (faulty). Bias refers to the intercept of the hyper plane that separates the two groups. RBF kernel has been used as a kernel function. Group names refer to the total data samples. Support vector indices refer to the training data that were selected as support vectors after the data was normalized. Shift refers to the negative of the mean across an observation in training while scale factor refers to 1 divided by the standard deviation of an observation in training. Based on the training data, the overall accuracy for SVM model was 99.4%.

4.3. ANN and SVM Models Performance

The ANN and SVM models were validated using 15% validation samples which were separated randomly from the original acquired data set. This method allows the fitted models to predict the valve condition from validation samples. The process was performed many times to obtain distribution of the predictive performance for each model. Thus, if the models classify the data correctly, then the usability of the model in other contexts can be assured. A lack of fit is possible if the model is unable to classify the data. In addition, receiver operating characteristic curves (ROC) is another comparable method to determine models classification ability [30]. The ROC curve is created by plotting the true positive rate (sensitivity) and false positive rate (one minus the specificity). The point on the curve that is nearest to the upper left corner corresponds to maximum sensitivity and specificity of model classification. The classification accuracy for ANN and SVM model are illustrated in Table 4 and Table 5 respectively. Fig. 11 shows the ROC curve for both models.

Table 4: ANN model classification based on validation samples.

Observed	Predicted		Total	Predicted Correctly
	Healthy	Faulty		
Healthy	6796	62	6862	99.1%
Faulty	73	14374	14350	99.5%
Total			21212	99.4%

Table 5: SVM model classification based on validation samples.

Observed	Predicted		Total	Predicted Correctly
	Healthy	Faulty		
Healthy	6842	97	6939	98.60%
Faulty	20	14253	14273	99.90%
Total			21212	99.4%

Table 4 and Table 5 clearly show that both ANN and SVM model have the same prediction accuracy of 99.4%. However, the models founds slightly different in term of sensitivity and specificity. By using the measure of percentage in the validation data that predicted correctly, ANN model could classify 99.1% from the healthy as healthy (true positive rate or sensitivity) and 99.5% from the faulty data as faulty (false positive rate or specificity) while the SVM model could classify 98.60% from the healthy as healthy and 99.90% from the faulty data as faulty. See Fig. 10.

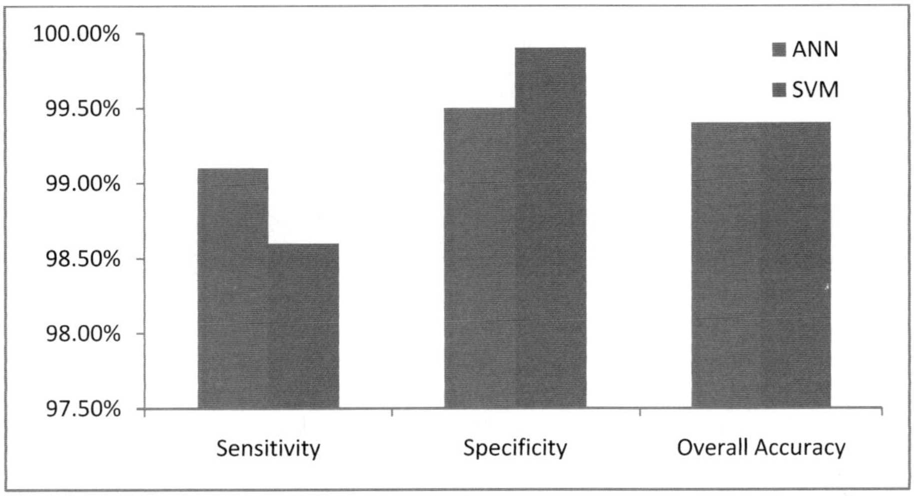

Fig. 10: A comparison between ANN and SVM performance.

Moreover, the ROC curve shows that both models were able to discriminate between healthy and faulty valve condition maximum sensitivity and specificity as both curves observed to be near to the upper left corner corresponding to maximum sensitivity and specificity of model classification.

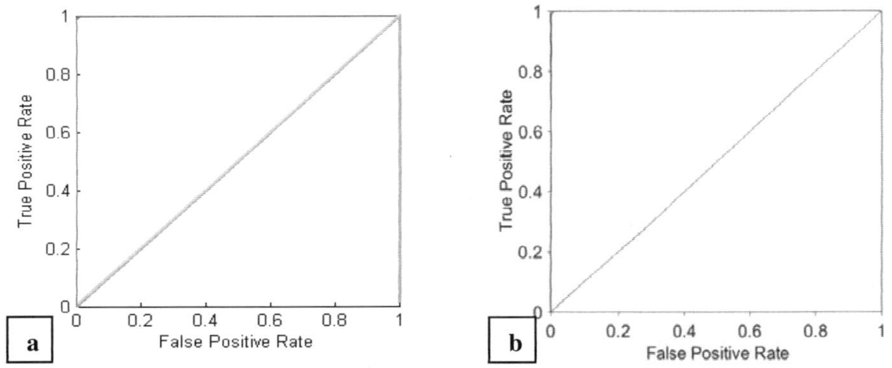

Fig. 11: ROC curve based on validation samples (a) ANN model and (b) SVM model.

5. Conclusion

The performance two machine learning approaches ANN and SVM models has been evaluated for detection valve fault in single stage reciprocating compressor based on AE signal. The experimental procedure was conducted by inducing two typical valve faults with different severity to acquire the combination of healthy/faulty AE signal during different speed and flow rate conditions. Same data sets, operational condition and AE parameters are used in constructing both models. ANN and SVM models have been train and evaluated based on hold and train method by using Matlab. In evaluating the results we consider the total accuracy of the detection and the ease of the model building. ANN and SVM models demonstrated slightly different in term of sensitivity and specificity while they have same ability to detect the valve condition with accuracy of 99.4%. Since, ANN accuracy is highly dependent on the neural networks structure such as number of nodes and hidden layers. While SVM has an excellent ability of handling a high number of input features with low sampling data sets [31]. SVM model is recommended for automate diagnose the valve condition in single stage reciprocating compressor.

Acknowledgment

The authors would like to extend their greatest gratitude to Institute of Noise and Vibration, University Technology Malaysia for funding the study under HICOE Grant Scheme (PY/2016/06784, PY/2016/07069 & PY/2016/07034) and Fundamental Research Grant Scheme (R.K130000.7840.4F653) financed by the Ministry of Higher Education, Malaysia.

References

[1] P. Reine, Industrial Air & Gas Compressors Report, (2014).
[2] D. Goebel, Insight: compressor monitoring. Technical Report, PROGNOST Systems GmbH, 2011.
[3] Q. Ren, X. Ma, G. Miao, Application of support vector machines in reciprocating compressor valve fault diagnosis, in: Proceedings of the International Conference on Natural Computation, 2005.
[4] V.T. Tran, F. AlThobiani, A. Ball, An approach to fault diagnosis of reciprocating compressor valves using Teager–Kaiser energy operator and deep belief networks, Expert Syst Appl. 41 (2014) 4113-4122.
[5] S. Manepatil, G. Yadava, B. Nakra, Modelling and computer simulation of reciprocating compressor with faults, (2000) 108-116.
[6] S.S. Manepatil, A. Tiwari, Fault diagnosis of reciprocating compressor using pressure pulsations, (2006).
[7] A. Tiwari, P. Yadav, Application of ANN in condition monitoring of a defective reciprocating air compressor, 38 (2008) 13-20.
[8] J. Gill, R. Douglas, Y. Neo, R. Reuben, J. Steel, Examination of plate valve behaviour in a small reciprocating compressor using acoustic emission, J. Acoust. Emission. 18 (2001) 96-101.
[9] M.H. El-Ghamry, R.L. Reuben, J.A. Steel, The development of automated pattern recognition and statistical feature isolation techniques for the diagnosis of reciprocating machinery faults using acoustic emission, Mech Syst Signal Pr. 17 (2003) 805-823.
[10] Y. Wang, X. Peng, Fault diagnosis of reciprocating compressor valve using acoustic emission, in: Proceedings of the ASME International Mechanical Engineering Congress and Exposition, Houston, Texas, USA, 2012.
[11] Y. Wang, C. Xue, X. Jia, X. Peng, Fault diagnosis of reciprocating compressor valve with the method integrating acoustic emission signal and simulated valve motion, Mech Syst Signal Pr. 56-57 (2015) 197-212.
[12] H. Sim, R. Ramli, A. Saifizul, M. Abdullah, Empirical investigation of acoustic emission signals for valve failure identification by using statistical method, Measurement. 58 (2014) 165-174.

[13] Y.H. Ali, R.A. Rahman, R.I.R. Hamzah, Acoustic emission signal analysis and artificial intelligence techniques in machine condition monitoring and fault diagnosis: a review, Jurnal Teknologi. 69 (2014) 121-126.
[14] Y.H. Ali, S.M. Ali, R.A. Rahman, R.I.R. Hamzah, Acoustic Emission and Artificial Intelligent Methods in Condition Monitoring of Rotating Machine–A Review, (2016) 212-219.
[15] S.M. Ali Al-Obaidi, Salman Leong, M., Raja Hamzah, R.I., Abdelrhman, A.M., Danaee, M. , Acoustic emission parameters evaluation in machinery condition monitoring by using the concept of multivariate analysis, ARPN Journal of Engineering and Applied Sciences. 11 (12) (2016) pp. 7507-7514.
[16] ASTM, Standard E1316: Standard terminology for nondestructive examinations, ASTM International, West Conshohocken, Pennsylvania, USA, 2006.
[17] S.M.A. Al-Obaidi, M.S. Leong, R. Hamzah, A.M. Abdelrhman, A review of acoustic emission technique for machinery condition monitoring: defects detection & diagnostic, Appl Mech Mater. 229 (2012) 1476-1480.
[18] K.H. Hui, M.H. Lim, M.S. Leong, S.M. Al-Obaidi, Dempster-Shafer evidence theory for multi-bearing faults diagnosis, 57 (2017) 160-170.
[19] A. Widodo, E.Y. Kim, J.-D. Son, B.-S. Yang, A.C. Tan, D.-S. Gu, B.-K. Choi, J. Mathew, Fault diagnosis of low speed bearing based on relevance vector machine and support vector machine, Expert Syst Appl. 36 (2009) 7252-7261.
[20] V. Vapnik, The nature of statistical learning theory, Springer-Verlag, New York, 2013.
[21] M.H.L. Kar Hoou Hui, Mohd Salman Leong, Salah Mahdi Al-Obaidi, Dempster-Shafer evidence theory for multi-bearing faults diagnosis, ENG APPL ARTIF INTEL. 57 (2016) 160–170.
[22] C.-W. Hsu, C.-C. Chang, C.-J. Lin, A practical guide to support vector classification. Technical Report, Department of Computer Science and Information Engineering, National Taiwan University, 2003.
[23] F. Chen, B. Tang, T. Song, L. Li, Multi-fault diagnosis study on roller bearing based on multi-kernel support vector machine with chaotic particle swarm optimization, Measurement. 47 (2014) 576-590.
[24] S.G. Barad, P. Ramaiah, R. Giridhar, G. Krishnaiah, Neural network approach for a combined performance and mechanical health monitoring of a gas turbine engine, Mech Syst Signal Pr. 27 (2012) 729-742.
[25] H.M. SOHEL RANA, S. K. SARKAR, Validation and performance analysis of the binary logistic regression model, in: Proceedings of the WSEAS International Conference on Environmental, Medicine and Health Sciences, Penang, Malaysia, 2010.
[26] Y.H. Ali, R. Abd Rahman, R.I.R. Hamzah, Artificial Neural Network Model for Monitoring Oil Film Regime in Spur Gear Based on Acoustic Emission Data, Shock and Vibration. (2015).
[27] A.K. Mahamad, S. Saon, T. Hiyama, Predicting remaining useful life of rotating machinery based artificial neural network, Comput Math Appl. 60 (2010) 1078-1087.
[28] M.H. Beale, M.T. Hagan, H.B. Demuth, Neural Network Toolbox 7, (2010).
[29] J. Yan, Y. Heng-hu, Y. Hong, Z. Feng, L. Zhen, W. Ping, Y. Yan, Nondestructive Detection of Valves Using Acoustic Emission Technique, Adv. Mater. Sci. Eng. 2015 (2015).
[30] J. Phillips, E. Cripps, J.W. Lau, M.R. Hodkiewicz, Classifying machinery condition using oil samples and binary logistic regression, Mech Syst Signal Pr. 60 (2015) 316-325.
[31] K.H. Hui, C.S. Ooi, M.H. Lim, M.S. Leong, A hybrid artificial neural network with Dempster-Shafer theory for automated bearing fault diagnosis, 18 (7) (2016).

Strategic Factors on Interpreting Remanufacturing Quality-Certifying Framework to Address Warranty Aftermarket for Malaysian Industry

N. Mohamed[1], M.Z.M. Saman[2], S. Sharif[2] and H.S. Hamzah[2]

[1] Department of Manufacturing and Industry, Faculty of Mechanical and Manufacturing, Universiti Tun Hussein Onn Malaysia, Batu Pahat- 86400 Johor, Malaysia,
[2] Faculty of Mechanical Engineering, Universiti Teknologi Malaysia, Skudai- 81300 Johor, Malaysia

Abstract. While the concept of remanufacturing is gaining popularity globally, literature and theory on strategic decision-making on certifying for warranty in this area remain limited. A strategic and establish concept flow is developed based on extensive literature review and surveys with experienced experts who are dealing with remanufactured, reconditioned, rebuilt and reused components. The remanufacturing research on evaluating quality assurance of remanufactured component targets macro-level parameters and the indicators which must be confirmed for evaluation. The strategic remanufacturing factors identified from the literature review are discussed in a brainstorming session with a number of remanufacturing researchers and academic experts. The study is further broadened by industrial surveys and case studies to justify the inputs on developing a framework to certify remanufactured components. Preliminary results have established the key factors of remanufacturing quality control that might lead to the strict quality assurance of remanufactured components. Later, the developed framework can be used as a benchmarking tool to certify remanufactured components and warranty issuance. The findings serve as the foundation for further research concerning Original Equipment Manufacturer (OEM) or Original Equipment Remanufacturer (OER) and Independent Equipment Remanufacturer (IER) in the Malaysian Remanufacturing Industry.

Keywords: remanufacturing quality-certifying framework; quality assurance; quality control; remanufacturing warranty, developing country

1. Remanufacturing scenario

Remanufacturing in Malaysia is nascent and has just begun to emerge. In practice, the terms recycling, reuse, rebuilding, and remanufacturing are confused from the perspective of Malaysian automotive sector. A clear framework should be established for a better aftermarket sector profile. Intensive study to perform a framework suits for Malaysian remanufacturing industry is a must for future industry. In the aftermarket business, a remanufacturer may be classified into two groups: Original Equipment Remanufacturer and Independent Equipment Remanufacturer.

Research done by [1] stated that a key problem for remanufacturers is how to balance the contradictory needs for low cost, high quality, and short lead time. In other studies by [2] indicate one of the core management issues in remanufacturing dealing industry is to effectively match demand and supply by dealing with uncertainty of the quality of returned products and of the market demand. In every aspect quality is extremely important and for remanufactured components quality is very challenging because the origin of each component has their own history during the life span and the ended of their useful life may vary. Continuous quality control is the best possible ways to achieve high quality remanufactured components as on OEM components that have appropriate development from the design concept through the manufacturing production, remanufactured components should have appropriate processes that merge with definition of continuous quality control.

Further, survey done by [4,5] listed nine Remanufacturing Decision-Making Factors (RDMF) for automotive industry companies engaged with OEM and IAM businesses included; financial impact of remanufacturing, core management, protection of intellectual property, green perception, OEM component specifications, government regulations, integrated organizational alignment, design for remanufacture (DfRem), and brand erosion. Out from the nine factors, core management and OEM component specification are two important factors in the remanufacturing processes in order to meet OEM sound conditions. From perspective of [5] criticized the issue of Original Equipment automotive suppliers which try to expand their portfolio of remanufactured components in the current state (without an upfront remanufacturing strategy), as the remaining service life requirements become only a key factor in their decisions to do remanufacturing. Not only the performance of component on remaining service life, remanufacturing need the re-issuance of the warranty period and coverage for aftermarket components in order to distinguish remanufacturing with other aftermarket businesses to secure its quality. Certification and trademarks of remanufactured component will help to counter this kind of issue in the grey market.

In this study, the quality of remanufactured components is the subject highlighted. Important factors or determinants which may reflect the quality outcome of remanufactured components is crucial to determine thorough the remanufacturing process, core management, quality management, and many other factors discussed in literature per summarized in Figure-1. A Remanufacturing-Quality Factors is important to perform in order to distinguish remanufacturing industry from other aftermarket businesses such as recycling, reusing, and reconditioning. Research is focusing on mechanical remanufactured parts, as applicable mostly in the Malaysian automotive, heavy machinery, marine equipment and aerospace industries.

2. Important factors for strategic quality interpretation

2.1 Core design

Used components or products going to be remanufactured are called core. Core selection for remanufacturing is very important as same as material selection for manufacturing processes to fabricate new products. Studies by [6] reported, the automotive industry in Malaysia, many have agreed that most components or parts intended for future remanufacturing should be designed in such a way that they would be able to undergo certain processes in making those components or parts usable to a certain extent, even after their originally planned life-cycle. Findings from [6] strengthen findings by [3,7] that the OEM/Supplier companies are making decisions on remanufacturing components late in the component life cycle, resulting in weakened remanufacturing results which makes DfRem is a key issue for remanufacturing. As extension of life-cycle process does not have a negative impact on component specification [8], or a negative impact on component performance after remanufacture, it may deteriorate the reliability and durability to a certain extent. Reliability is among the eight dimensional models for product quality [9]. Hence, it is important that such a design concept (Design for Disassembly, Design for Remanufacturing, Design for Reuse and so on) is incorporated before the components or parts are manufactured. The issue here is that three-quarter of remanufactured component are not designed for disassembly and this has a significant impact on the remanufacturing processes [10]. Parts without a reengineering concept may be prone to damage during assembly; desired final specifications are not achieved and this may increase the scrap rate.

2.2 Production layout

Does non-existence of proper production layout jeopardize the quality of remanufacturing processes? This question is only answerable by an authorized entity or government body by rules or regulations as set up. Flexible or standard remanufacturing looks significant to represent the remanufacturing processes, from Entrance Diagnosis to the end of Final Component Testing. To prove that remanufacturing is a sustainable business and not just a backyard junk business, here it must be emphasized that a planned layout is requirement. As volume of core is an issue in remanufacturing business, flexible production plan is the best alternative to set up.

In forward manufacturing industry, working manual is kind of Standard of Procedure (SOP) need to be issue to the production plan. Without a production layout, SOP is hard to document and it is not easy to ensure that remanufacturing processes are as efficient as possible. According to [11], organization and layout of the production system and network are enablers of, and have effects on, the efficiency of production processes. This is applicable to forward manufacturing and this rule should be applied to reverse manufacturing or remanufacturing as well.

Since there are no well-known strategies for evaluating the remanufacturing processes, methods for evaluation of ordinary manufacturing process were used as indicate in research by [6, 7], as facility layout is important to an aftermarket process organization. Through a literature study, [13] summarized Flexible Manufacturing Systems, Process Sequence Cell Layouts, Flow Processing and Mixed Model Flow Processing which can directly respond to facilitate flow processing within the remanufacturing environment. As long as a remanufacturer can justify their production layout, it shows that they understand the necessity of processes to be carried out and for high-quality production not just doing local workshop repairing work.

2.3 Remanufacturing processes

Previous study indicates, six processes should involve in the remanufacturing components; disassembly, cleaning, inspection and sorting, reengineering, reassembly and final testing [14]. To reach the same quality, warranty, reliability safety, effectiveness, lifetime and safety standards as a new component, adequate quality assurance measures are applied in every step. In particular, the final test of every component to be sold is not an independent "sixth step", but an integral part of the whole remanufacturing process chain [14]. The testing, measurement, and quality control methods used should similar to those in original manufacturing. The only difference is that remanufacturing requires 100% inspection because in remanufacture all parts are presumed faulty until proven otherwise [15].

Existing claimed remanufacturing companies operate their processes according to varying strategies based on the characteristics of component they remanufacture, because they have no developed theory on how remanufacturing processes should be done. Some companies may have diagnosis entrance system for the components; some may knock down components up to single part and check for the condition and; some may just remanufacture the components according to needs based on their manual judgment. The lack of standard procedure on how remanufacture should be performed makes it harder to bring remanufacturing up to the ultimate level of end-of-life cycle processes. All strategies, theories, and processes in remanufacturing business which can bring out the high quality of remanufactured components have not yet been integrated in order to issue remanufactured components that same as new OEM. Warranties to match new OEM components are a must to complete the remanufacturing definition.

Analyses by [12] on remanufacturing facilities, for household appliances and automotive parts revealed that the cleaning and repairing steps are most critical in the remanufacturing process. In a remanufacturing study by [15], three key areas of inspection during the remanufacturing process were found to be very important: core inspection/testing; part inspection/testing; and finished component inspection/testing. This was embraced by [17] where agreed that remanufacturing always requires 100% inspection rate at one or more stages during the remanufacturing process. This is a quite detailed process but makes sense to determine quality of remanufactured components, as testing on final output only seems unreliable for remanufactured components.

2.4 Data measurement/ Documentation

'Lack of documentation of remanufacturing procedure' is one of remanufacturing problem considered under study done by [15]. However, only 2 out of 5 companies under the case study find that documentation significant for their businesses. In addition, [18] revealed that many companies do not have an ISO9001 or ISO14001 certifications, since the management does not see the value of having these since there is no customer demand [18] nor enforcement from government body for having it.

Findings by [18] contradict to latest finding by [19], who found that the end-of-life vehicle (ELV) management industry is well established in North America and the processing technologies are generally understood, while specifics about each stage or unit operation of the ELV management process are not well documented. This issue in the remanufacturing industry means it operates like a black market business without any auditing. Business and operations related compliance, documentation and reporting must be performed to standardize remanufactured output as same as a new OEM part or else it just a reused component without any quality assurance and warranty given.

However, in recent years, remanufacturing is developing as an industry for future business. In order to declare a company as remanufacturer and provide a warranty for remanufactured components, they must run their business as any other manufacturing company. Everything must be clear and documented for auditing purposes. Remanufacturing is not a backyard or lame industry but a new business for future industry.

2.5 Management expert and skill workers

As revealed by [15] in a study, a company stated that the most important task of remanufacturers is obtaining the correct blend of technical skill, component knowledge and component history and scarcity of skilled was cited as the top concern among USA remanufacturer [20]. These criteria reflect the needs of expert and skill worker in remanufacturing sector. Skilled workers here represent experienced workers or skilled person with technical flexibility. Experienced technical workers are a great need for flexibility in remanufacturing businesses because this can help to increase worker productivity.

Repairing, rebuilding, reconditioning or other restoration procedures need a lot of experience and skill. For an example, local mechanics repairing all types and brands of vehicle without knowing the original condition and details of each part of component may fix most vehicle problems. It takes a great deal of experience as learning processes to cater all problems but basically, the fundamental element of repairing vehicle parts must be mastered. Skilful persons are crucial and the documented procedure and judgment criteria should be documented in a database as a systematic approach. Practically, most remanufacturer and reuse company only depend on traditional inspection to either rejected or remanufactured/ reused the components [21].

2.6 Organizational alignment

An integrated organizational alignment between OEM and independent remanufacturer is important to remanufacture a component. Compatibility of remanufacturing concept with current organization's policy and direction is one of the main point researched by [6] for Malaysia Automotive Remanufacturing Sector. Manufacturing data such as design tolerance, component specifications, and testing are required to return components to their original condition or perhaps even upgrade them to a better level. However, the sources of data are not easily to be access and it is confidential for most manufacturers. Hence, integrated organizational alignments supported by regulated bodies are important to perform in order to assist the independent remanufacturer to get support from the OEM in order to establish remanufacturing businesses.

By practices, some remanufacturers are depending reverse engineering strategy make profiles for the ELV components. For an example 3D scanner is used to develop the cores profiles, how far it can take place for a legal business is thing to be find out. Until present, based on report by Asia-Pacific Economic Cooperation, Malaysia does not have a single regulation on Remanufacturing Businesses. This is a strong argument for Malaysia to come out with legislation and regulation for Remanufacturing industry for a clear organizational integration.

2.7 Serviceable and aesthetic

One thing is certain; high quality means pleasing the consumer, not just protecting them from annoyances [9]. That quote reflects our developing remanufacturing industry in this decade. From [9] the eight quality dimensions that can serve as a framework for strategic analysis: performance, features, reliability, conformance, durability, serviceability, aesthetics, and perceived quality. From literature performance, serviceability, aesthetic and reliability and conformance are the elements can be studied under remanufacture components. Capturing original OEM performance is the goal for remanufactured components.

During the warranty period, a component usually is applicable for service or and inspection from the authorized dealer or maker. In remanufacturing case this issue is not been highlighted, maybe because the industry itself is not established and run on a big scale. However, remanufactured components are applicable for same warranty as the New OEM and this issue should be discussed further to makes it a fair business for forward manufacturing players and aftermarket players. Furthermore, warranty and services are criteria for insurance claim purposes. In the other hand, originality in term of shape, surface finish and is might be important to retain in remanufactured components but goods surface finish sound like an additional cost for remanufacturer to bear.

2.8 Complementary performance

Complementary performance here is trying to highlight the factors to be investigated before decided to remanufacture component, whether the component is reliable to undergo remanufacturing processes. This includes the condition of component return and total life span the component can serve the end market. Safety and reliability issues still play a key role in serious and often controversial discussions about the future of remanufacturing and reuse of parts [17]. Reliability is the most important quality dimension to be understood to make sure remanufactured components can serve without failing within the warranty period.

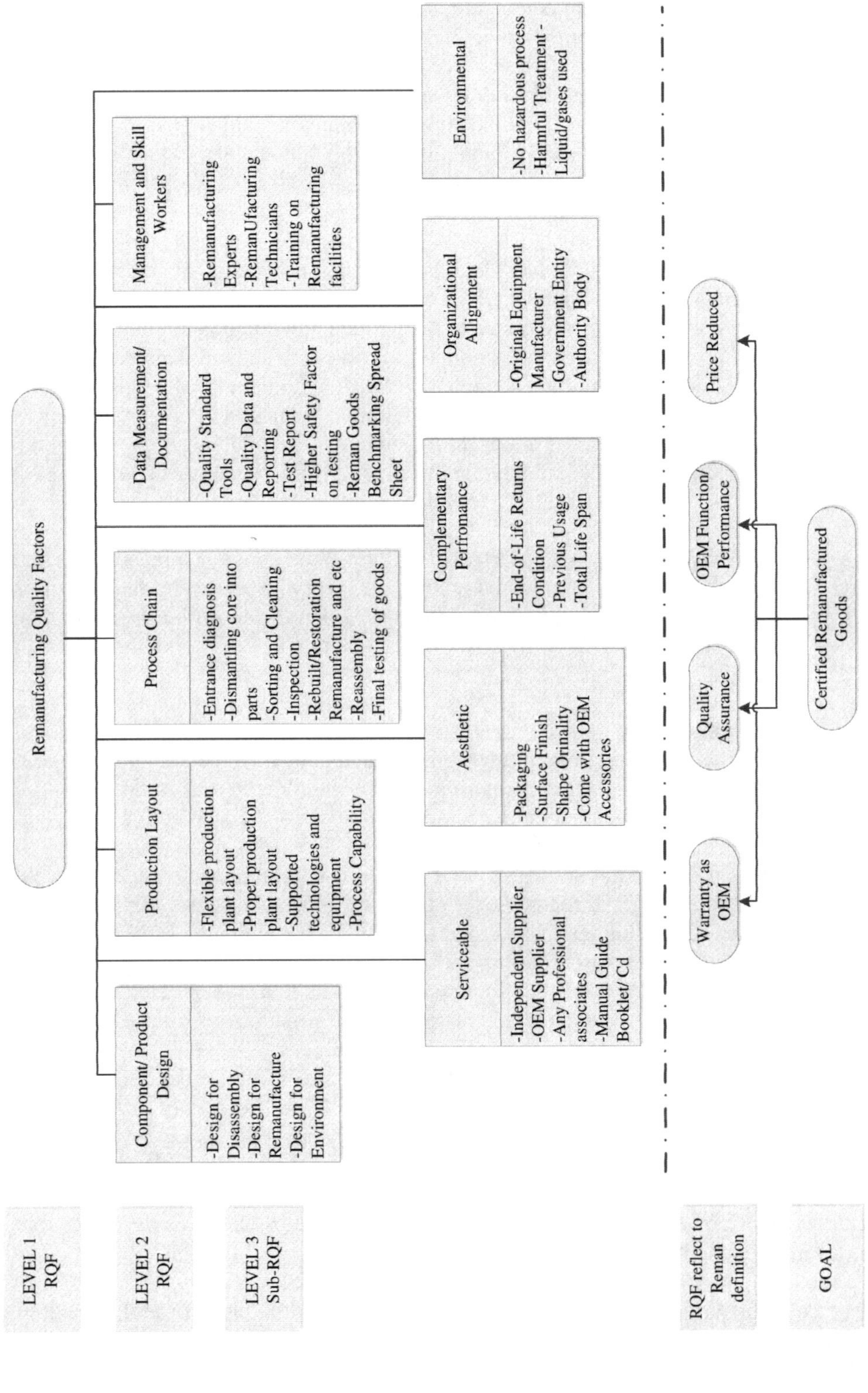

Figure 1. Remanufacturing-Quality Criteria for Remanufacturing Industry

IOP Conf. Series: Materials Science and Engineering **328** (2018) 012033 doi:10.1088/1757-899X/328/1/012033

2.9 Environmental issue

6Rs is a greening approach for a sustainable future. The goals of 6Rs close-loop lifecycle are to keep the better environment and sustain nature and besides preserving mineral and natural resources. With this reason, remanufacturing process must be practiced in cleaner and non-hazardous environment. For an example, no extreme chemical should be used through cleaning out the excessive oil and liquid from the used components. From surveys in Malaysia the excessive oil which is pumped out from used component are not well managed, where most of the components are transporting from outside Malaysia.

3. Methodology

A framework is developed by literature study and expert opinion because of limitation of industrial players in remanufacturing sectors in Malaysia. A comprehensive literature study is done to find the important criteria in remanufacturing area which may reflect to component quality and warranty issuance. Brainstorming sessions with researchers involved in remanufacturing and recycling activities are conducted in the preliminary stage to make sure the future developed framework is significant and reliable. Further, feedback has been gathered to finalize and construct the preliminary framework base on researchers and academics points of view.

This methodology needs a joint effort and commitment from government and related bodies, aftermarket manufacturers (remanufacturer/ recycler/rebuilder and so on) and researchers to standardize a system in order planning and evaluated quality remanufactured of components in perspective of industrial expert and academic expert to fill up the gap. After intensive literature study, brainstorming session and interview conducted with a few recyclers/ remanufacturer in Malaysia resulted as Figure- 1 where items are grouped into categories. Final frameworks have been constructed by an interview with representative from Ministry of International Trade and Industries (MITI) in Malaysia for initial validation of research work.

4. Finding and discussion

From preliminary survey it shown that remanufacturing is mixed up with recycling, reusing and rebuilt industry. With no regulations on remanufacturing, Malaysia already has total of 39 registered remanufacturing businesses in five different sectors: aerospace (1), automotive/heavy duty (10), electronic (19), heavy machinery (5), marine equipment (1) and plant machinery (8). Each remanufacturing sectors has different strategies for bringing used components at par with new OEM components. Practitioners remanufactured cores depending on consumer demand and acceptance.

In term of recyclers and rebuilders population, Malaysia have hundred them under different association. However, from their valuable experience and feedback, author tuned it to academic components and finds that recycling and rebuilding can be upgraded to remanufacturing business with additional of processes and quality management system. All information from surveys is gathered and categorized in Figure-1 based on themes selected from literature.

On current situation it is promising and shown positive sign of future remanufacturing industry in Malaysia as industrial player has interest in this business and certainly profitable. To help remanufacturing industry growth in Malaysia remanufacturer and the government/ authorities body need to cooperated and negotiated with the existing industrial players to fulfil the basic needs of having qualified remanufacturing sector. There is no regulation or standard procedure performs in order to acknowledge remanufactured component/ product is better and reliable rather than repair, refurbished, reconditioned, reused, rebuilt or other execution mode.

The methodology presented in this paper provides a comprehensive evaluation of RQF, covering stages of the remanufacturing processes, management system, and component early designing stages which reflect good remanufacturing practices for quality and warranty issuance. Through these stages, the authors have found that remanufactured components have issues with reliability and the reliability should been understood before components being dismantled. Component history and original design is a must to be understood. Reliability study is the best indicator for remanufacturer to gain confident on their remanufactured quality and finally provide a warranty to their remanufactured components.

5. Conclusion

From the 10 elements of Remanufacturing Quality Factors (RQF) proposed above, process chain and expert management/ skill worker are the two elements practices by most remanufacturers in Malaysia regardless they are OER or IER. For recyclers and rebuilders the most important to on running their business is expert management/ skill worker only. However, to accredit component/ product of remanufactured good, every aspect of component/ product must fulfilling the requirement as per remanufacturing definition (refer GOAL in Figure-1). Further study will be conducted from the perspective of quality, safety, and component warranty in order to come out with a RQF that helps as a blue print for Malaysia remanufacturing industry. Which and how factors will bring the warranty as same as OEM new component/ product warranty will be finalized further.

Further RQF be applied by both OER and IER to justify their remanufactured components are meets the quality for second life use through reliability point of view. Besides helping recyclers and rebuilders upgrading their business, RQF can be auditing benchmark criteria for a consortium or government body to examine the quality of remanufacturer outputs. Since there are no standard strategies for evaluating the remanufacturing procedure in Malaysia, factors for evaluation in Figure-1 is the basic need for IER and OER to work as independent/ authorized remanufacturer.

6. Acknowledgements

The authors express appreciation to the scholars, Universiti Tun Hussein Onn Malaysia and Ministry of Higher Education Malaysia on their sponsorship for this research and also to Universiti Teknologi Malaysia for funding this publication under registered grant scheme no. 4F570.

7. References

[1] W. L. Ijomah, C. a. McMahon, G. P. Hammond, and S. T. Newman. Development of robust design-for-remanufacturing guidelines to further the aims of sustainable development. *Int. J. Prod. Res.* 2007; 45: 4513–4536.

[2] S. K. Mukhopadhyay and H. Ma. Joint procurement and production in decisions remanufacturing under quality and demand uncertainty. *Int. J. Prod. Econ.* 2009; 120; 5–17.

[3] R. Subramoniam, D. Huisingh, R. B. Chinnam, and S. Subramoniam,. Remanufacturing Decision-Making Framework (RDMF): Research validation using the analytical hierarchical process. *J. Clean. Prod.* 2013; 40 212–220.

[4]] R. Subramoniam, D. Huisingh, and R. B. Chinnam. Aftermarket remanufacturing strategic planning decision-making framework: Theory & practice. *J. Clean. Prod.* 2010; 18: 1575–1586.

[5] R. Subramoniam, D. Huisingh, and R. B. Chinnam. Remanufacturing for the automotive aftermarket-strategic factors: literature review and future research needs. *J. Clean. Prod.* 2009; 17: 1163–1174.

[6] N. M. Yusop, D. A. Wahab, and N. Saibani. Realising the automotive remanufacturing roadmap in Malaysia: Challenges and the way forward. *J. Clean. Prod.* 2016; 112: 1910–1919.

[7] S. S. Yang, S. K. Ong, and A. Y. C. Nee. Towards implementation of DfRem into the product development process. *Procedia CIRP*. 2015; 26; 565–570.

[8] N. M. Yusop, D. A. Wahab, and N. Saibani. Analysis of remanufacturing practices in the malaysian automotive industry. *J. Teknol. ;Sciences Eng*. 2012;59:77–80.

[9] D. A.Gavin. Competing on the eight dimentions of quality. *J. for Profession Manager*, Harvard Business Review 1987;65:101-108.

[10] V. D. R. G. Jr and A. Õ. Forbes. Production planning and control for remanufacturing: industry practice and research needs. *J. Operations Management* 2000; 18: 467-483.

[11] M. Koho, M. Tapaninaho, J. Heilala, and S. Torvinen. Towards a concept for realizing sustainability in the manufacturing industry. *J. Ind. Prod. Eng*. 2015; 32: 12–22.

[12] E. Sundin and B. Bras. Making functional sales environmentally and economically beneficial through product remanufacturing. *J. Clean. Prod*. 2005;13: 913–925.

[13] S. Khalifa. *Implementing Flow Processing with Product End of Life Remanufacturing*. De Montfort University; 2013.

[14] R. Steinhilper, B. Rosemann, S. Freiberger, and R. Technology. Product And Process Assessment For Remanufacturing Of Computer Controlled Automotive Components. *Proceeding 13th CIRP Int. Conf. Life Cycle Eng*. 2006; 441–445.

[15] W. L. Ijomah. *A model-based definition of the generic remanufacturing business process*.University of Plymouth; 2002.

[16] M. Errington. *Business processes and strategic framework for inspection in remanufacturing*. University of Exeter; 2009.

[17] A. C. Brent and R. Steinhilper. Opportunities for remanufactured electronic products from developing countries: Hypotheses to characterise the perspectives of a global remanufacturing industry. *SAIEE Africa Res. J*. 2005; 96: 196–201.

[18] E. Sundin. Product and process design for successful remanufacturing. *Institute of Technology, Link¨opings Universitet* ; 2004.

[19] S. S. Sawyer-Beaulieu and E. K. L. Tam. Regulation of end-of-life vehicle (ELV) retirement in the US compared to Canada. *Int. J. Environ. Stud*. 2006; 63,: 473–486.

[20] I. A. Williamson, D. R. Pearson, S. L. Aranoff, and D. a Pinkert. Remanufactured Goods : An Overview of the U . S . and Global Industries , Markets , and Trade. *Technical Report of United States International Trade Commission*. 2012; 332; 284, 2012.

[21] A. M. Yazid, J. K. Rijal, M. S. Awaluddin, and E. Sari. Pattern recognition on remanufacturing automotive component as support decision making using Mahalanobis-Taguchi system. *Procedia CIRP*. 2015; 26; 258–263.

Institute of Physics
Dirac House, Temple Back
Bristol BS1 6BE UK

ISSN: 1757-8981
ISBN 978-1-5108-5985-2

9 781510 859852

CURRAN ASSOCIATES INC.
proceedings.com